Mass Transfer and Kinetics of Ion Exchange

NATO ASI Series

Advanced Science Institutes Series

A series presenting the results of activities sponsored by the NATO Science Committee, which aims at the dissemination of advanced scientific and technological knowledge, with a view to strengthening links between scientific communities

The series is published by an international board of publishers in conjunction with NATO Scientific Affairs Division

A	**Life Sciences**	Plenum Publishing Corporation
B	**Physics**	London and New York
C	**Mathematical and Physical Sciences**	D. Reidel Publishing Company Dordrecht and Boston
D	**Behavioural and Social Sciences**	Martinus Nijhoff Publishers Boston/The Hague/Dordrecht/Lancaster
E	**Applied Sciences**	
F	**Computer and Systems Sciences**	Springer Verlag Berlin/Heidelberg/New York
G	**Ecological Sciences**	

Series E: Applied Sciences – No. 71

Mass Transfer and Kinetics of Ion Exchange

edited by

Lorenzo Liberti

Senior Researcher at the Water Research Institute
of the National Research Council of Italy
Associate Professor of Industrial Chemistry
at the Faculty of Engineering
University of Bari, Italy

and

Friedrich G. Helfferich

Professor of Chemical Engineering
Pennsylvania State University
University Park, PA 16802, USA

1983 **Martinus Nijhoff Publishers**
The Hague / Boston / Lancaster
Published in cooperation with NATO Scientific Affairs Division

Proceedings of the NATO Advanced Study Institute on
Mass Transfer and Kinetics of Ion Exchange,
Maratea, Italy, May 31 - June 11, 1982

Library of Congress Cataloging in Publication Data

NATO Advanced Study Institute on Mass Transfer and
Kinetics of Ion Exchange (1982 : Maratea, Italy)
Mass transfer and kinetics of ion exchange.

(NATO ASI series. Series E, Applied sciences ;
no. 71)
"Published in cooperation with NATO Scientific Affairs
Division."
"Proceedings of the NATO Advanced Study Institute
on Mass Transfer and Kinetics of Ion Exchange, Maratea,
Italy, May 31-June 11, 1982"--T.p. verso.
1. Mass transfer--Congresses. 2. Ion exchange--
Congresses. I. Liberti, Lorenzo. II. Helfferich,
Friedrich G., 1922- . III. North Atlantic Treaty
Organization. Scientific Affairs Division. IV. Title.
V. Series: NATO advanced science institutes series.
Series E, Applied sciences ; no. 71.
TP156.M3N37 1982 660.2'8423 83-13191
ISBN 90-247-2861-4

ISBN 90-247-2861-4 (this volume)

Distributors for the United States and Canada: Kluwer Boston, Inc., 190 Old Derby Street, Hingham, MA 02043, USA

Distributors for all other countries: Kluwer Academic Publishers Group, Distribution Center, P.O. Box 322, 3300 AH Dordrecht, The Netherlands

Printed in The Netherlands

PREFACE

Ion exchange, its theory and its applications in laboratory and industry, can look back on a long and distinguished history and many view the 1950's as its high point. Indeed, it was at that time that most of the current uses of ion exchange were developed and reduced to practice, and the foundations laid for the theory as we know it today. However, in recent years there has been a resurgence of interest in ion exchange as one of the tools to overcome the many increasingly difficult problems posed by our growing need to recover valuable materials from waste and to prevent or correct damage to the environment. Therefore this is a good time to re-examine theory, established applications, and as yet unrealized potentials of ion exchange.

Like many others the technology of ion exchange relies heavily on three elements: synthesis, for suitable materials; thermodynamics, for equilibrium properties; kinetics, for dynamic performance. Of these three elements, synthesis, in a thoroughly matured field, builds largely on industrial know-how and proprietary information, leaving little room for input from the scientific community at large. Thermodynamics has been studied extensively for many years and brought to a high degree of perfection. The same cannot be said of kinetics, however, which still poses unsolved problems once we wish to probe beyond the simple "ideal" case of binary ion exchange of strong electrolytes with strong-acid or strong-base resins. It is from this background the idea of an international study meeting devoted to kinetics of ion exchange arose and the principal lectures given at that meeting have been collected in this volume.

No specialty thrives in isolation. The scope of the meeting was therefore broadened to include input from other fields of ion exchange -- from thermodynamic theory to industrial processes — while kinetics and mass transfer remained the focal point. This gave the meeting the spice of contributions, formal and informal, from experts of many diverse backgrounds and nationalities, and helped to secure the broad-based support from NATO, the Italian Consiglio Nazionale delle Ricerche and Cassa per il Mezzogiorno, and industrial companies that made the conference possible in the first place.

It is not for the organizers and lecturers to say that the meeting was a success. This must be judged from future perspectives by those who derived benefit from attending the meeting or from reading the contributions presented there. We are glad, however, to express our deep appreciation for the untiring work of the other members of the Scientific Board of this meeting: Prof. G. Dickel, Prof. J.A. Marinsky, Dr. J.R. Millar, Prof. R. Passino, and Prof. V. Soldatov, and for the support of all who helped us provide a stimulating climate, both technical and social for a true meeting of minds.

Lorenzo Liberti   Friedrich G. Helfferich

TABLE OF CONTENTS

Preface V

1. John R. Millar
On the Synthesis of Ion-Exchange Resins 1
On the Structure of Ion-Exchange Resins 23

2. Robert Kunin
The Nature and Properties of Acrylic Anion Exchange Resins 45

3. Jacob A. Marinsky
Selectivity and Ion Speciation in Cation-Exchange Resins 75

4. George Eisenman
The Molecular Basis of Ionic Selectivity in Macroscopic Systems 121

5. Friedrich G. Helfferich
Ion Exchange Kinetics -- Evolution of a Theory 157

6. Lorenzo Liberti
Planning and Interpreting Kinetic Investigations 181

7. Reinhard W. Schlögl
Non-Equilibrium Thermodynamics - a General Framework to Describe Transport and Kinetics in Ion Exchange 207

8. Gerhard Klein
Column Design for Sorption Processes 213

9. Alirio E. Rodrigues
Dynamics of Ion Exchange Processes 259

10. Roberto Passino
Simplified Approach to Design of
Fixed Bed Adsorbers 313

11. Patrick Meares
Ion-Exchange Membranes 329

12. G. Dickel
The Nernst-Planck Equation in
Thermodynamic Transport Theories 367

13. A. Berg, T.S. Brun, A. Schmitt and K.S. Spiegler
Water and Salt Transport in
Two Cation-Exchange Membranes 395

ON THE SYNTHESIS OF ION-EXCHANGE RESINS

John R. Millar

(formerly Senior Research Scientist
Duolite International, Inc.)

Introduction:

Ion exchange is a phenomenon which has been around on Earth from very early times, perhaps something of the order of 10^9 years, and in the Galaxy probably for a trifle longer. Recognition of the phenomenon (in soil) is conventionally attributed to Way, and Thompson, about 130 years ago (1); though it may well have been used by Moses on the bitter waters of Marah several millenia before (2), this was on Divine recommendation (3) and inadequately documented.

All practical ion-exchange materials, inorganic and organic, are essentially insoluble matrix structures containing mobile ions capable of reversible exchange with ions of similar charge in a surrounding solution. The processes of such exchange, and their rates, form the subject of later sections of the present symposium.

The early materials, naturally occurring greensands or zeolites, were used for base exchange, i.e. the replacement of one cation in solution for the cation present in the ion-exchanger. Apart from water-softening (4), (replacement by sodium of the magnesium or calcium ions which formed precipitates with, and thus interfered with the detergent properties of, soap) zeolites were also used in sugar treatment to reduce molasses formation (5).

Synthetic alumino-silicates were later made, and naturally occurring polymers like coal (6) or cellulose (7) were chemically modified, but it was not until the mid-thirties of the present century that a completely synthetic organic material capable of base exchange was reported. The pioneering work was that of Adams and Holmes of the Chemical Research Laboratory at Teddington, near

London. They prepared crosslinked polymeric materials by the condensation of various polyhydric phenols with formaldehyde and demonstrated their capability for cation exchange on the (very weakly ionized) phenolic groups. The extension was soon made to the condensation of aromatic polyamines with formaldehyde to give materials on which strong mineral acids could be taken up, and in January 1935 their paper was published in the Transactions of the Society of Chemical Industry (8). From their presentation it is clear that they had foreseen almost every major application of ion-exchange; all that was missing were the practical materials to achieve them.

Adams and Holmes' first materials had partially-ionized functional groups, and thus both their ion-exchange capacity and their exchange rates were pH-dependent. Within ten years strongly-acidic (9) and -basic (10) functional groups had been incorporated in stepwise condensation polymers, and by 1950 these materials were being overtaken in the market by chain-growth (addition) polymer materials based on poly(styrene) (11, 12).

The decade of the fifties was one of tremendous activity in the ion-exchange field, particularly in the area of synthesis, and a staggering variety of materials became available for experiment. Initially, the emphasis was on the functional group, with crosslinked poly(styrene) being the first choice for modification. Later other addition polymers, such as the acrylics, received attention. Towards the end of the decade, as the result of the pioneering research by Pepper and his co-workers in Britain, and Gregor, Boyd, Bonner and others in the U.S., the modification of ion-exchange properties as a function of structure began to be understood (13). This paved the way for the introduction of macroporous matrices, in which structural modification was introduced at a supra-molecular level (14).

The sixties commenced with the feeling that ion-exchange was at the height of its powers, and nothing was impossible. The chelating resins dreamed of by Griessbach, Skogseid, and Mellor, and hopefully synthesized by so many workers in the fifties (15, 16), became a symbol of the sixties. Their exchange rates, slow because of their relatively low dissociation, were enhanced somewhat by the use of macroporosity, but as a result of the careful investigative work by a number of workers it became clear that the ion-specific resin, the "philosophers stone of the 50's", was indeed a myth. Many of the other cheerful predictions of the enthusiasts were weighed in the economic balance and found wanting, and, as a colleague of mine succinctly put it, the sixties could be regarded as ion-exchange's Decade of Disillusion (17). A great deal of very high quality work, both on synthesis and application, was published; nevertheless the overall tone was less optimistic than during the Golden Age.

The seventies were, perforce, more pragmatic. The limitations of ion-exchange were now recognized, and the economic requirements of any process were better understood, as a result of the global economic situation. In particular energy and the environment became important consideration, which led to the exploitation of the ion exchange process to the full with the tools already available. As clean water, and the penalties for throwing water away in a contaminated state, became more expensive, so did the efficient use of a low energy cost means of recycling it become more attractive. When the contaminant itself was at a low-level, yet intrinsically of some value, removal and concentration by ion exchange could go some way towards subsidizing environmental protection.

Today, ion-exchange is a mature technology, but it is as true today as it was twenty years ago that the development of ion exchange resins and their applications progresses more rapidly than the theory.

Despite a present (1982) slowdown in several of the industries which historically have used large amounts of ion-exchange resins, it is predicted that present usage will have increased by a factor of some 3½ times before the end of this century (see Table I). For this reason alone, the necessity for state-of-the-art reviews such as those of the Society of Chemical Industry (18), and the present meeting, is obvious.

Characteristics of a Practical Ion-Exchanger

The requirements for such a material have been categorized by Wheaton and Hatch as "quantitative performance, quality, and economics" (19). Naturally the ideal exchanger will combine high ion-exchange capacity, appropriate selectivity, rapid kinetics, and low price. It will probably surprise few readers to be told that this ideal does not exist, and that all practical materials are compromises.

Certain characteristics, however, can be expected. The resin (to use a conventional abbreviation) should be insoluble in, and unattacked by the medium in which the exchanging ions are dissolved. Its working life should be long enough to make the process cost-effective, and it should not give rise to foreign materials (other than the counter-ion) in the process stream. The resin should show the appropriate polarity, and selectivity. For the exchange of cations, for example, it should possess on its matrix negatively charged groups, and show under the the exchange conditions a distribution coefficient sufficiently in favor of the ingoing ion that the process is feasible. As to its performance, it should show a sufficiently high ion-exchange capacity on a volume basis, and an adequate rate of exchange under the process conditions, such that equilibrium is established without the ion-exchange installation being grossly oversized, nor the leakage of the ingoing ion unacceptably high.

Table I

Estimated World Ion-Exchange Production
(1967-2000)

(Figures are given in thousands of cubic meters)

AREA	YEAR						
	1967	1970	1974	1977	1981	1984	2000
N. AMERICA	35	40	78	50	61	90	250
W. EUROPE	16	19	35	30	32	45	100
E. EUROPE & U.S.S.R.	4	5	9	16	20	30	60
JAPAN & REST OF ASIA	3	4	10	7	17	35	50
OTHER	2	2	6	3	10	15	20
TOTAL	60	70	138	125	140	205	500

NOTE: THESE FIGURES ARE TAKEN FROM A NUMBER OF SOURCES AND OFFER ONLY GUESSTIMATES OF THE QUANTITIES INVOLVED, AND THEIR GEOGRAPHICAL DISTRIBUTION. SEE, E.G.,

REPORTS ON PROGRESS OF APPLIED CHEMISTRY, 55, 66-76, (1970)

REPORTS OF PROGRESS OF APPLIED CHEMISTRY, 59, 235-41, (1974)

CHEMICALS, 8-11, 12-18, (1981)

Table II

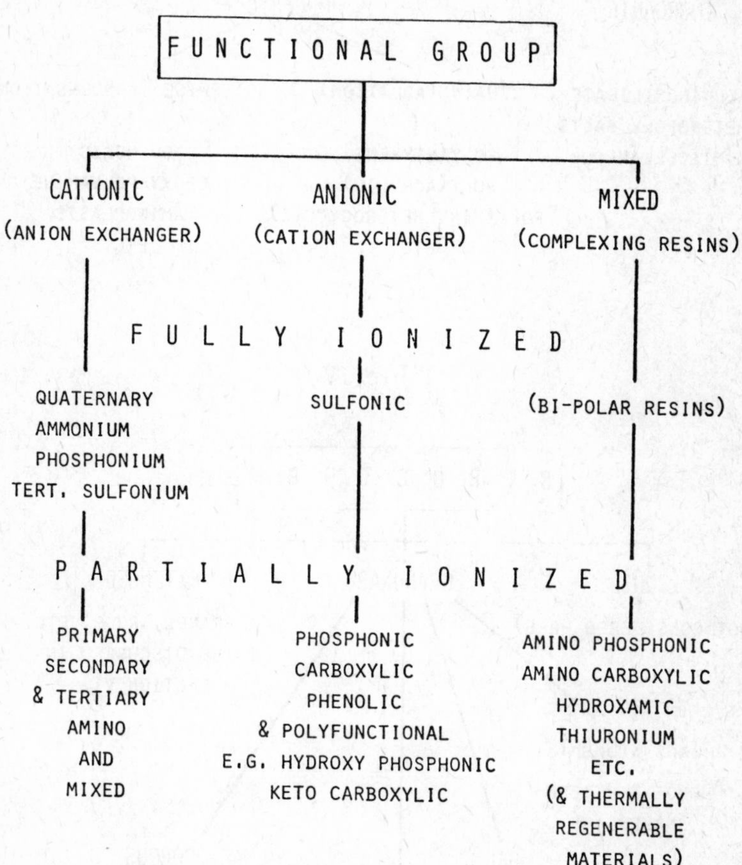
FUNCTIONAL GROUP
CATIONIC
(ANION EXCHANGER)
ANIONIC
(CATION EXCHANGER)
MIXED
(COMPLEXING RESINS)
FULLY IONIZED
QUATERNARY
AMMONIUM
PHOSPHONIUM
TERT. SULFONIUM
SULFONIC
(BI-POLAR RESINS)
PARTIALLY IONIZED
PRIMARY
SECONDARY
& TERTIARY
AMINO
AND
MIXED
PHOSPHONIC
CARBOXYLIC
PHENOLIC
& POLYFUNCTIONAL
E.G. HYDROXY PHOSPHONIC
KETO CARBOXYLIC
AMINO PHOSPHONIC
AMINO CARBOXYLIC
HYDROXAMIC
THIURONIUM
ETC.
(& THERMALLY
REGENERABLE
MATERIALS)

Table III

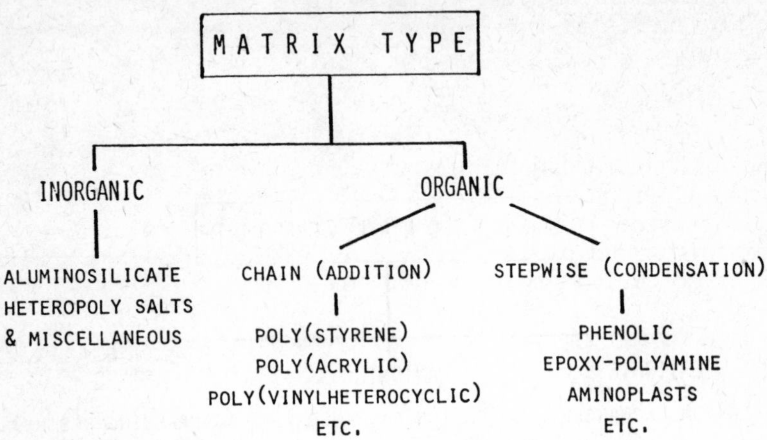

Table IV

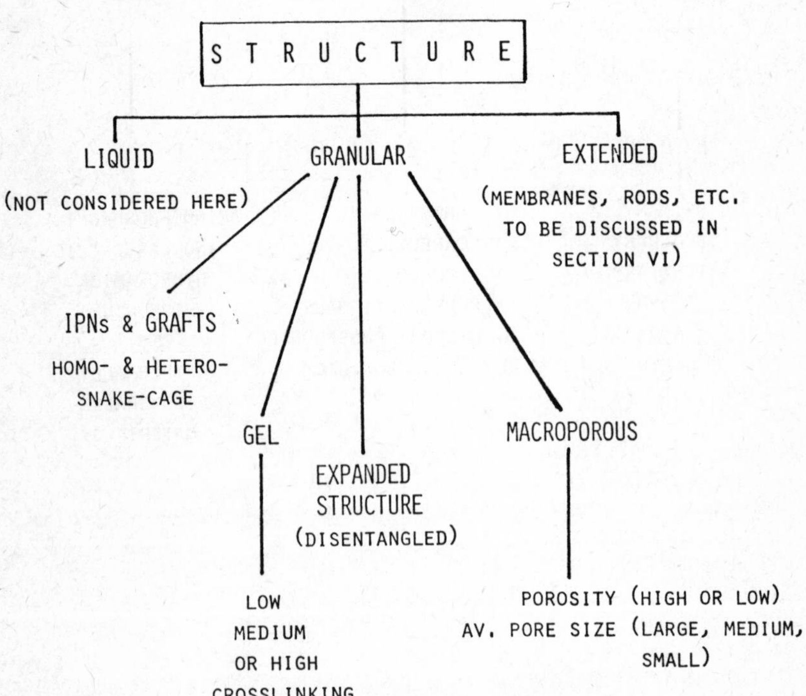

Economically, it is necessary that the original installation cost of the resin be relatively modest, its running costs moderate, and its replacement (when eventually required) a sufficiently small fraction of the total cost of the project, making any alternative more expensive.

Synthesis

Ion-exchangers can conveniently be categorized in several ways; by functional group, by matrix, and by structure (see accompanying Tables). In order to maintain a perspective on what appears to be an impossibly broad picture it is as well to recall that of the 140,000 or so cubic meters of synthetic organic ion-exchangers which were sold last year, well over 120,000 were poly(styrene)-based strong-acid or strong-base gel resins in bead form, substantially the same as those developed in the early fifties.

Chemically, the synthesis of organic polymer-based ion exchangers can follow two main routes (see Figure 1). Of these, the direct polymerization of functional monomers is the least common, for a number of reasons. The requirement of insolubility, with its usual involvement of a crosslinking monomer, makes the appropriate choice of the latter difficult, especially if suspension polymerization methods are to be used. The monomers themselves are specialty chemicals and correspondingly expensive. They also tend to be reactive in ways other than those to be employed in the synthesis, for example by hydrolysis or oxidation.

More often, a cheap commodity monomer is prepared in crosslinked form, and appropriate chemical reactions are performed to introduce the required functional group. This brings up a fundamental limitation on such syntheses, that incompleteness of reaction, or the occurrence of side reactions, cannot be avoided by the conventional purification techniques of organic chemistry. Thus, only clean-cut, essentially quantitative reactions are suitable for this route, and the most operationally economical synthetic scheme is also usually the cheapest in practice.

As can be seen from Table V, the selection of an organic matrix for anything other than a very exotic or high-priced material is essentially limited to addition polymers based on styrene, or acrylic monomers, and condensation products of phenols or amines with formaldehyde or epichlorhydrin. Some of the reactions which may be carried out on these matrices are illustrated in Figures 2, 3, and 4. None of these is new, and all the products have been described in earlier reviews (19, 20).

The phenol-formaldehyde matrices which are still in use are macroporous in nature, and are mainly used in specialized sorption processes (21), for which they are well suited. The styrene and

Figure 1

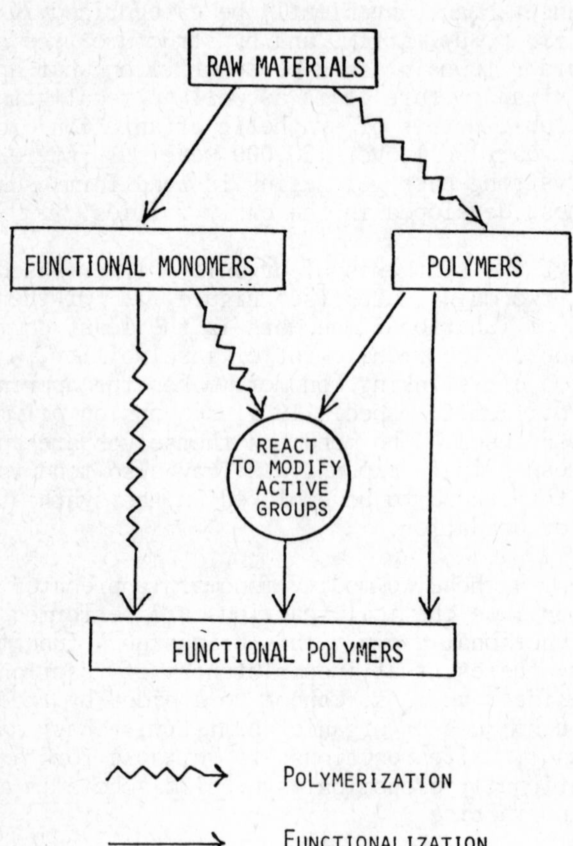
RAW MATERIALS
FUNCTIONAL MONOMERS
POLYMERS
REACT TO MODIFY ACTIVE GROUPS
FUNCTIONAL POLYMERS
POLYMERIZATION
FUNCTIONALIZATION

Table V

Comparative[a] Bulk Monomer Prices

(U.S.A., April 1982)

ADDITION		CONDENSATION	
Vinyl Chloride	1.0	Formaldehyde[b]	1.1
Styrene	1.5	Phenol	1.4
Acrylonitrile	1.8	Epichlorhydrin	1.5
Ethyl Acrylate	2.3	Melamine	2.0
Vinyl Toluene	2.6	Triethylenetetramine	2.5
Methacrylic Acid	3.1	Bisphenol A	2.6
Butyl Methacrylate	3.5	Tetraethylene Pentamine	3.0
TMPTMA[c]	6.4		
EGDMA[d]	6.8	m-Phenylene Diamine	7.6
Divinylbenzene[b]	9.0		
2-Vinyl Pyridine	13.2	Resorcinol	17.3

a) Ethylene = 1.0 (at 55¢ per kilo)

b) As 100% Active Product (44% HCHO; 55% DVB)

c) Trimethylolpropane Trimethacrylate (crosslinker)

d) Ethylene Glycol Dimethacrylate (crosslinker)

Figure 2

Polystyrene Resin
(Post Treatment)

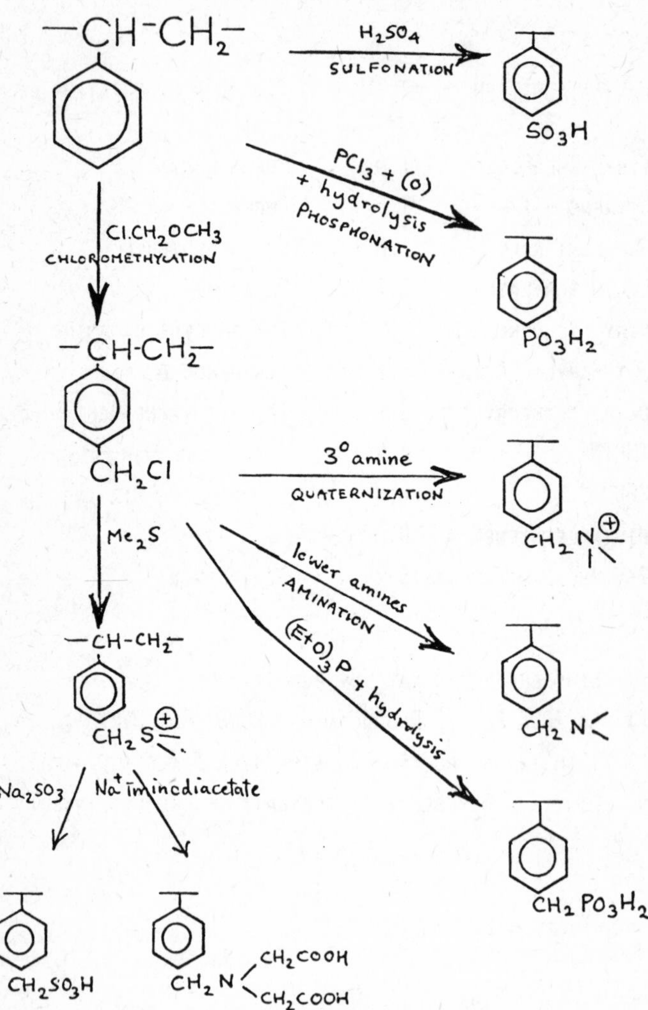

Figure 3

Phenol-Formaldehyde Resin
(Post Treatment)

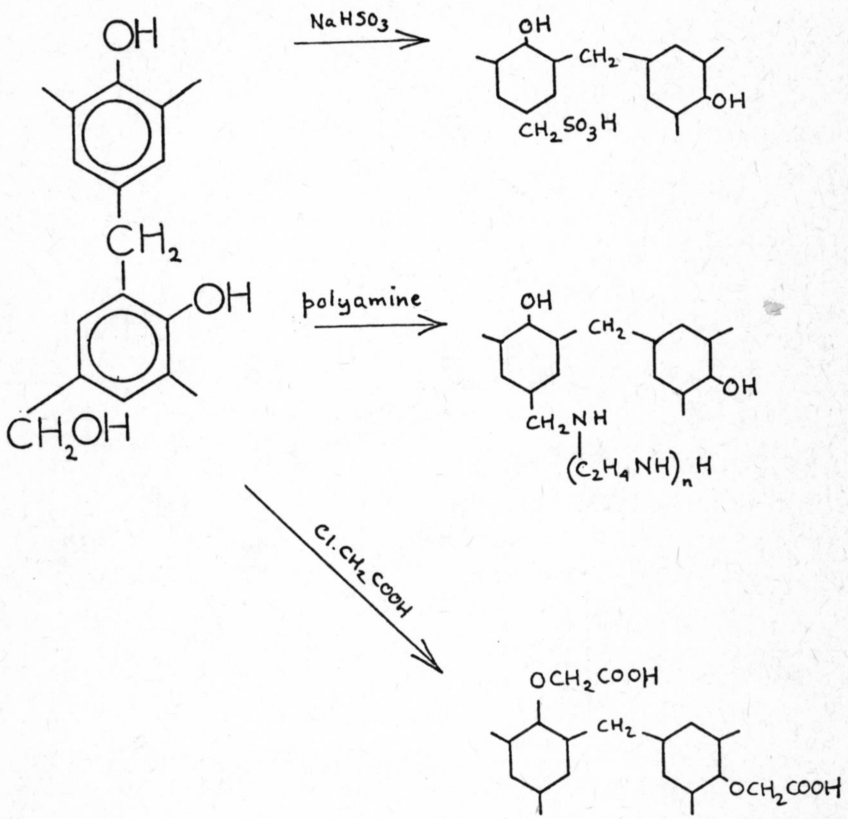

OH
CH2
OH
CH2OH
NaHSO3
OH
CH2
OH
CH2SO3H
polyamine
OH
CH2
OH
CH2NH
(C2H4NH)nH
Cl.CH2COOH
OCH2COOH
CH2
OCH2COOH

Figure 4

Acrylonitrile Resin
(Post Treatment)

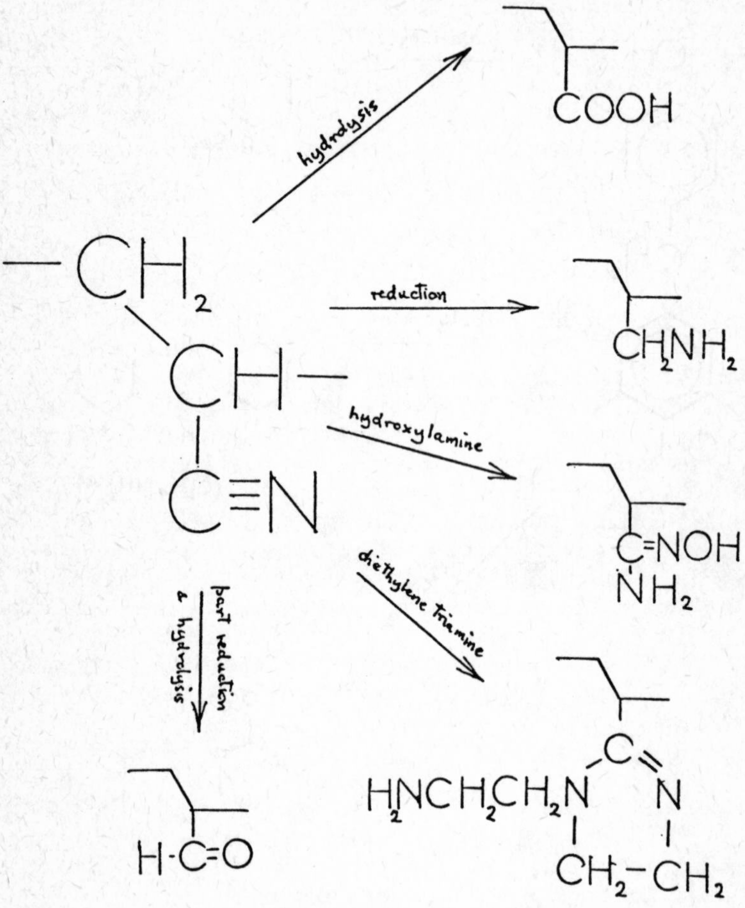

$$-CH_2-CH(C{\equiv}N)-$$

hydrolysis → $-CH_2-CH(COOH)-$

reduction → $-CH_2-CH(CH_2NH_2)-$

hydroxylamine → $-CH_2-CH(C(=NOH)NH_2)-$

diethylene triamine → $-CH_2-CH-$ substituted with a ring C that is double-bonded to N, with $N-CH_2-CH_2-N$ closing the ring; the other ring N bears $H_2NCH_2CH_2$

part reduction & hydrolysis → $-CH_2-CH(H{\cdot}C{=}O)-$

acrylic polymers are normally made by conventional suspension polymerization techniques. Here a mixture of the monomer with the chosen amount of crosslinker and a free-radical initiator is dispersed in droplets in an aqueous suspension medium. This medium contains stabilizers (inorganic and/or organic colloids) to ensure the formation of spherical beads of the desired size range; the modal bead size is dependent on the interfacial tension between monomer droplet and suspension medium, as well as on the rate of shear imparted to the droplets by stirring (Figure 5). The temperature of polymerization is usually in the range 60-85°C, and is a function of the monomers used and the particular initiator system.

Polymerization is often completed at a higher temperature to ensure maximum conversion and a strong product. Polymerization recipes are usually proprietary, since they play a large part in determining the mechanical properties of the final product (22, 23). The crosslinking of the matrix, and its macroporosity if present, also play a large part in determining the physical and some of the chemical properties. This will be discussed in a later section.

Sulfonation of Crosslinked Polystyrene

This is a typical aromatic electrophilic substitution reaction, in which the sulfonating agent attacks at the position of highest electron-availability. Since the styrene polymer is an alkyl-substituted benzene, the main products would be o- and p-substituted. Steric hindrance of the polymer backbone at the ortho position results in a mainly para-substituted product. A tetra-functionally polymerized divinylbenzene unit (a chemical cross-link) is unlikely to substitute under normal conditions, but the accompanying ethylstyrene units in the polymer may give disulfonated 1:2:4 trisubstituted benzene rings. Overall, the practical limit in normally crosslinked matrices is just under a (statistical) monosubstitution (24).

Special techniques have been used to prepare more highly substituted materials (25) but the maximum extent of reaction has been about 1.5 SO_3H groups per ring. This extra substitution gives no advantage in total capacity per unit volume, as the swelling of the product in water parallels the increased charge density of the dry product, and only second-order differences in behaviour are observed.

The sulfonation reaction results in elimination of H_2O from the reactants. In the presence of heavy metals which act as catalysts, and particularly where free SO_3 exists in the sulfonating medium, there is a tendency to eliminate a further molecule of water from a sulfonic acid group and a neighboring aromatic hydrogen atom, giving a sulfone cross-link. This also occurs to an increasing extent at higher temperatures when chlorosulfonic acid is used as the sulfonating reagent. Apart from this, which results in a loss of potential

Figure 5

Suspension Polymerization

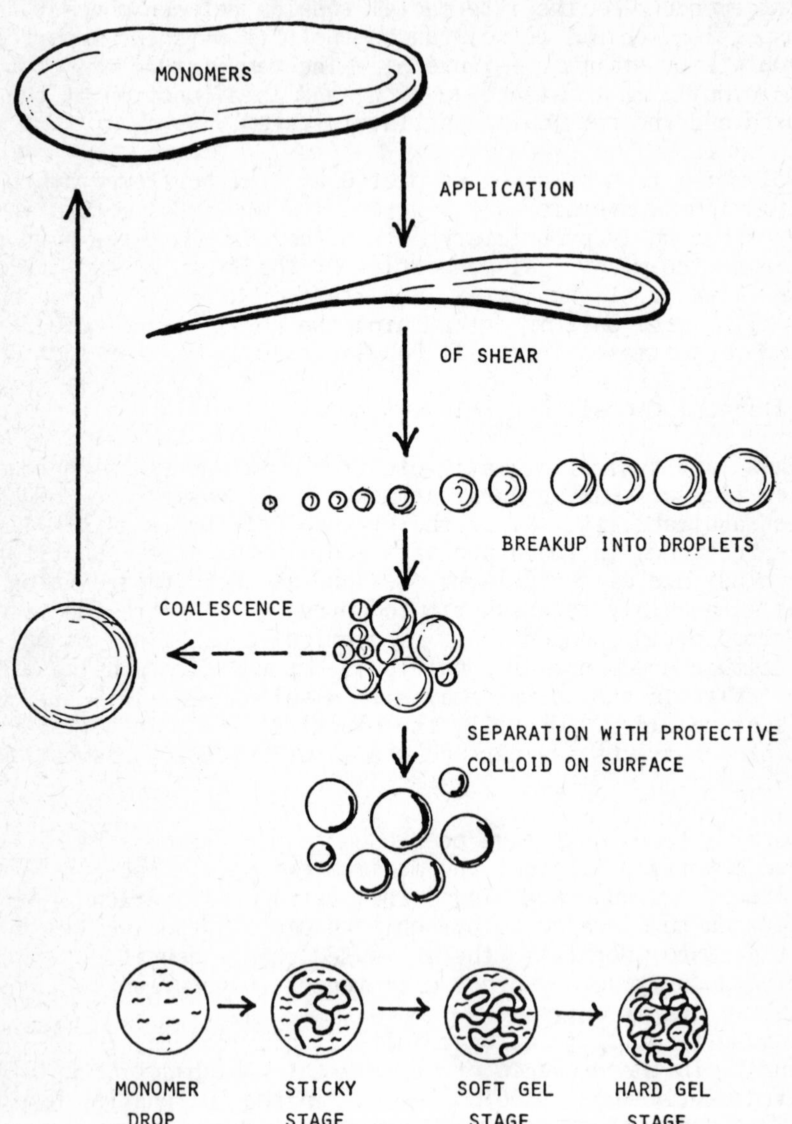

sulfonation sites, there are few side reactions, and the reaction is a smooth one. It is, however, diffusion-limited; as a result, even with a pre-swollen bead, the reaction is shell-progressive (26), and incomplete sulfonation tends to leave an unreacted and hence hydrophobic core. Stresses set up on hydration will tend to destroy such a bead, and most if not all commercial products are essentially fully-substituted.

One of the advantages of macroporous resins is that reactions of this sort proceed throughout the bead, often with an enhanced overall rate, without the necessity for solvent-swelling. Subsequent removal of the sulfonating agent, and washing, may be carried out under much less rigorously-controlled conditions without fragmentation of the product.

Chloromethylation of Crosslinked Polystyrene

This is also an electrophilic substitution reaction, and is catalyzed by Lewis acids, such as ferric, stannic, aluminum, zinc, titanium and similar chlorides, BF_3, etc. The active reagent depends on the actual catalyst used, but is generally the $^{+}CH_2C\ell$ carbonium ion. This may be derived from a number of reaction mixtures, but all share the unfortunate circumstance that a regular accompaniment of all these mixtures is the compound bis(chloromethyl) ether. This material is a double-ended alkylating agent, and a potent carcinogen (27). Since this was discovered, extreme precautions in the use and handling of bis-CME and mixtures containing it have been mandatory in the U.S., and essential anywhere.

A number of alternative routes to the final products previously made by reaction of the chloromethylated polymeric intermediate have been developed, as Dr. Kunin will mention later. However, modern techniques of material handling have enabled several major manufacturers to design plant which will handle the containment problems, and monitor potential exposure at the sub-ppb level.

The reaction, like sulfonation, yields o- and p-substituted derivatives, and as before the major product is the para-chloromethylstyrene unit. The $CH_2C\ell$ group deactivates the ring and there is little or no di(chloromethylation). A major side reaction is the catalyzed Friedel-Crafts alkylation of one aromatic ring by a substituted benzyl chloride unit nearby. Analogously to the sulfone formation previously mentioned, a methylene bridge which acts as a chemical crosslink is formed. The side reaction occurs readily when substitution goes beyond about 80%, according to Pepper, Paisley and Young (28). The order of effective crosslinking catalysis by some common Lewis acids is $H_2SO_4 > A\ell_2C\ell_6 \simeq FeC\ell_3 > SnC\ell_4 > ZnC\ell_2$.

Some measure of the extent of this crosslinking may be gained by a comparison of the water uptake per milliequivalent of capacity or specific water regain (29) in a trimethylamine-aminated quaternary resin with that of a sulfonated resin made from the same matrix. The specific water regains for early quaternary materials with essentially no methylene bridging were identical with those of the sulfonated resins of the same nominal divinylbenzene content (30). Although materials such as these, and their more recent acrylic-based analogs, will be dealt with in more detail by Dr. Kunin, the chemistry of amination itself requires a brief mention at this point.

Amination of Chloromethylated Polystyrene Matrices

The combination of a chloromethylated polystyrene matrix with an amine is a typical nucleophilic reaction. For the tertiary amines normally employed commercially in the synthesis of strong base resins – trimethylamine (or TMA) for Type I, dimethylethanolamine (or DMEA) for Type II – the reaction is smooth, mildly exothermic, quantitative and generally uncomplicated by side reactions. With a small excess of 40% aqueous reagent, TMA aminations are typically complete in one or two hours at ambient temperatures. DMEA aminations, using at least sufficient water to swell the final product, typically require four or five hours and higher temperatures (35-40°C) for complete conversion. The final quaternary ammonium resin remains in the chloride salt form, since the aqueous amine is a weaker base than the quaternary ammonium hydroxide. Thus only near-stoichiometric amounts of the tertiary amines are necessary.

The Type II resins, with a 2-hydroxyethyl group on the nitrogen, are somewhat weaker quaternaries than the Type I and in consequence show better regeneration efficiency with alkaline regenerants. They also show somewhat lower chemical stability, with a tendency to undergo a sigmatropic elimination of 2-hydroxyethyl and leaving a tertiary benzyl dimethylamine group (31). Such a group is protonated only under low pH conditions; its salt form is hydrolyzed to the free base by water.

Weak base resins, in which essentially all the groups are of this type, can be made by aminating with secondary or primary amines, or less usefully with ammonia itself. However, although the reaction is still smooth, the exotherm is quite marked, and there is the complication that more than one chloromethyl group may react with each amino nitrogen. This in turn may bridge two polymer chains, and give rise to added crosslinking. In the case of DMA amination of the chloromethylated polystyrene, the main reaction products are the free base form of the substituted benzyl dimethylamine, and the hydrochloride salt of dimethylamine, since in this instance the secondary amine is a stronger-base than the benzyl tertiary. Any double-ended reaction, however, results in dibenzyl dimethyl quaternary groups which, like those in the conventional quaternary resins, are in the chloride salt form. In the absence of any other means of fixing $HC\ell$, such as added alkali (32), the reaction requires at least 2 moles of amine per chloromethyl group.

Compared with the conventional benzyl trimethyl quaternary groups, these dibenzyl quaternaries are more labile, but like them they are ionized over almost the full range of pH. They therefore contribute to the swelling of the alkali-regenerated product, which in the free base form is otherwise virtually unionized and hence only marginally hydrophilic. At the same time, they therefore reduce the so-called "breathing volume" of the resin, the change in swelling on going from the salt to the free base form and back again. The observed increase in breathing volume and decrease in strongly-basic capacity of a number of weak base resins on long-time use or on prolonged cycling between alkali and acid solutions is related to this lower stability of the dibenzyl quaternaries, since in the salt form there is even less restraint on swelling when these crosslinks are broken.

Other Reactions Involving Chloromethylated Polystyrene Matrices

A number of other products beside anion exchange resins have been prepared from the chloromethylated styrene matrix. Some of these were shown in Figure 2. Since the chloromethylated matrix is essentially hydrophobic, it is difficult to introduce nucleophiles like sodium sulfite, or disodium iminodiacetate, but the conversion to a relatively unstable "onium" salt with dimethyl sulfide has been used to facilitate the reactions (33). This technique can be used to introduce primary or secondary amino groups without the cross-linking side reaction, but the primary group can also be introduced by amination with hexamethylenetetramine, followed by hydrolysis (34). As an alternative to the formation of methylene thiol groups by a similar technique in which thiourea is used and the thiuronium compound hydrolyzed (35), this "sulfonium" technique may be used with sodium hydrosulfide as nucleophile.

Recently, a variant of this technique for introducing incompatible reagents has been described, under the name of phase transfer catalysis (36). If the nucleophile is provided with a suitable organic cation the ion pair formed can diffuse into, and react readily with, the hydrophobic (organophilic) matrix. The classic example is the reaction of n-octyl chloride with cyanide ion. With the hydrated sodium ion as counterion, virtually no reaction occurs, but when, e.g., tetra-n-butylammonium ion is present reaction is reasonably rapid, smooth and essentially complete. Frechet and his co-workers have used this technique to prepare a number of modified crosslinked polystyrene resins (37), but so far no commercial resin is reported as having been made in this way.

Poly(vinylbenzene phosphonic acid) resins were originally made by direct phosphonylation of crosslinked polystyrene using the Clay-Kinnear-Perren reaction (38). Phosphonic resins are often made nowadays by the Arbuzov reaction of e.g. triethyl phosphite with the chloromethylated polystyrene matrix (39). The phosphonic ester so formed is hydrolyzed to the free acid. The Arbuzov procedure is preferred, since it leads to a cleaner and better defined product.

Phosphonic acid resins are, of course, chelating resins in their own right, giving 4-ring phosphonato chelates with e.g. uranyl ion. However, the incorporation of an amino group, as in the comparatively recent aminomethylene phosphonic acid resins (34), gives chelating resins with selectivities analogous to those of the iminodiacetic acid resins, but with a more highly dissociated primary cation-exchange group. The synthesis is from the chloromethylated polystyrene matrix via a primary amino derivative which is reacted with, e.g., a mixture of trioxymethylene & $PC\ell_3$. These resins have recently proven valuable for the removal of small amounts of Ca^{++} and Mg^{++} from saturated $NaC\ell$ brines used for electrolytic chlorine. A fuller discussion of chelating resins will be given in a subsequent paper by Dr. Hudson.

Carboxylic Resins

Although not normally classified as chelating resins, the carboxylic ion-exchangers do appear to form chelate complexes of the -ato type with the uranyl ion (40) and possibly also with some of the lanthanide elements (41). The main uses of carboxylic acid resins, however, derive from their incompletely-ionized state. Under neutral conditions, more than 99% of the carboxylic groups of a hydrogen-form carboxylic resin in contact with pure water exist as covalent -C(O)OH. Under alkaline conditions quite high capacity for cations is shown (for conventional addition-polymer carboxylic resins, the exchange capacity is of the order of 10 meq/g of hydrogen-form resin), but regeneration by low pH solutions, even in the presence of monovalent cations, is essentially quantitative.

The earliest carboxylic exchangers were the sulfonated coals (6) in which oxidation of the coal structure had occurred. However the first monofunctional carboxylic exchangers were derived from maleic anhydride (42) or various acrylic or α-substituted acrylic acid compounds (43) copolymerized with an appropriate crosslinking agent such as divinylbenzene, or a polyfunctional methacrylate ester. Copolymers of e.g. methacrylic acid with ethylene glycol dimethacrylate as a crosslinker, however, were found to autohydrolyze somewhat (with loss of crosslinking) on storage in the moist hydrogen form, and today most current commercial materials are divinylbenzene crosslinked.

There are two main types of carboxylic acid resin which differ in the chemistry of their functional group. Those deriving from methacrylic acid are appreciably less dissociated than those from acrylic acid. The apparent pKa of a resin is of course dependent on the extent of crosslinking (and substitution), but the thermodynamic pKa values (corrected for internal concentration and ionic strength) are almost one pK unit apart, with the acrylic acid resins at about 4.8 ± 0.1, and the methacrylic resins at about 5.7 ± 0.2 (44). The use of the acrylic acid resins permits the significant uptake of "bicarbonate alkalinity" in water treatment (45).

Several manufacturers have used acrylonitrile as the precursor for their acrylic acid resins (46), while others use acrylate esters, with or without some added methacrylic acid which facilitates the introduction of alkali for the final hydrolysis step. Provided hydrolysis is essentially complete, the products are equivalent, and the major differences between such materials lie in their internal structure, either in terms of gel crosslinking or in macroporosity. Nowadays, every major ion exchange manufacturer offers macroporous carboxylic resins, in addition to the older gel type. The implications of the internal structure of ion exchange resins for mass transfer and ion exchange kinetics will be discussed in the second part of this presentation.

REFERENCES

1a) Thompson, H.S. J. Roy. Agric. Soc. Engl., 11 (1850), 68.
 b) Way, J.T. ibid., 11 (1850) 313 & 13 (1852) 123.
2) Kunin, R. Ion Exchange Resins 2nd ed., p.1, (N.Y. Krieger 1972).
3) Anon. The Book of Exodus XV.23.
4) Gans, R. Jahrb. preuss. geol. Landesanstalt (Berlin), 26(1905). 179 & 27 (1906) 63.
5a) Gans, R. Ver. deut. Zucker-Ind. 57(1907) 206.
 b) idem, Ger. Pat. 174,097 (1915).
6a) --- Brit. Pats. 450,179 & 450,540 to N.N. Octrooien Mij. Activit.
 b) Liebknecht, O. Ger. Pat. 763,936.
 & idem, Brit. Pats. 450,574-5.
 c) Furness, R. Brit. Pat. 455,374 to Jos. Crosfield & Sons.
 d) --- Brit. Pat. 478,134 to Permutit A.G.
7) Kullgren, C. Svensk. Kem. Tid. 43(1931) 99.
8) Adams, B.A. & E.L. Holmes. J. Soc. Chem. Ind. 54(1935) 1T.
9a) Wassenegger, H. Ger. Pat. 733,679 (1943) to I.G. Farben.
 b) Holmes, E.L. and L.E. Holmes. Brit. Pat. 588,380 (1947) to The Permutit Co. Ltd.
10) Kressman, T.R.E. Brit. Pat. 660,130 (1949) to The Permutit Co. Ltd.
11) D'Alelio, G.F. U.S. Pat. 2,366,007 (1944) to General Electric Co.
12) McBurney, C.H. Brit. Pat. 654,706 (1948), U.S. Pats. 2,591,573-4 (1952) to Rohm & Haas Co.
13) Kitchener, J.A. Physical Chemistry of Ion Exchange Resins, Ch.2 of Modern Aspects of Electrochemistry, J. O'M. Bockris (Ed.) (London, Butterworths 1959).
14) Seidl, J., K. Dusek, J. Marinsky, & W. Heitz. Advances in Polymer Science 5 (1967) 113-213.
15) Millar, J.R. Chem. & Ind. (London) (1957) 606.
16) Nickless, G. and G.R. Marshall. Chromatog. Rev. 6 (1964) 154.
17) Anderson, R.E. personal communication.
18a) Ion Exchange & Its Applications (5-7 April 1954)(Soc. Chem. Ind. London 1955).
 b) Ion Exchange in the Process Industries (16-18 July 1969) (ibid. London 1970).
 c) The Theory & Practice of Ion Exchange (26-30 July 1976) (ibid. London 1977).
19) Wheaton, R.M. and M.J. Hatch. Ion Exchange vol.2, J. Marinsky (Ed.) pp. 191-234 (N.Y. Dekker 1969).
20) Helfferich, F.G. Ion Exchange pp. 26-71 (N.Y. McGraw-Hill 1962).
21) Abrams, I.M. Ind. & Eng. Chem. Prod. Res. & Dev. 14(1975)108.
22) Dales, M.J. U.S. Pat. 4,192,921 (1980) to Rohm & Haas Co.
23) Howell, T.J. U.S. Pat. 4,283,499 (1981) to Rohm & Haas Co.
24) Pepper, K.W. J. Appl. Chem. 1 (1951) 124.
25) Pirs, M. and D. Dolar. Vestnik Slov. Kemi. Drustva 13 (1966) 13.

26) Schmuckler, G. and S. Goldstein. Ion Exchange vol. 7, J. Marinsky (Ed.) pp. 1-28 (N.Y. Dekker 1977).

27) Van Duuren, B.L., B.M. Goldschmidt, C. Katz, L. Langseth, G. Mercado and A. Sivak. Arch. Environ. Health 16 (1968) 472-476.

28) Pepper, K.W., H.M. Paisley and M.A. Young. J. Chem. Soc. Lond. (1953) 4097.

29) Kressman, T.R.E. and J.R. Millar. Chem. & Ind. (London)(1961) 1833-4.

30) Wheaton, R.M. and W.C. Bauman. Ind. Eng. Chem. 43 (1951) 1092.

31) Baumann, E.W. J. Chem. Eng. Data 5 (1960) 376.

32) Seifert, H., W. Richter and M. Quaedvlieg. Brit. Pat. 1,133,920 (1968) to Bayer A.G.

33a) Hatch, M.J. U.S. Pat. 3,300,416 (1967) to Dow Chem. Co.

b) Hwa, J.C. U.S. Pats. 2,874,131 & 2,895,925 (1959) to Rohm and Haas Co.

34) Carbonel, J., P.D.A. Grammont, and J.E.E. Herbin. U.S. Pat. 4,002,564 (1977) to Diamond Shamrock Corporation.

35) Cerny, J. and O. Wichterle. J. Polymer Sci. 30 (1958) 505. c.f. Urquhart, G. et al. in Org. Synth. Coll. vol.3 p.363 (N.Y. John Wiley 1955).

36) Starks, C.M. & C. Liotta. Phase Transfer Catalysis — Principles & Techniques (N.Y. Academic Press 1978).

37a) Frechet, J-M.J., M.D. deSmet, and M.J. Farrall. J. Org. Chem. 44 (1979) 1774-9.

b) Frechet, J-M.J. Tetrahedron 37 (1981) 663-83.

38) Bregman, J.I. and Y. Murata. J. Am. Chem. Soc. 74 (1952) 1867.

39) McMaster, E.L. and W.K. Glesner. U.S. Pat. 2,980,721 (1961) to Dow Chem. Co.

40) Kressman, T.R.E. and J.R. Millar. Brit. Pat. 812,815 (1959) to The Permutit Co. Ltd.

41) Arnold, R. and L.B. Son Hing. J. Chem. Soc. (London) A (1967) 306-8.

42) D'Alelio, G. U.S. Pat. 2,340,110 (1944) to General Electric Co.

43) D'Alelio, G. U.S. Pat. 2,340,111 (1944) to General Electric Co.

44) Kunin, R. and S. Fisher. J. Phys. Chem. 66 (1962) 2275.

45) Anderson, R.E. Weak-Acid Cation Exchange Resins in Water-Treatment Tech. Publ. DS-FC7-404 (Redwood City, Diamond Shamrock Corp., 1977).

46a) Tavani, L. and M. Morini. U.S. Pat. 2,885,371 (1959) to Montecatini.

b) Carbonnel, J., P.D. Grammont and L.E. Werotte. U.S. Pat. 3,674,728 (1972) to Diamond Shamrock Corporation.

ON THE STRUCTURE OF ION-EXCHANGE RESINS

John R. Millar

(formerly Senior Research Scientist
Duolite International, Inc.)

Introduction

The ideal case of an ion-exchange resin is that of a homogeneous isotropically-swelling gel with a regular distribution of charged functional groups throughout the particle. For convenience, the particle geometry should be spherical, since this requires only one parameter for its definition, and of course each particle should have the same radius. The charged groups should each have a single counterion, and the Donnan membrane effect should permit virtually no invasion of external electrolyte. When considering the exchange of ions in such a system, both the original counterion and the ion for which it exchanges should be of the same size, hydration, and valency, and should have identical diffusion coefficients both in the external solution, and inside the homogeneous gel phase. Again for convenience we should wish the resin to show no preference for one ion rather than the other.

Life, however, is not like that! (47)

Since the validity of a number of the assumptions which are made in simplifying kinetic expressions are dependent on structure, the second part of this presentation is concerned with the non-ideality of real materials.

Crosslinking and Entanglement

To ensure that the resin is insoluble in the solutions it is designed to treat, the polymeric chains bearing the charged functional groups must form part of a three-dimensional network. This can be assured by polyfunctional condensation reactions, as in the

Table VI

Some Crosslinking Monomers

DIVINYLBENZENE
ETHYLENEGLYCOL DIMETHACRYLATE
TRIMETHYLOLPROPANE TRIMETHACRYLATE
DIVINYL KETONE
METHYLENE BIS-ACRYLAMIDE
DIVINYL PYRIDINE
ETHYLENE GLYCOL DIVINYL ETHER
DI-ISOPROPENYL BENZENE
TRIALLYLAMINE
TRIALLYLPHOSPHATE
TRIVINYLBENZENE
DIVINYLSULFONE
VINYL ACRYLATE
PENTAERYTHRITOL TRIACRYLATE
TETRAETHYLENE GLYCOL DIMETHACRYLATE

phenolic resins, by addition copolymerization of bifunctional monomers such as styrene or acrylates with tetra- or higher-functional comonomers (see Table VI), and by post-polymerization reactions of various types such as those already described.

The swelling of crosslinked networks has been studied for many years, and it is well-known that while crosslinks restrain swelling, the nature and extent of crosslinkage and the state of the network at its original formation all influence the effectiveness of this restraint (48).

Conventionally, a crosslinked polystyrene network has for long been represented as a two-dimensional structural formula of the type shown in Figure 6 (49). Topologically speaking this is a 4-connected plane net. Obviously a more reasonable representation would be a 4-connected three-dimensional net, corresponding to the tetrahedral carbon atoms, in fact a diamond-like structure. This however is still an idealized structure because the network is not created but grown, and while it is growing parts of the network can grow through and interpenetrate the earlier-formed parts. The end result is a multiply-connected system of considerable complexity.

This network entanglement can be deliberately enhanced, as in the interpenetrating network materials (50), known nowadays as IPN's (51), or reduced by polymerization in the presence of an inert, swelling, diluent (52). In conventional materials, viscoelastic measurements on unmodified styrene-divinylbenzene copolymers have given values for the effective intra-crosslink chain length which are much less than those calculated from the divinylbenzene content (53). Table VII gives a summary of the results, in which the theoretical values of M_c are calculated assuming that only 50% of the divinylbenzene is effective as a crosslinker, (in line with published estimates by Gordon & Roe (54), Haward & Simpson (55) and Dusek (56)). This is a very rapid increase, analogous to the increase with N of the number of entanglement isomers in linked catenanes with N rings (Table VIII).

A major effect of crosslinking (or its equivalent) on the functionalized matrix is of course on the internal average molality of the active groups. This is illustrated in Table IX which shows the total ion exchange capacity (in equivalents per litre of bed volume) of typical polystyrene sulfonate resins as a function of their divinylbenzene content. It affects a number of other properties which are related to mass transfer and kinetics of exchange, such as the diffusion coefficients of ions in the resin phase (57) and the extent of electrolyte invasion (58), as well as the equilibrium properties which are involved in determining the extent of exchange in limited-bath kinetic measurements (see Table X).

Figure 6

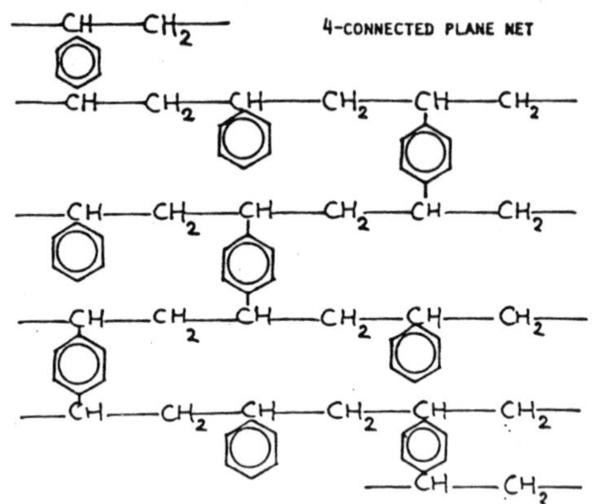

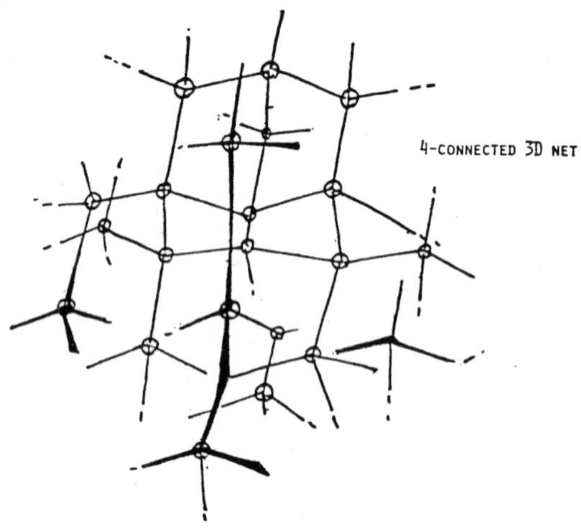

Table VII

Contribution of Entanglement to Crosslinking

% DVB	Intra-crosslink mol. wt. (M_c)		Extra Crosslinking (as % DVB) Due to Entanglement
	Theoretical	from Elasticity	
1	12,900	10,500	0.2
2	6,500	4,500	0.9
4	3,200	1,800	3.1
7	1,800	800	8.8
10	1,400	600	13.3
15	790	230	36.5

Table VIII

Entanglement Isomers of Linked Catenanes with N Rings

N	1	2	3	4	5	6
Entanglement Isomers	---	1	2	6	19	68

Table IX

Moisture Retention & Wet Volume Capacity

(Figures are for fully-substituted RSO_3H resins in the hydrogen-ion form)

DVB %	WR (G/G)	$\overline{\text{S.W.R.}}$ (G/MEQ)	WET VOLUME CAPACITY (EQUIV./LITER)
1	10.5	1.90	0.31
2	3.6	0.680	0.78
5	1.5	0.275	1.60
10	0.83	0.155	2.26
15	0.60	0.115	2.67
25	0.38	0.078	2.99

Table X

Physicochemical Properties of Crosslinked Sodium Polystyrene Sulfonate Resins as a Function of Divinylbenzene Content

DVB %	ELECTROLYTE INVASION $\tilde{\tilde{M}}/M$ @ 0.1 M (REF. 58)	SELECTIVITY COEFFS. K_H^{Na} @ $\bar{X}$ = 0.5 (REF. 60)	SELF-DIFFUSION COEFFS. $10^7\ \bar{D}_{Na}$ * (SEE FOOTNOTE)
1	0.027	~ 1	---
2	0.018	1.10	~ 22
5	0.008	1.55	~ 15
10	0.002	1.94	~ 8
15	0.0006	2.13	~ 3.5
25	VERY SMALL	2.40	~ 1

NOTE: THE ISOTOPIC SELF DIFFUSION COEFFICIENTS ARE ESTIMATES BASED ON THE WORK OF BOYD & SOLDANO, MODIFIED TO TAKE INTO ACCOUNT LATER CRITICISM (59), AND ADJUSTED ON THE BASIS OF INTER-DIFFUSION COEFFICIENTS REPORTED BY MILLAR ET AL. (60).

Some early work on polystyrene sulfonate resins with deliberately enhanced entanglement (50) indicated that not all these properties were affected to the same extent. Thus, the selectivity coefficients were markedly increased and ionic diffusion rates appeared to be higher than appropriate to a conventional resin with the reduced swelling achieved in the IPN by using a second interpenetrating network (60).

The reverse effect, reduction in entanglement for a given amount of crosslinker, is observed when the copolymerization occurs in the presence of an inert diluent which solvates the polymer, such as toluene. Here, for moderate amounts of crosslinker and not too exaggerated dilutions, electrolyte invasion was enhanced, selectivity coefficients reduced, and diffusion coefficients increased in comparison with those of a conventional material of the same divinylbenzene content (61). They were essentially similar to those of a resin of the same average internal molality.

This should be recalled when comparing the early work of Pepper and his colleagues, or that of Gregor and others, with later publications. In the manufacture of commercial DVB, the divinylbenzene isomers are obtained as a mixture with their precursors in the original feedstock, and are separated by fractional distillation. Over the years, the techniques have improved to the extent that normal commercial DVB today contains 55-60% of divinylbenzene isomers the remainder consisting almost entirely of the corresponding ethylstyrenes, while concentrates of up to 88% divinylbenzene isomers are regularly available. In the late forties, however, the crosslinker content was only 25-35% and the ethylstyrene content was similar. The remainder in those days was mainly the original diethylbenzene feedstock, in quite significant amounts. Thus the higher DVB content materials of those days were in fact solvent-modified, with the resultant physicochemical differences already described.

Pores and Heterogeneity

Arising from the work on solvent-modification, which had already been used empirically by Clarke for ion-exchange membranes (62), and by Mikes (63) for ion-exchange resins, it was found that with high crosslinker contents and substantial dilution the solvent-free matrix was opalescent in appearance and demonstrably porous in character (61). The existence in the gel phase of micropores has generally been accepted, and recent work by Freeman & Schram using inverse gel permeation chromatography (64) has given a measure of the micropore structure in the THF-swollen DVB-crosslinked polystyrene matrix. Their figures are interesting to compare with average pore diameters calculated by Grubhofer (65) for the water-swollen sulfonates (Table XI).

Table XI

Average Swollen Pore Diameters (Å)

DVB %	HC Matrix (Ref. 64)	Sulfonated Copolymer (Ref. 65)
1	77	343
2	54	151
4	37	58
8	14	30
16	13	15

In the macroporous copolymers there is a much wider range of pore sizes and the existence of pores of 10,000 Ångstroms diameter, or larger, has been demonstrated by electron-micrography and mercury porosimetry in a number of instances. These pores, unlike the gel porosity, persist in the unswollen state. They are usually continuously interconnected, and can accommodate liquids which would not normally be taken up by the resin, making it possible to carry out exchange in non-aqueous liquids at rates not achievable in conventional materials. The porosity also cushions the material against damage by osmotic shock, and permits extensive electrolyte invasion which facilitates exchange in partly-ionized resins. Most weakly-basic resins at the present time are synthesized on macroporous matrices.

The pore structure (and indeed the surrounding gel structure) of a macroporous resin is critically dependent on a number of factors. We have already seen that for solvating diluents crosslinker content and dilution are very important. The nature of the diluent itself is also a determining factor. In order to achieve a macroporous product, phase separation must occur during the formation of the matrix. If this is achieved by the use of a poorly-solvating or non-solvating diluent (66), phase separation occurs early on, and a high crosslinker content is not essential. In the ultimate case where the diluent is polymeric (67), phase separation is very early and the pore sizes achieved are large.

The characteristics of the three classic types of macroporous matrices are summed up in Table XII. An excellent early review on synthesis and structure of macroporous styrene-divinylbenzene copolymers is given by Seidl, Malinsky, Dusek and Heitz (68).

More recently, the need to tailor matrices to obtain optimum performance has occasioned a great deal of work, most of which is proprietary and published only in patent form, on the judicious blending of SOL and NONSOL porogens to achieve special results. The most recent and most comprehensive of the patent disclosures on the subject of mixed porogens (69) are those from Asahi (70).

The solvents used as components of the mixed porogen are divided into three groups:

i) solvents for all homopolymers of the monomers used
ii) non-solvents for all homopolymers of the monomers used
iii) solvents for some, non-solvents for other of the homopolymers of the monomers used.

(In the case of polyvinyl monomers, "good swellant" is understood to replace the term "solvent".)

Table XII

Characteristics of Macroporous Matrices

POROGEN TYPE	MINIMUM CROSSLINKER REQUIREMENT	TYPICAL SIZE OF MACROPORES	STRUCTURE OF GEL FRACTION
SOL (SOLVATING DILUENT)	REL. HIGH	~ 200 Å	DISENTANGLED < NORMAL CROSSLINKING
NONSOL (POORLY SOLVATING DILUENT)	LOW-MEDIUM	~ 500 Å	ENTANGLED > NORMAL CROSSLINKING
POL (POLYMERIC DILUENT)	VERY LOW	~ 1500 Å	NORMAL CROSSLINKING

The mixed porogen (here termed a modifier) is made up of a number of solvents chosen from among the three groups. Five types of modifier can be distinguished, the simplest of which is a single solvent from group iii. Addition of group ii solvents results in an increase in pore diameter, while addition of group i solvents reduces pore size. A mixture of group i and group ii solvents gives a wide variation in pore structure as the i:ii ratio is varied, but the finest control is obtained by using an appropriately chosen mixture of solvents from group iii.

Clearly much depends on the types of monomers being copolymerized. Unlike the classic DVB/Styrene copolymers which are of low polarity and whose components all have very similar solubility parameters (8.8-9.1), the choice in the case of polar co-polymers is often much more difficult.

Six examples of types of polar monomer mixes are given. In each case at least one crosslinker, one polar monomer, and one comonomer (whose polarity range is that from butadiene to 2-vinylpyridine) are involved. Table XIII lists the (approximate) values of solubility parameter for these, and the Examples concerned. For each of these Examples I-VI suitable solvents are given, and Table XIV indicates these solvents with approximate values of solubility parameter and the groups into which they fall for each Example.

The type of porosity can thus be adjusted for any given copolymer. However, the pore volume is dependent on the volume of modifier and in the Asahi patents the amount of modifier appropriate to a given crosslinking is defined, curiously, in terms of the square root of the % crosslinking. The scope of the patent is given, in the usual manner, by bracketing, giving first a range within which the invention will work, secondly a preferred range, and thirdly a recommended range. The percentage by weight of modifier (D) referred to the total weight of monomers is given as a function of the percentage by weight of crosslinking agent (X) on the same basis by the following relations:

typical	$7\sqrt{X} \leq D \leq 500\sqrt{X}$
preferred	$20\sqrt{X} \leq D \leq 200\sqrt{X}$
recommended	$34\sqrt{X} \leq D \leq 150\sqrt{X}$

It can be seen that there is a fairly wide scope for variation, and that at least in the S/DVB case the recommended limits cover most of the useable range for NONSOL-modified materials.

The large range of polarities and porosities which may be required, and the empirical nature of "solubility parameters" for insoluble polymers, make it difficult to sum this information up

Table XIII

Examples of Polar Monomer Mixtures from Asahi Patents

Crosslinker[a-f]	Polar Monomer	Comonomer	δ
divinylbenzene		1:3 butadiene[a,c,d]	8.3
			8.9
		ethylstyrene[a,b,c,d,e,f]	∼ 9
		styrene[a,b,c,d,e,f]	9.1
ethylene glycol dimethacrylate			9.3
		methyl methacrylate[a,c,d,e,f]	9.5
	methacrylonitrile[a]	methyl acrylate[a,c,f]	∼ 10
	methacrylonitrile[a]		10.7
	acrylonitrile[a]		12.5
divinylpyridine	N-vinylcarbazole[b]		∼ 13
	c——2-methyl-5-vinyl pyridine——a,d		13.1
	c——2-vinyl pyridine——a,d,f		13.5
	4-vinyl pyridine[d]		13.5
	2-vinylimidazole[d]		14
	2-methyl-N-vinyl imidazole		14
	acrylamide[e]		14.5
	N-vinylpyrrolidone[f]		15

a) Example III
b) Example IV
c) Example II
d) Example I
e) Example VI (ethylene bis-acrylamide can also be included among x-linkers)
f) Example V

Table XIV

Classification of Porogens or "Modifiers" from Asahi Patents

Solvent	Polarity	I	II	III	IV	V	VI	δ
hexanes	---	ii	ii	ii	ii	ii		∿ 7
didodecyl phthalate	small							7.2
heptanes	---	?	ii	ii	?	ii		7.5
octanes	---	ii	ii	ii	ii	ii		7-7.5
decanes	---	ii	?	ii	?	ii		7.5-8
dioctyl phthalate	small							7.9
cyclohexane	---	ii	ii	ii		ii		8
di-isopropyl ketone	+					iii		8
ethyl benzoate	+	iii	i	iii				8.2
ethyl propionate	+	iii	i				iii	8.4
methyl isobutyl ketone	+	iii		[iii]	iii			8.4
benzonitrile	(+)	i	i					8.4
butyl acetates	+	iii					iii	8.3-8.5
n-butyl propionate	+	iii		iii		iii		8.8
xylenes	---	iii	i	iii	i		iii	8.8
ethylbenzenes	---	iii		iii	i	iii	iii	8.8
toluene	---	iii	i	iii	i	iii	iii	8.9
methyl propionate	+			iii	iii			8.9
dibutyl adipate	+	iii		iii				8.9
ethyl acetate	+	iii		iii	iii	iii		9.1
benzene	---	iii	i	iii		iii	iii	9.2
chloroform	(+)				i			9.3
methyl ethyl ketone	+	iii	i	[iii]	iii			9.3
tetralin	---	iii			i	iii	iii	9.5
chlorobenzene	(+)	iii	i					9.5
tetrachloroethane	(+)	i			i			9.7
2-nitropropane	(+)	i	i	[iii]				9.9
cyclohexanone	+	i	i	[iii]				9.9
tetrahydrofuran	+				i			9.9
o-dichlorobenzene	(+)	iii	i		i			10
dioxan	+				i			10
diethyl phthalate	+	iii		iii				10
amyl alcohols	+	iii	iii	ii				10-11
octanols	+	iii	iii	ii	ii			∿ 10.5
n-butyronitrile	(+)			[iii]				10.5
methyl benzoate	+	iii	i	iii			iii	10.5
acetophenone	+	i						10.6
propionitrile	(+)			[iii]				10.7
pyridine	+							10.7
dimethyl phthalate	+	iii	i					10.7
butanols	+	iii	iii					10.5-11.5
hexanols	+	iii	iii	ii	ii			∿ 10.7
nitroethane	(+)	i		[iii]				11.1
cyclohexanol	+	iii	iii	ii	ii			11.4
benzyl alcohol	+	i						12.1
nitromethane	(+)							12.7

( ) = poor hydrogen bonding [ ] = in acrylonitrile copolymers only

more concisely.

Pore Size Measurements

Up until now we have used the concept of porosity as though everyone were familiar with it (which they usually are) and understood it (which they frequently do not). Dictionary definitions of porosity are usually cyclic and unhelpful. If one uses the term "porosity" to mean a local absence of matrix, the state of having holes in the medium, then it becomes as unquantifiable as "friability" or "happiness". What is often referred to as the porosity of a solid material is the percentage (or fraction) of voids within the porous solid. This can be defined by the apparent density, if the normal matrix density is known.

Thus, a silicate glass with density 2.8 g/mL when foamed to 80% porosity will give a product every mL of which will contain 0.2 mL or 0.56 g of glass, and 0.8 mL of air. The apparent density will be 0.56, and provided the porosity is closed, i.e. liquids cannot enter, the product will float on water.

If the pores are open and interconnected, the water will penetrate and the material will behave as a composite with an average density of 1.36, sinking in the water. To calculate the apparent density, out of water, you would have to take the external volume, and the net weight, and correct for the amount of water held internally.

With macroporous organic sorbents or ion exchange resins, the situation is complicated by the swelling of the matrix. Unlike the foamed glass in the previous example, where a hole is a hole is a hole, a hole in an elastic matrix will increase in size as its boundaries increase. If the swelling of the polymer is isotropic, a volume change of x% in the polymer implies a corresponding volume change in the pore, and usually this is so for the materials we have been considering.

Unfortunately, electron micrographic, mercury porosimetric and BET/N_2 adsorption techniques, which have found widespread application to the measurement of pore-size distribution in macroporous organic polymers, require dry out-gassed samples of the polymer in question. In consequence neither give an adequate description of the porosity of the sample under normal operating conditions. Indeed, unless appropriate precautions are taken in the preparation of the dry out-gassed samples, the figures obtained even on these are open to considerable criticism (71), and may indeed be meaningless.

One method which permits measurement of surface areas in swollen resins is the sorption of p-nitrophenol (PNP) from aqueous or organic solution, introduced some twenty years ago by Giles and Nakhwa (72), and used by Fang and Golownia on macroporous organic polymer matrices (73). Some comparisons of BET/N_2 and PNP figures were given in a recent review (74). The method is apparently little used, probably because it is not easy to automate.

For most people, seeing is believing, and there is something very convincing about an electron micrograph. Since, most ion exchange chemists are aware of the possibility of artefacts, such pictures as have been published have survived extensive review. However, uncertainties arise in mercury porosimetry, and in surface area and pore size distribution measurements by adsorption methods, which are not so well appreciated outside the specialized surface chemistry laboratory. Taking the BET technique first, the most obvious is that the surface area is measured in terms of an amount of sorbate, and the amount of sorbate per unit area is not accurately known. The best known of the permanent gases used in the BET technique is nitrogen, with a molar cross-sectional area of 16.2 $Å^2$, and relatively unambiguous isotherms on most surfaces. For many of the other sorbates, the apparent cross-sectional area A_m is dependent on the system or the sorbent, and wherever possible calibration should be made against a N_2 value. Krypton is a case in point – values from 17-22 have been used at various times in various circumstances, giving a possible variation in surface of ± 13%. Similar problems arise in sorption from solution; p-nitrophenol has an A_m value which ranges from about 15 on metal oxides in hydrocarbon solvents to 53 in polar solvents on carbon.

Secondly, the sorbate will absorb only when it can approach the surface, and it will desorb only if it has first adsorbed. Consequently, the probing of micropores with larger sorbates is pointless, while restricted porosity or the so-called ink-bottle pores will give rise to hysteresis effects on the adsorption curve which require interpretation. In any case, there is evidence that the Kelvin equation, which is the basis for pore size calculations based on sorption, becomes invalid at relative pressures below p/p_o values of $\sim$ 0.2, i.e in the 20 Å range, despite the lack of steric hindrance to the diffusion of N_2 (with a rotational diameter of 3.2 Å).

With the mercury intrusion technique, the interfacial tension between the mercury and the polymer surface is the factor which controls the penetration of mercury at a given pressure. The pressure required (Table XV) is dependent on the radius of the capillary, and under normal conditions (unwetted surface, clean mercury, open pore structure) there is a natural limitation of around 20 Å radius, at which the pressure required is approximately 3600 Kg. cm^{-2}, about 3,500 atmospheres). Pressures as low as 2000 Kg. cm^{-2} can result in irreversible damage to the pore structure, and if the newer scanning porosimeters are used there may be uncertainty about what is actually

Table XV

Range of Pore Size Determination

PORE DIA Å	POROSIMETER PRESSURE (KG. CM^{-2})	BET/N_2 (p/p_o)
20	--	0.16
50	2742	0.57
100	1300	0.78
200	598	0.89
500	253	0.95
1000	112	--
5000	26.4	--
10000	12.7	--

being measured. The ink-bottle pores in a structure, again, give rise to problems. What is actually measured is a volume penetrated, and if some pore volume of 1000 Å diameter is only accessible through a 100 Å orifice, it will register only when 1300 Kg. cm^{-2} pressure is applied, and will be associated therefore with a 100 Å diameter pore. It will also not empty when the pressure is reduced, and the variation in sequential volume/pressure curves shown on a single sample of Diaion HP 20 (Figure 7) can reflect either retained mercury in ink-bottle pores or small pore wall breakdown.

Surface areas calculated from mercury porosimetry are dependent on assumptions of pore geometry, and usually cover the range of pore size from about 50-100,000 Å diameter, so that pores below 50 Å which can contribute appreciably to internal surface are neglected. By that same token, pore volumes derived from the BET adsorption measurements are indirectly estimated also, and cover a range of pore size from about 25 Å diameter to about 250 Å above which there is no condensation. Thus the very large macropores which can contribute significantly to the total volume in an actual pore size distribution can be omitted from consideration. Clearly, even for the limited objective of obtaining as full as possible pore size distributions on dry, unswollen materials, both types of measurement are essential, and they should only be combined with circumspection and with consistent geometrical assumptions.

From the foregoing discussion, it will be evident that parameters like "mean effective pore diameter" are essentially meaningless, even as a means of empirical comparision, if the actual distributions are not already known to be similar. Similar remarks apply to integrated values like "total surface area" or "pore volume", even in the rare instances when the measurement techniques are clearly defined.

Effects of Structural Heterogeneity on Mass Transfer and Kinetics

While this will be dealt with in later sections of this Advanced Study Institute, one or two points should be made in the light of the ideas presented in this section.

It has already been pointed out that electrolyte invasion into the macropore regions of macroporous resins is extensive. There is however a significant difference between invasion in SOL- and NONSOL modified materials resulting from the difference in detailed structure, the ratio of internal to external electrolyte concentration being higher for the NONSOL-modified materials of comparable average internal molality (66).

It may be assumed that because of the existence of macropores, the thickness of the unconvected Nernst film, which is (under low concentration conditions) a limiting factor in the so-called film-

Figure 7. Repeated Hg Scanning Porosimetry on Diaion HP-20

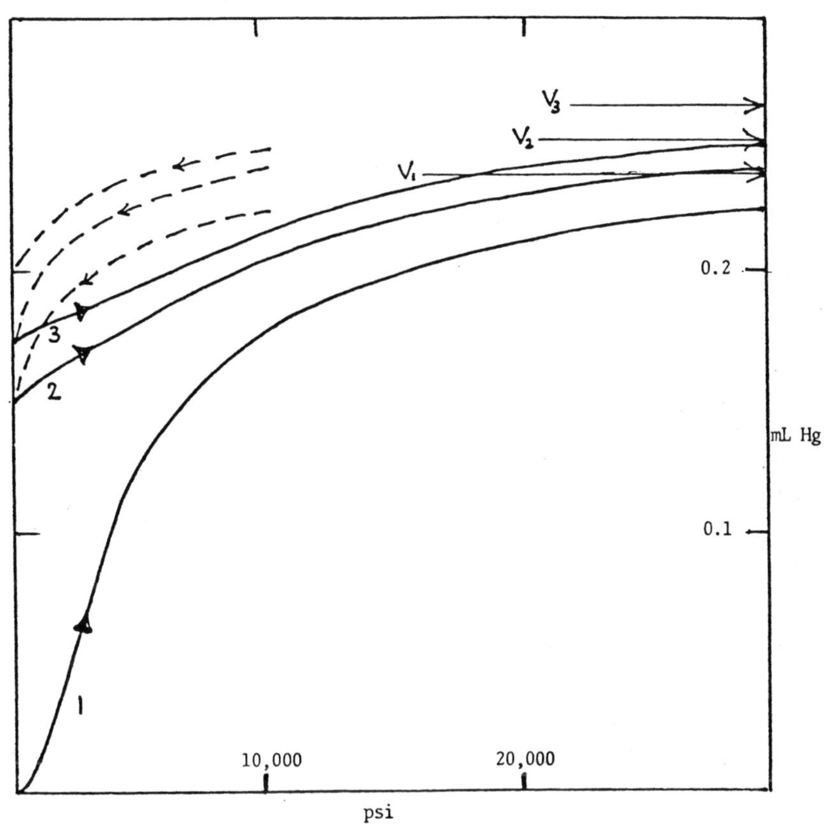

(Sample size, 0.2189 g)

V_1 : 1.10 mL/g

V_2 : 1.19 mL/g

V_3 : 1.23 mL/g

diffusion kinetics (75), is somewhat higher than for corresponding conditions on a conventional gel-type exchanger. Since the stirred film thicknesses are of the order of microns, however, this is unlikely to be more than a second-order effect. The effect in unstirred systems, such as the immobilized biocatalysts discussed by Kasche and Kuhlmann (76), is negligible.

The effect of macroporosity on the titration curves of ion-exchange resins is on the apparent dissociation constant only, and the thermodynamic dissociation constant of the group is essentially unaffected. Some second-order differences may exist as a result of the distribution of charged groups which is demonstrably more heterogeneous than in conventional gel resins.

One area where clear-cut differences have been reported is in the diffusion coefficients within the resin for relatively large ions (61b). The physical constraints on self-diffusion within an exchanger involve excluded volume effects, tortuosity and anisotropy of pore structure. The presence of a macropore system in the resin phase will tend to reduce such constraints, and although the measured diffusion coefficients were indeed higher, the major effect was on the Arrhenius activation energies for exchange interdiffusion, which were essentially those for diffusion in free aqueous solution. However, since the large ions in this case were appreciably excluded from the gel region, this may be no more than a reflection of this fact, an artefact of the kinetic theory employed (77).

Much more work is needed to obtain the data which will enable us to understand and quantify the kinetics of ion exchange in such complex systems. In the meantime, let us be on our guard against taking the mathematical consequences of our plausible assumptions as a reflection of what some of us like to regard as the "real world".

REFERENCES

47) Murphy's Law. See The Scientist Speculates: an Anthology of Partly-Baked Ideas I.J. Good (ed) p. 212 (N.Y. Putnam 1965).
48a) Flory, P.J. Principles of Polymer Chemistry (N.Y. Cornell U. Press 1953).
b) Mark, J.E. J. Amer. Chem. Soc. 92 (1970) 7252-57.
49) Houwink, R. (ed.) Elastomers & Plastomers (Amsterdam Elsevier 1949).
50) Millar, J.R. J. Chem. Soc. Lond. (1960) 1311-17.
51) Sperling, L.H. Interpenetrating Polymer Networks and Related Materials (N.Y. Plenum Press 1981).
52) Millar, J.R., D.G. Smith, W.E. Marr & T.R.E. Kressman. J. Chem. Soc. Lond. (1963) 218-25.
53) Dusek, K. Coll. Czech. Chem. Commun. 27 (1962) 2841-53.
54) Gordon, M. and R-J Roe. J. Polymer Sci. 21 (1956) 27-90.
55) Haward, R.N. & W. Simpson. J. Polymer Sci. 18 (1955) 440.
56) Dusek, K. Coll. Czech. Chem. Commun. 32 (1967) 1182-89.
57) Boyd, G.E. and B.A. Soldano. J. Amer. Chem. Soc. 75 (1954) 6091-99.
58) Attridge, C.J. and J.R. Millar. J. Chem. Soc. Lond. Supp. 2 (1965) 6053-60.
59) Kitchener, J.A. in Modern Aspects of Electrochemistry 2. J.O'M. Bockris (ed) p. 146 (London Butterworths Scientific Publications 1959).
60) Millar, J.R. D.G. Smith and W.E. Marr. J. Chem. Soc. Lond. (1962) 1789-94.
61a) Millar, J.R., D.G. Smith, W.E. Marr and T.R.E. Kressman, J. Chem. Soc. Lond. (1963) 218-25.
b) Idem, ibid. (1963) 2779-84.
c) Idem, ibid. (1964) 2740-46.
62) Clarke, J.T. U.S. Pats. 2,730,768 (1951)
2,731,408 (1956)
2,731,411 (1956) to Ionics Inc..
63a) Mikes, J. Magyar Kem. Lapja 6 (1964) 303-8.
b) Idem J. Polymer Sci. 27 (1958) 587 30 (1958) 615-623.
c) Vandor, J and J. Mikes. Magyar Szab. VA-348 (1951).
64) Freeman, D.H. and S.B. Schram. Anal. Chem. 53 (1981) 1235-8.
65) Grubhofer, N. Makromol. Chem. 30 (1959) 96-108.
66) Millar, J.R., D.G. Smith, W.E. Marr, and T.R.E. Kressman. J. Chem. Soc. Lond. (1965) 304-10.
67) Roubinek, L. and A.G. Wilson. U.S. Pat. 3,509,078 (1970).
68) Seidl, J., J. Malinsky, K. Dusek and W. Heitz. Adv. in Polymer Science 5 (1967) 113-213.
69a) Walters, H.A. U.S. Pat. 3,275,548 (1966) to Dow Chem. Co.
b) Gustavson, R.L. U.S. Pat. 3,531,463 (1970) to Rohm & Haas Co.
c) Haupke, W., R. Hauptmann, K. Stelzner and E. Roesel. Bit. Pat. 1,274,361 (1972) to VEB Chemie Kombinat Bitterfeld.
d) Jung, E., V. Vossius, and J. Schindewahn, D.B.R. Offen. 2,121,448 (1972) to Akzo N.Y.

70a) Ikeda, A., K. Imamura, T. Miyaka and K. Takeda. U.S. Pat. 4,093,570 (1978) to Asahi Kasei Kogyo K.K.

b) Miyake, T., T. Kunihiko, A. Ikeda and K. Imamura. U.S. Pat. 4,154,917 (1979) to Asahi Kasei Kogyo K.K.

c) Abe, T., A. Ikeda, and T. Sakurai. U.S. Pat. 4,202,775 (1980) to Asahi Kasei Kogyo K.K.

71) Hilgen, H., G.J. de Jong, and W.L. Sederel. J. Appl. Polymer Sci. 19 (1975) 2647-54.

72) Giles, C.H. and J.N. Nakhwa. J. Appl. Chem. 12 (1962) 266.

73) Fang, F.T. and R.F. Golownia. A.C.S. Polymer Preprints 8 (1967) 374.

74) Millar, J.R. J. Polymer Sci. (Symposium) 68 (1980) 167-77.

75) Boyd, G.E., A.W. Adamson and L.S. Myers, Jr. J. Amer. Chem. Soc. 69 (1947) 2836.

76) Kasche, V. and G. Kuhlmann. Enzym. Microb. Technol. 2 (1980) 309-12.

77) Reichenberg, D. J. Amer. Chem. Soc. 75 (1953) 589.

THE NATURE AND PROPERTIES OF ACRYLIC ANION EXCHANGE RESINS

Robert Kunin
Yardley, Pennsylvania USA

INTRODUCTION

The discussion of a topic such as the acrylic-based ion exchange resins before a group such as the NATO/CNR course requires some background indoctrination in order that a proper perspective be developed and maintained.

Ion exchange is a unique technology since it occupies a special place in at least three other scientific disciplines i.e., polymer chemistry, polyelectrolytes and adsorption. Interestingly, those interested in the synthesis of ion exchange resins focus their attention on the polymer chemistry discipline. Those involved in the theory of the ion exchange process itself appear to focus their attention on the relationships with the general theory of polyelectrolytes. Finally, those primarily interested in the use or application of ion exchange are primarily attracted to its role in the general area of adsorption technology which encompasses the technology of ion exchange.

In studying an area as complex as ion exchange, most "students" find it of importance to develop some model and implant this model in their minds. The model need not be an accurate one as long as it serves a useful purpose; i.e., enabling one to predict or explain various phenomena associated with ion exchange systems. A model that I have used for years is based upon a young man preparing his first meal of pasta or spaghetti. As an inexperienced chef, our young man makes a series of errors. His first error is not having sufficient water. His second error is the omission of the fast cold water rinse to remove soluble starches. His final error involves not having prepared the sauce on time and hence his cooked spaghetti has cooled and congealed on the platter. At this point, the mass of spaghetti can be picked up by our young man, platter and all, by merely holding on to the uppermost strands. This "blob" or mass of improperly prepared pasta is a replica of the internal structure of half of a crosslinked copolymer bead or ion exchange resin intermediate or ion exchange resin itself.

If we now focus our attention on this structure, we will find randomly intertwinned strands of spaghetti glued together by the soluble starch at various points of contact. In essence, we are looking into the internal structure of a crosslined copolymer bead. The spaghetti strands are the polymer chains and the starch-glue points binding together the spaghetti strands represent the points of crosslinking which form a three dimensional polymer structure.

If we focus our attention more closely, it becomes evident that the density of spaghetti strands is not uniform nor is the density of starch glue points. There are dense areas of spaghetti strands and also areas in which there are but a few strands glued together. In other words, we see that there is a heterogeneity of agglomeration of spaghetti strands. In other words, our copolymer model exhibits a heterogeneity of polymer chains and crosslinks.

To complete our model of an ion exchange resin particle we merely have to attach fixed or immobile ionic groups with an equivalent number of mobile ionic moieties of opposite change to satisfy the Law of Electroneutrality. For the model of the cation exchange resin, the fixed ionic groups are anionic structures such as the sulfonic acid ($-SO_3^{(-)}$) or carboxylic acid ($-COO^{(-)}$) groups and the mobile charges are such ions as the sodium (Na^+) and hydrogen (H^+) ions. The model of the anion exchange resin has the fixed or immobile group of positively charged group such as the quaternary ammonium ($-CH_2\overset{CH_3}{\underset{CH_3}{N}}CH_3^{(+)}$) structure with the mobile charges being the hydroxide (OH^-), chloride (Cl^-), etc. anions.

It is important to realize that this model pictures an unequal distribution of polymer chains, crosslinks, fixed and mobile ionic charges, and even an unequal distribution of water of hydration. As homogeneous as an ion exchange resin bead may appear to be visually, it is definitely a heterogeneous mass. This degree of heterogeneity will vary, of course, with the chemistry of the copolymerization. Evidence for this heterogeneity is quite substantial. The nature of the polymerization and crosslinking reactions is such that heterogeneity must be inferred. Rarely is the rate of polymerization equal to the rate of crosslinking. Hence, the ratio of monomer (styrene, for example) to crosslinking agent (divinylbenzene) varies due to the course of the copolymerization reaction. Penetration of large and colored ionic species into various ion exchange resin beads confirm this heterogeneous state.

The situation is even more dramatic if we realize that there also exists a particle to particle degree of heterogeneity particularly in case of vinyl polymerizations where the particles are prepared by suspension or dispersion polymerization which results in a mass of copolymer beads having a statistical distribution of particles of varying size. Since there is a large exotherm during vinyl polymerization, this heat must be dissipated in order to avoid any thermal degradation. The water of the dispersion medium serves to remove this heat. However, the heat formed during the polymerization of a small bead is more readily dissipated than in the case of a larger bead. Since the polymerization and crosslinking reaction rates are temperature dependent, but not equally so, one must expect some basic differences amongst beads of a single batch, whether produced in the laboratory or commercially. Finally, these differences also reflect differences within single beads or between beads during the subsequent sulfonation, chloromethylation, ammination, etc. reactions.

The foregoing model merely attests to the heterogeneous state of ion exchange resins which also reflects the disordered state of these substances. It is interesting to compare these structures with the crystalline, homogeneous and ordered zeolite minerals. Figures 1 and 2 compare these structures in a simplistic manner.

The foregoing model is commonly referred to as gel-type ion exchange resins. If we now consider the case of the macroreticular structures, the case becomes somewhat more complicatd because of the added heterogeneity due to the macroreticular pores (Figures 3 and 4).

CURRENT STATUS OF SYNTHESIS

In many respects, the current status of the synthesis of ion exchange resins is one of refinement. The past two decades have not witnessed any profound changes in basic chemistry. The polymer chemist at the present time has the ability to introduce almost any known functional group onto many crosslinked copolymer structures. Much of the recent effort has been directed towards the following:

1. Improvement of physical stability.
2. Production of particle uniformity of small and large particles in hugh yields.
3. Development of unique pore structures such as the macroreticular structures.

THE MACRORETICULAR STRUCTURE

The physical structures just described refer to what is commonly called gel-type and macroreticular structures. The author shall not dwell upon the semantic problems associated with the terms macroreticular and macroporous except to discuss the term porosity as it applies to ion exchange resins. As applied to ion exchange resins, the author many years ago used the term to describe that portion of the swollen or hydrated structure occupied by water. In essence, the pores were the spaces or distances between the polymer chains and crosslinks. On a statistical basis, one could calculate the pore sizes of such pores which could be substantiated using large organic ionic species (1). Of course, the sizes of these pores would vary with the degree of crosslinking as well as the ionic strength or osmotic pressure of the solution in which the ion exchange resin was immersed. If the ion exchange resin was dehydrated, these pores would actually disappear.

To distinguish the pores of the gel-type ion exchange resin from the extra-gelular pores of the macroreticular structure, the author and his associates coined the term, macroreticular. Based upon size, the "pores" of the gel-type ion exchange resins are much smaller than those of the macroreticular structure but encompass the pore size range that has been called micro -, transitional -, and macro-pores by those in the general fields of colloid chemistry and heterogeneous catalysis. To avoid this dilema, the term macroreticular was chosen. The development of these unique structures has given a new dimension to ion exchange resins and adsorption technology. However, it gives rise to an added complication to the basic theme of this conference since we now have to consider true pore diffusion as well as particle and film diffusional process in our kinetic models.

THE STYRENIC STRUCTURES

The technology of ion exchange was truly launced into a new era with the development of the anion exchange resins. Until this development occurred, the usefulness of ion exchange was quite limited. Although the polymer chemists found several routes to the creation of such ion exchangers, the major breakthrough occurred with the discovery that the styrene-divinylbenzene crosslinked copolymer could be chloromethylated and then aminated to form many new and useful products. This chloromethylated intermediate is a most useful and important intermediate in ion exchange resins. It can be prepared from both gel-type and macroreticular crosslinked styrene-DVB copolymer beads from also zero % to almost 100% DVB contents. Figures 5, 6, 7 and 8 describe the chemistry for the chloromethylation reaction and for the formation of both weakly basic and strongly basic anion exchange resins as well as such unique products as the boron-specific and chelating products. These latter two resins are available commercially and are being used in a number of applications.

It is most interesting to note that it is quite difficult to synthesize resins based upon bulky functional groups because of the steric problems in introducing such groups into gel-type beads without straining and fracturing the beads. The unique physical structure of the macroreticular beads makes it possible to synthesize such products with a satisfactory degree of bead integrity.

It is of importance to note at this point that there are several basic reasons why one elects to consider a macroreticular structure. These include the following:

1. Improved physical strength towards osmotic and physical stresses.
2. Ease of introducing bulky functional groups.
3. Development of pore structure that enables one to operate under anhydrous conditions and as catalysts.
4. Development of pore structure that permits removal of colloids.
5 Production of unique adsorbents.

THE ACRYLIC STRUCTURES

In the model described earlier, the actual polymer chains or the basic structure is usually considered as being based upon styrene. However, the structure may be based upon phenol, vinyl pyridine or even various acrylic-based monomers. I have selected the acrylic structure as the topic for this lecture since it represents a class of ion exchange resins that have been available and used commercially throughout the world for many years, but they represent a class of ion exchange resins about which little has been written and which has been absent in most lectures on ion exchange. Whereas there may be justification for a lack of interest in vinyl pyridine polymers or ion exchangeresins based upon some unique nomomers or crosslinking agents because of a lack of availability or high cost of these reagents, these situations are not applicable in the case of the acrylic-based ion exchange resins. Raw materials for their synthesis are readily available. Even though these ion exchange resins are now widely used, data on their basic properties and structure have been lacking and studies on their kinetic and equilibrium behavior have also been lacking.

Except for a strongly acid cation exchange resin analogue, all basic types of ion exchange resins based upon acrylic copolymers are available commercially throughout the world. These include the weakly acid cation and weakly basic anion exchange resins, as well as strongly basic anion exchange resins. In fact, they have been prepared in the form of gel-type as well as macroreticular structures.

The chemical structures of these acrylic-based ion exchange resins are described in Figures 9 and 10. They may, of course, be compared with comparable ion exchange resins based upon the styrene-DVB copolymers. All of these structures may be prepared on the basis of gel-type as well as macroreticular copolymer bead structures.

The basic difference between the acrylic-and the styrenic-based is, of course, the nature of sketal structure. The latter is aromatic and the former is aliphatic. Although most studies in ion exchange have ignored the contribution of the skeletal or polymer with respect to both the kinetics and the equilibria of ion exchange. Particularly with respect to the anion exchange resins, the nature of the polymer or skeletal structure plays a most important role. This factor becomes even more important as the size of the anionic species increases. Studies (2) over the past decade have demonstrated that the kinetics and equilibria of anion exchange have become much more favorable as the skeletal structure is changed from the hydrophobic, aromatic styrenic structure to hydrophylic, aliphatic structure. For example, the recovery of various pharmaceuticals, the removal of humic matter from water supplies, and the decolorization of sugar syrups are examples of this relationship.

Interestingly, this structural relationship has been known for at least a decade; however, industry is only making use of it slowly.

Although the major purpose of this lecture is to be devoted to the acrylic-based anion exchange resins, some mention must be made of the acrylic-based weakly acidic cation exchange resins. The structures of these are described in Figure 9. It is of interest to note that the methacrylic acid-derived weakly acidic cation exchange resin which has been used worldwide for the recovery and purification of vitamins, antibiotics, enzymes and many other biologicals is prepared in a single step dispersion polymerization using methacrylic acid and divinylbenzene. The other and more acid weakly acid cation exchange resin is more conveniently prepared in two steps by hydrolysis of the cross-linked ethyl or methyl acrylate copolymer (first step) with either acid or alkali (second step). These weakly acid cation exchange resins occupy a unique place in the water, pharmaceutical and sugar industries.

The weakly basic anion exchange resin described in Figure 9 is an unusual anion exchange resin in that it is the most basic of the weakly basic anion exchange resins. It is sufficiently basic to form a salt with CO_2 (carbonic acid) but not with SiO_2 (silica acid). It is now being used in water treatment because of its ability to remove humic matter reversibly during demineralization or deionization.

The strongly basic anion exchange resin described in Figure 10 may be derived from the weakly basic anion exchange resin of Figure by methylation with either methyl sulfate or methyl chloride. These strongly basic anion exchange resins are essentially as basic as their styrene-based analogues; however, their acrylic skeletal structure renders them more hydrophylic than the styrene-based resins. The fact that these anion exchange resins are essentially aliphatic and hydrophylic renders them to remove humic matter from water supplies and colorants from sugar juices and syrups with a degree of reversibility orders of magnitude greater than those of the styrene-based anion exchange resins. Hence, the acrylic-based anion exchange resins foul to a much lesser degree than the styrenics. In fact, in most instances. their fouling due to humic matter and sugar colorants is virtually nil. They have found a profound degree of acceptance in these areas.

The chief or major disadvantage of the acrylic-based anion exchange resins is that these resins do not have the thermal stability exhibited by the styrenic analogues; however, in the areas indicated the thermal stability factor is not of significance. These areas include the decolorization and deashing of sugar juices and the deionization of surface water supplies-situations in which one is concerned with the removal of high molecular weight anionic moieties which must be removed and in which their removal results in the fouling of the styrenic anion exchange resins.

CONCLUSIONS

The synthesis of ion exchange resins has entered a new era. The polymer and physical chemists are now focusing their attention on improving (1) physical strength, (2) particle uniformity and (3) pore structure. In some respects, the development of the acrylic-based ion exchange resins represents one of the few major developments that has occurred recently in the chemistry of ion exchange resin synthesis.

REFERENCES

(1) Kunin, R. Ion Exchange Resins. Robt. E. Krieger Pub. Co. Huntington, N.Y., (1972).
(2) Gustafson, R.L. and Lirio, J.A. I&EC Prod. Res. & Dev. 7, 116 (1968).
(3) Gustafson, R.L., Filluis, H.F., and Kunin, R. I&EC Prod. Res. & Dev. 7, (1968).

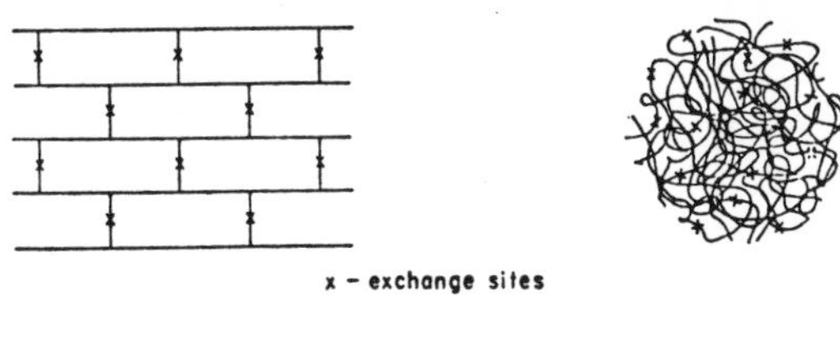

Figures 1 & 2 Comparison of hypothetical oriented and unoriented cross-linked ion-exchange structures.

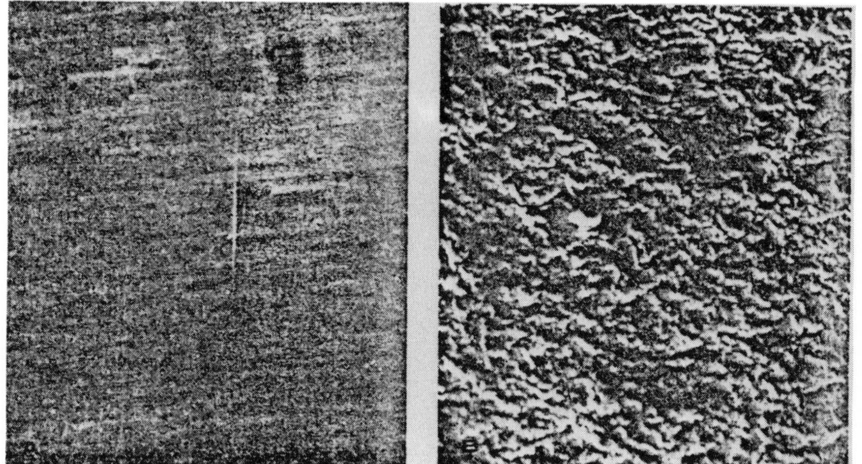

Figures 3 & 4 Electron micrograph of (A) conventional gel and (B) macroreticular structures

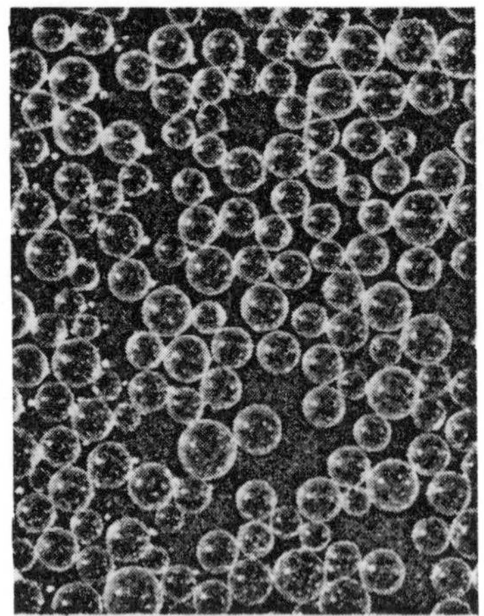

Figure 3A

Gelular ion exchange resin

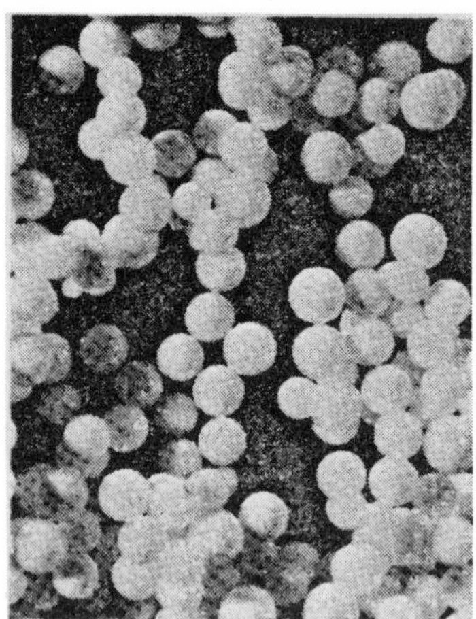

Figure 4A

Macroreticular ion exchange resin

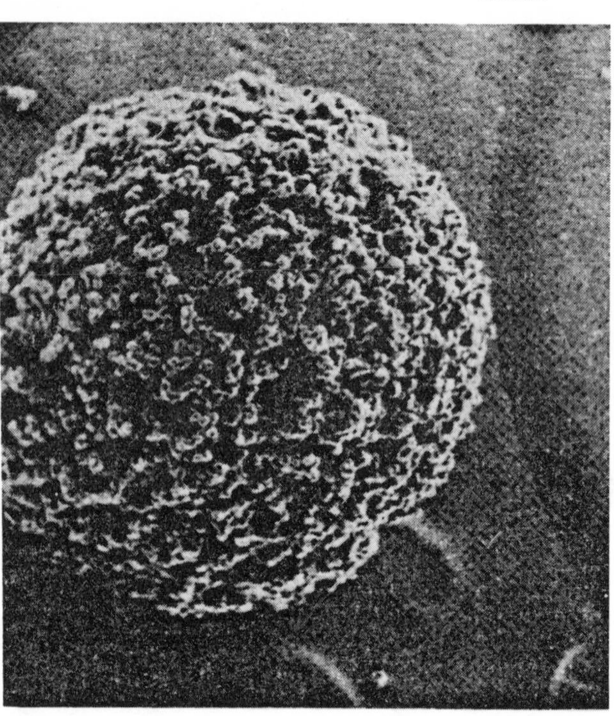

Figure 5 SEM of single macroreticular resin bead

Figure 6 Synthesis of the copolymer

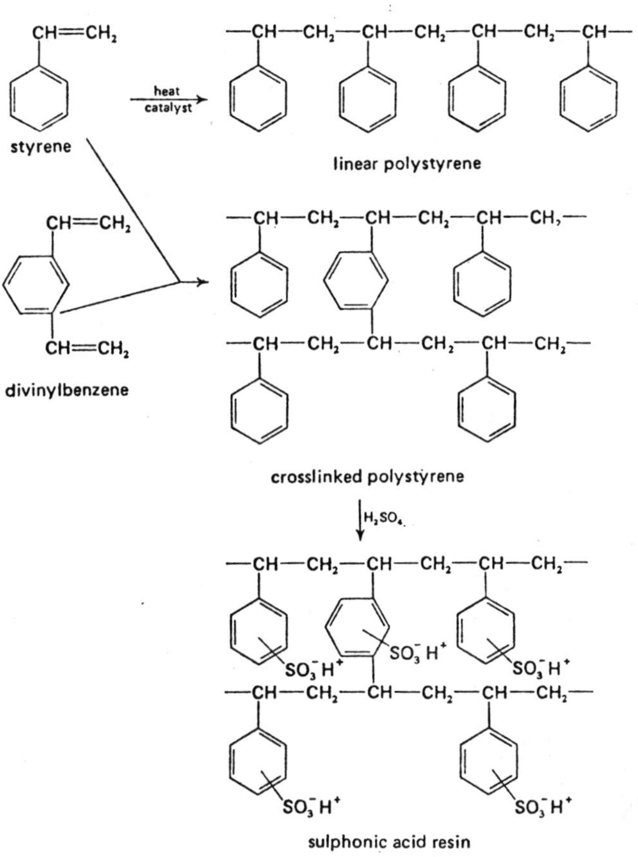
CH=CH2
heat
catalyst
styrene
—CH—CH2—CH—CH2—CH—CH2—CH—
linear polystyrene
CH=CH2
CH=CH2
divinylbenzene
—CH—CH2—CH—CH2—CH—CH2—
—CH—CH2—CH—CH2—CH—CH2—
crosslinked polystyrene
H2SO4
SO3- H+
sulphonic acid resin

Figure 7 Structure of Boron Selective Resin

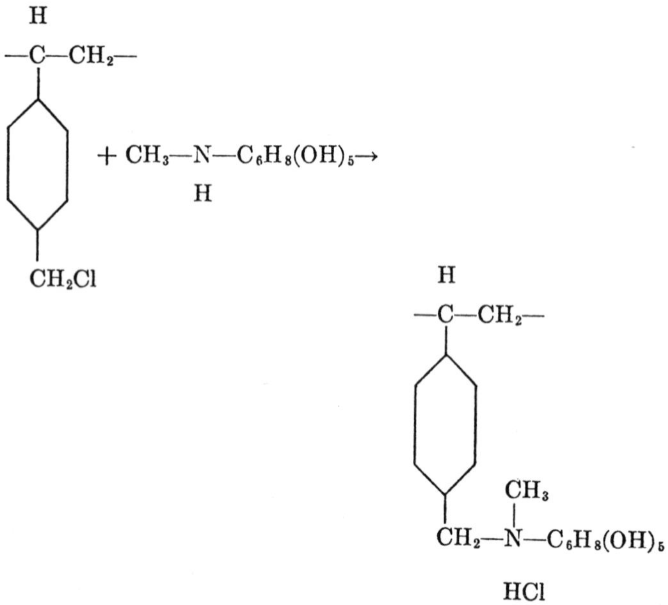

$$-\overset{H}{C}-CH_2- \text{ (ring) } -CH_2Cl + CH_3-\underset{H}{N}-C_6H_8(OH)_5 \rightarrow -\overset{H}{C}-CH_2- \text{ (ring) } -CH_2-\overset{CH_3}{\underset{HCl}{N}}-C_6H_8(OH)_5$$

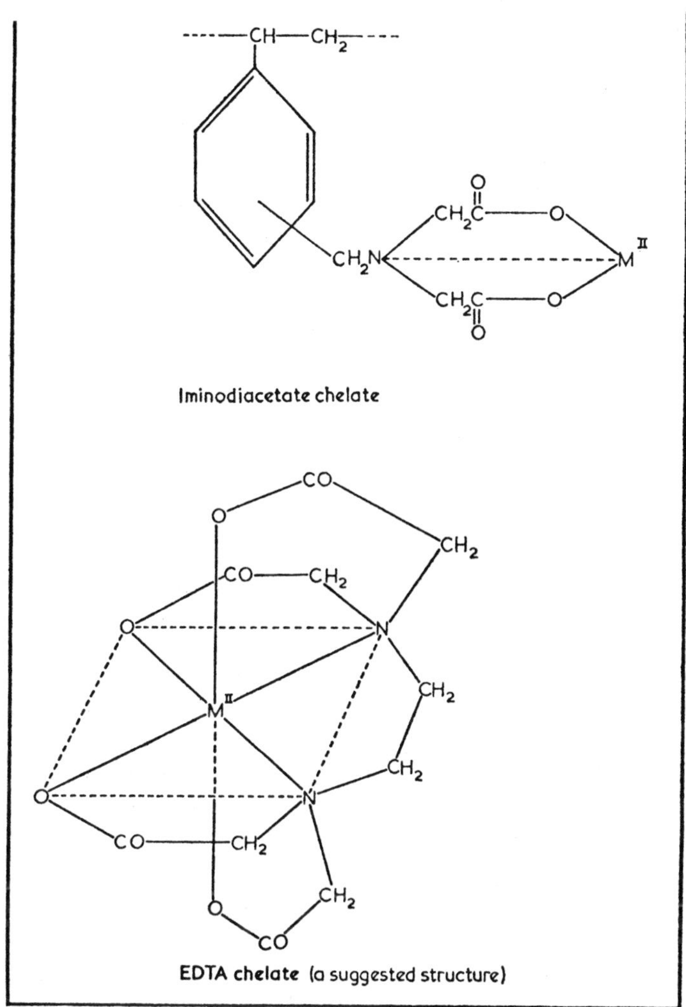

FIG. 8. An example of a specific ion-exchange resin and its relation to an analytical reagent.

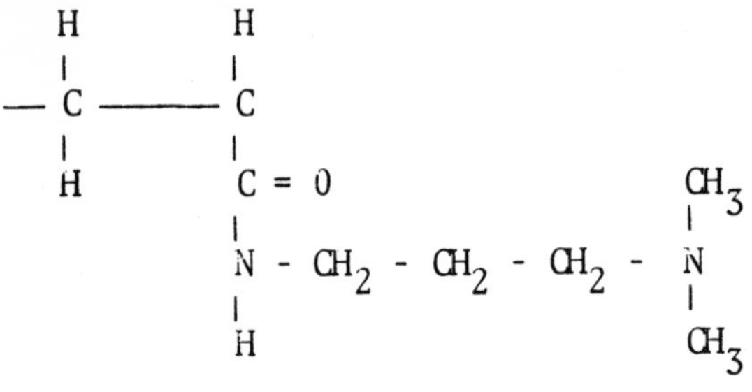

```
  H         H
  |         |
— C ——————— C
  |         |
  H         C = O                         CH3
            |                             |
            N - CH2 - CH2 - CH2 -  N
            |                             |
            H                             CH3
```

Figure 9 - Weakly Basic Acrylic Anion Exchange Resin

```
  H         H
  |         |
— C ——————— C
  |         |                                    +
  H         C = O                   CH3
            |                       |                       -
            N - CH2 - CH2 - CH2 - N - CH3          Cl
            |                       |
            H                       CH3
```

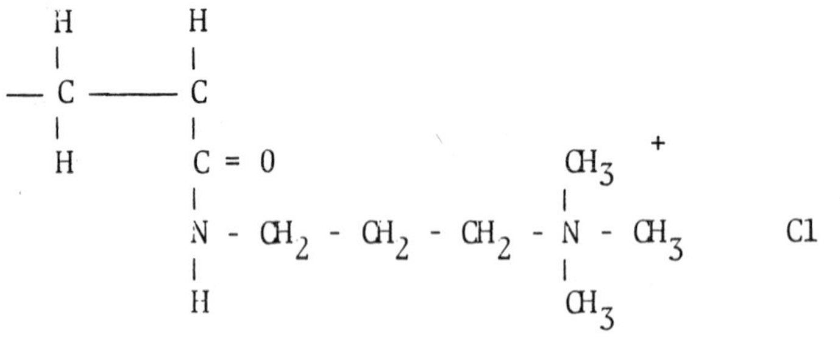

Figure 10 - Strongly Basic Acrylic Anion Exchange Resin

TABLE 1

COMPARISON OF SELECTIVITIES OF ACRYLIC AND STYRENIC ANION EXCHANGE RESINS FOR NAPTHALENE SULFONIC ACID [(1)]

	K_A^o	
Θ	Acrylic	Styrenic
0.22	5.6	
0.31		67
0.44	9.9	
0.49		51
0.63		45
0.72	17.5	

(1) Versus Chloride Ion.

TABLE 2

COMPARISON OF KINETICS OF ACRYLIC AND STYRENIC ANION EXCHANGE RESINS FOR ANTHRAQUINONE SULFONIC ACID

θ[1]	Acrylic	Styrenic
	(min)	
0.10		4.0
0.13		8.0
0.23	2.1	
0.26	4.0	
0.65	9.8	
0.77		20.0

[1] Fractional Attachment of Equilibrium

TABLE 3

GENERAL PROPERTIES OF TWO ACRYLIC-BASED ANION EXCHANGE RESINS

(Typical Values)

	Amberlite IRA-458	Amberlite IRA-68
Appearance	Clear, colorless beads	Clear, colorless beads
Shipping Weight	45 lbs/ft^3	45 lbs/ft^3
Moisture Content	60%	60%
Total Anion Exchange Capacity	1.2 meq/ml	1.6 meq/ml
Typical Regeneration Level	4-6 lbs. NaOH/ft^3	4 lbs. NaOH/ft^3
Typical Regenerant Concentration	4%	4%

The Nature and Properties of Acrylic Anion Exchange Resins

APPENDIX I

AMBERLITE® ACRYLIC ANION EXCHANGE RESINS

- AVAILABLE IN STRONG AND WEAK BASE, MACRORETICULAR AND GELULAR
- HIGH PHYSICAL STRENGTH AND RESISTANCE TO ATTRITION
- HIGHER OPERATING CAPACITY THAN HOMOLOGOUS STYRENIC RESINS
- SUPERIOR ADSORPTION AND ELUTION OF ORGANICS

STRONG BASE ANION RESINS

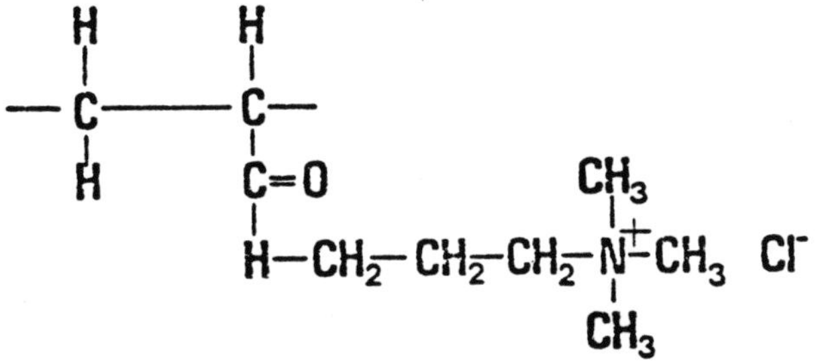

$$-\overset{\overset{H}{|}}{\underset{\underset{H}{|}}{C}}-\overset{\overset{H}{|}}{\underset{\underset{\underset{H-CH_2-CH_2-CH_2-\overset{\overset{CH_3}{|}}{\underset{\underset{CH_3}{|}}{N^{+}}}-CH_3 \quad Cl^{-}}{|}}{C=O}}{|}}{C}}-$$

Acrylic quaternary amine (type 1)

$$-\overset{\overset{H}{|}}{\underset{\underset{H}{|}}{C}}-\overset{\overset{H}{|}}{\underset{|}{C}}-$$

$$C_6H_4$$

$$H_3C-\overset{\overset{CH_2}{|}}{\underset{\underset{CH_3}{|}}{N^{+}}}-CH_3 \quad Cl^{-}$$

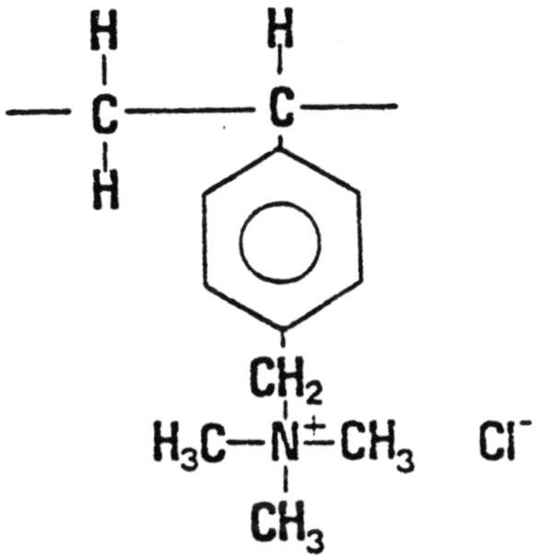

Styrene quaternary amine (type 1)

Amberlite IRA-458

- Gelular, strong base acrylic resin
- Generally used in industrial water treatment applications
- High capacity
- Efficient use of caustic
- Organic fouling resistance superior to styrenics

GENERAL PROPERTIES OF TWO ACRYLIC-BASED ANION EXCHANGE RESINS

(Typical Values)

	Amberlite IRA-458	Amberlite IRA-68
Appearance	Clear, colorless beads	Clear, colorless beads
Shipping Weight	45 lbs/ft^3	45 lbs/ft^3
Moisture Content	60%	60%
Total Anion Exchange Capacity	1.2 meq/ml	1.6 meq/ml
Typical Regeneration Level	4-6 lbs. $NaOH/ft^3$	4 lbs. $NaOH/ft^3$
Typical Regenerant Concentration	4%	4%

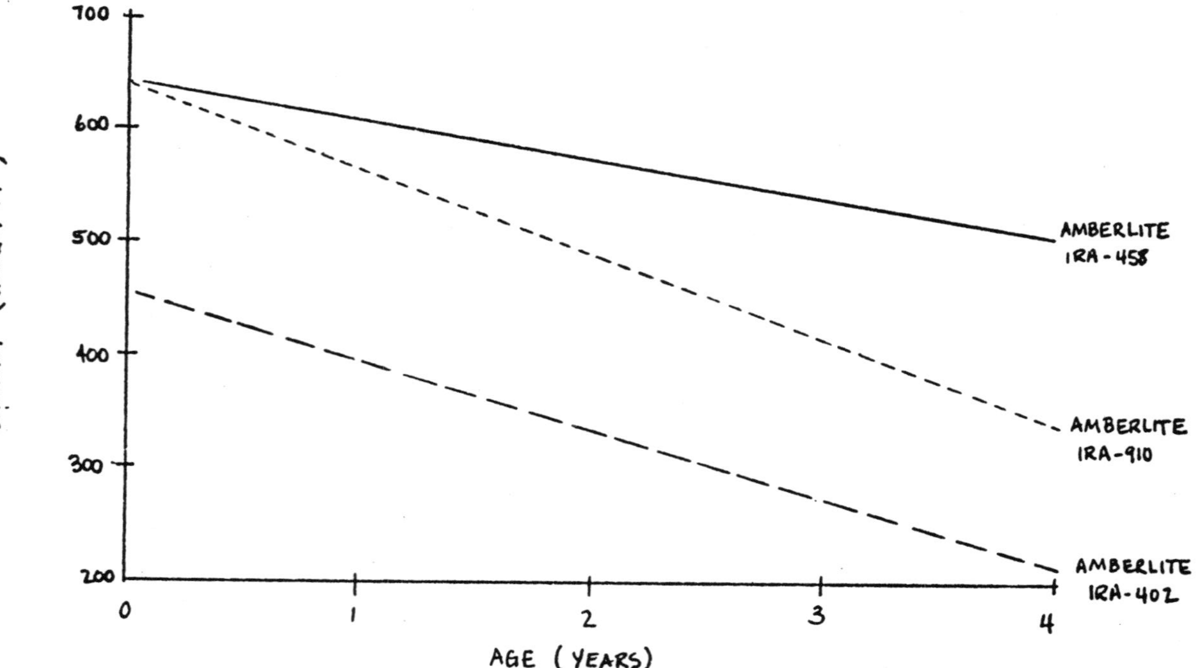
PLANT EXPERIENCE - STRONG BASE ANION EXCHANGE CAPACITY
CAPACITY (G.EQ./M³)
700
600
500
400
300
200
0
1
2
3
4
AGE (YEARS)
AMBERLITE IRA-458
AMBERLITE IRA-910
AMBERLITE IRA-402
FROM: "THE ASSESSMENT OF ANION EXCHANGE RESIN CAPACITY ..." BY G.H.MANSFIELD
AT THE THEORY AND PRACTICE OF ION EXCHANGE CONFERENCE, CAMBRIDGE, 1976

humic acid structure proposed by Dragunov

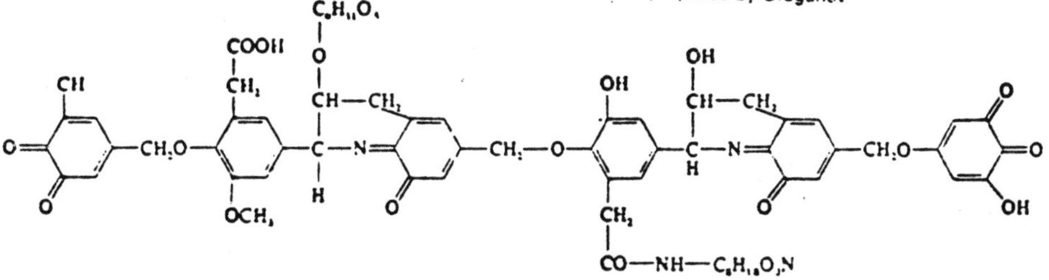

Structure of humic acid proposed by Kleinhempel. Note the multifunctional nature of the compound and the presence of iron as a complexing heavy metal

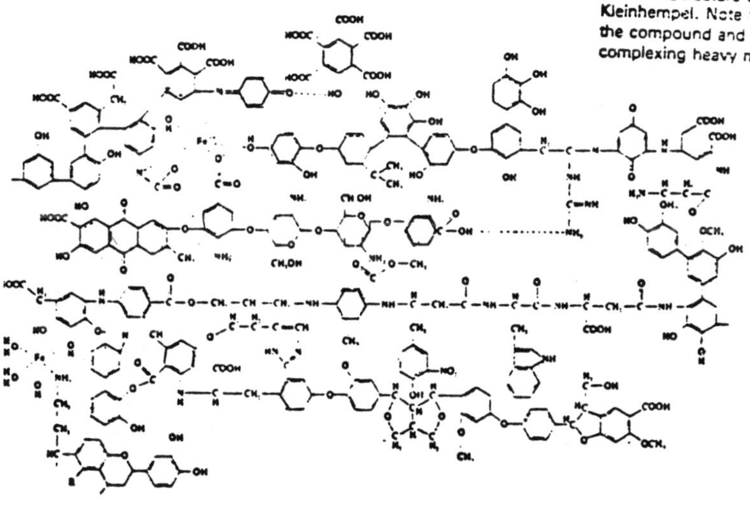

LOADING AND ELUTION OF ORGANICS

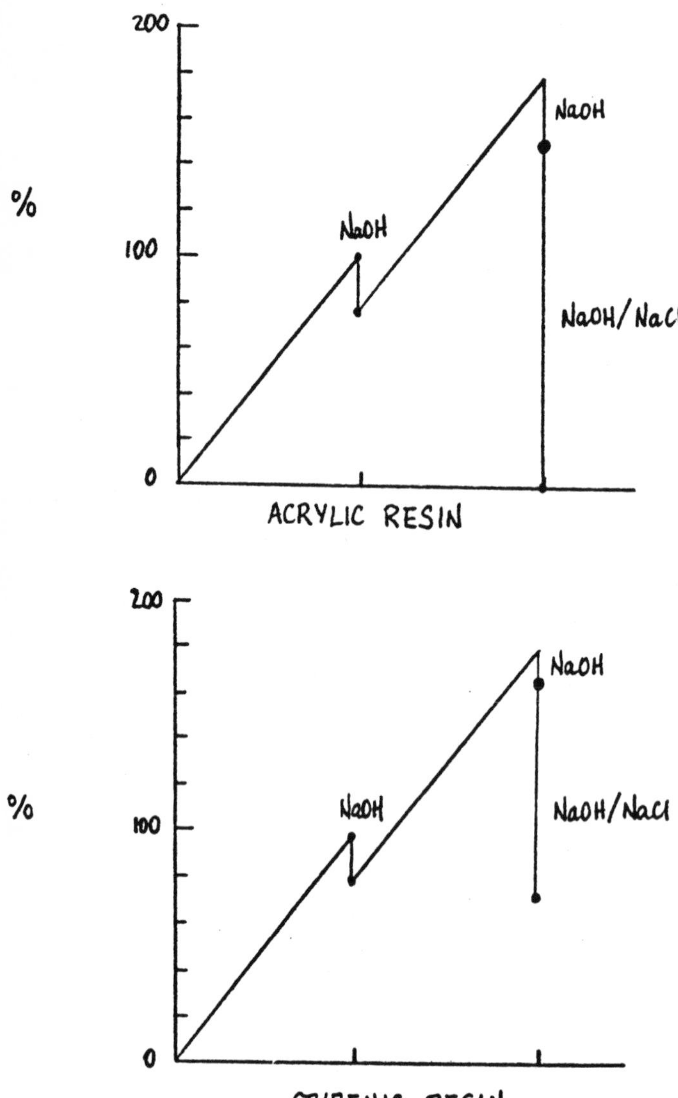

FROM: L.G. OVERMAN, H.W. VANDERBOSCH, BEHAVIOR OF ORGANIC MATTER ON POLYACRYLATE AND POLYSTYRENE - TYPE ANION EXCHANGERS, VGB CONFERENCE "POWER STATION CHEMISTRY 1980"

AMBERLITE IRA-958

- MACRORETICULAR, STRONG BASE ACRYLIC RESIN
- MAINLY USED IN DEIONIZATION/DECOLORIZATION OF SUGAR SYRUPS
- SUPERIOR PERFORMANCE AS ORGANIC SCAVENGER

Amberlite IRA-68

- Gelular, weak base acrylic resin
- Generally used in industrial water treatment applications
- Approved by FDA for use in food processing applications
- High operating capacity
- Unique ability to remove CO_2

CAPACITY CHANGE OF WEAK BASE RESINS

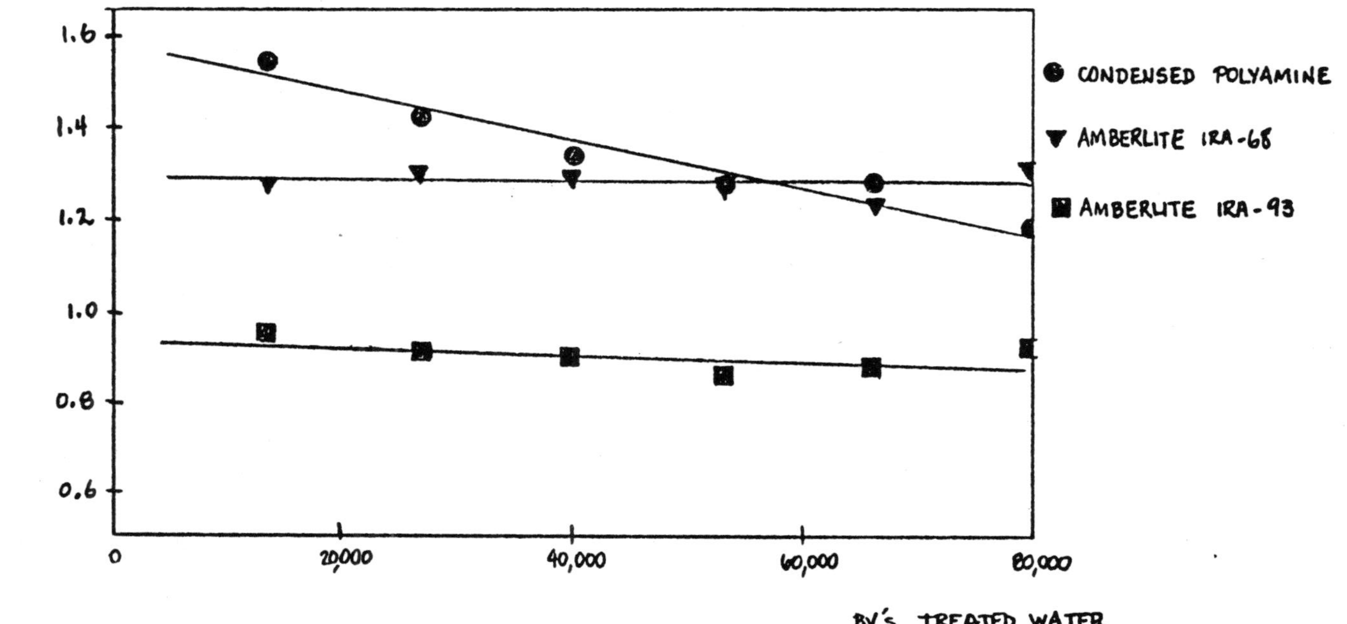

FROM: ROHM AND HAAS TECHNICAL REPORT 36-1627

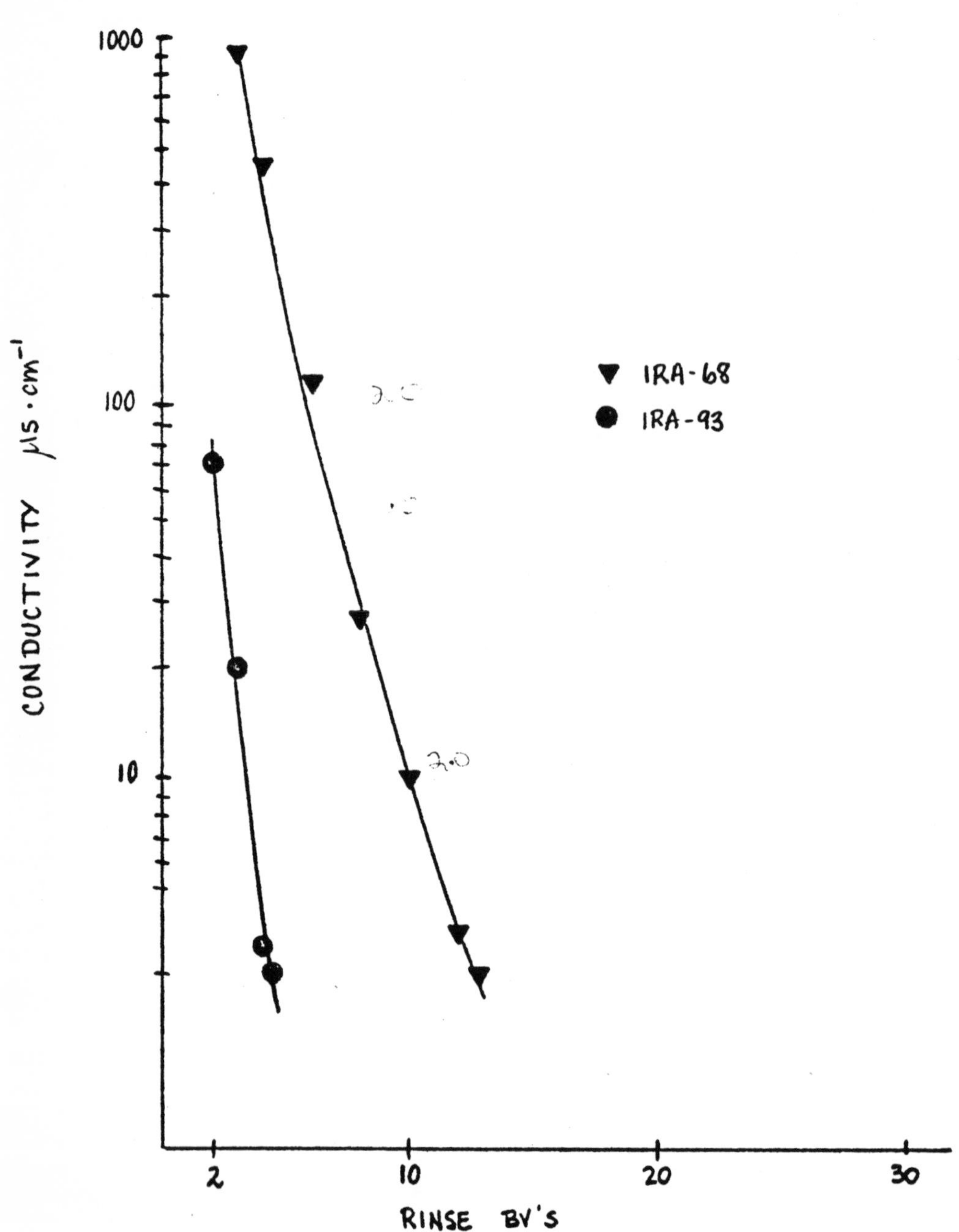
TYPICAL RINSE PROFILE OF
WEAK BASE RESINS.
1000
100
10
CONDUCTIVITY µS · cm⁻¹
IRA-68
IRA-93
2
10
20
30
RINSE BV'S

AMBERLITE IRA-35

- MACRORETICULAR, WEAK BASE ACRYLIC RESIN
- USED IN CHEMICAL PROCESSING APPLICATIONS, (CYANIDE AND SILVER REMOVAL)

SELECTIVITY AND ION SPECIATION IN CATION-EXCHANGE RESINS

Jacob A. Marinsky

Chemistry Department, State University of New York at Buffalo, Buffalo, New York 14214

SUMMARY

The development of a model for the prediction of ion-exchange selectivity has been described in detail. Cation-exchange selectivity in aqueous media is shown to be determined by the difference in interaction of various cations with water. The difference in selectivity is resolvable through the analysis of osmotic coefficient data obtained for the pure ionic forms of the linear polyelectrolyte analogue of the cross-linked cation exchange resin. The utility of this approach has been tested and is shown to be applicable to a weakly dissociated cation-exchange resin as well as to the fully dissociated one. The insight gained from this study is believed to be useful to the analysis of mass transport phenomena in cation-exchange materials.

INTRODUCTION

Appropriate consideration of the mass transport properties of an ion-exchange resin in particle or membrane form requires an accurate assessment of speciation in the gel phase. Diffusion processes are usually described in terms of Fick's first law

$$J_i = -D \text{ grad } C_i$$

where J_i is the flux in moles per unit time and unit cross section of the diffusing species i, C_i is its concentration in moles per unit volume and D is the diffusion coefficient. This equation, of course, applies only to diffusion of one species where there is no possibility of interference from the coupling effects of another simultaneous transport process. An example of this ideal situation, represented by the above equation, is provided by isotopic diffusion in a system at equilibrium except for the isotope distribution. For proper assessment of the nature of diffusion processes which are coupled, such as those which determine the rate of ion exchange, the identity and relative concentrations of the various species encountered in these systems as well as the factors which influence their distribution need to be known. Such information is essential for accurate, educated estimates of the important variables which affect the diffusion coefficients in these systems.

For this purpose a complete analysis of the cation-exchange phenomenon will be developed in this presentation. First, the model employed to achieve this objective will be examined and fully tested. Examples of its validity will be eventually demonstrated by the successful prediction of ion-exchange selectivity patterns. Finally a detailed examination of the method developed from the basic model for the achievement of our ultimate goal, the unambiguous elucidation of speciation in gels whose repeating functional unit is a potential complexer of metals, will be presented. With accurate assessment of metal ion immobilization by its attachment to the matrix functional unit repeated in the macromolecule the correct concentration of free mobile counter ion in the gel becomes available for use in the rate equations.

The Gibbs-Donnan Model.

Numerous attempts to interpret ion-exchange equilibria have been reported[1-12] and a rather complete discussion of the spectrum of theoretical approaches and models that have been employed for this purpose has been presented by Helfferich[13]. One of the most successful of these is the Gibbs-Donnan[14a,b] model. In this model the charged matrix of a resin is considered to constitute a membrane that is permeable to simple electrolyte and solvent. When equilibrated with solvent the solvent is transported into the resin phase. Operationally, the flexible matrix is thought of as a network of elastic springs which exert pressure on the inner pore liquid of the resin phase until the chemical potential, μ, of solvent in both phases is equal. For simplicity, it is assumed that the chemical potential in isothermal systems can be split into two additive terms, one of which depends only on composition; the other only on pressure.

With this representation

$$\bar{\mu}_s = \mu_s \tag{1}$$

and

$$RT \ln \bar{a}_s + \bar{P}\,\bar{V}_s = RT \ln a_s + P\,V_s \tag{2a}$$

where a_s and V_s represent the activity and partial molal volume of the solvent; P, the pressure exerted on the solvent and the bar placed above the symbols is used to differentiate the resin phase from the liquid phase. Since V can be assumed to be independent of pressure without introduction of serious error eq. 2a can be restated as

$$RT \ln \bar{a}_s + \pi V_s = RT \ln a_s \tag{2b}$$

where π is the difference in pressure between the interior of the resin phase and the external solution.

With this thermodynamic description of each component in the ion-exchange reaction.

$$\overline{MX} + NX \rightleftarrows \overline{NX} + MX \tag{3}$$

the following expression is derived for the concentration distribution of uni-univalent ions at equilibrium.

$$RT \ln \frac{\bar{m}_N m_M}{\bar{m}_M m_N} = RT \ln K_M^N = \pi(V_M - V_N) + RT \ln \frac{\bar{\gamma}_M}{\bar{\gamma}_N} - 2RT \ln \frac{\gamma_{\pm MX}}{\gamma_{\pm NX}} \tag{4a}$$

For the univalent-divalent ($NX + MX_2$) and divalent-divalent ($NX_2 + MX_2$) systems, the equations are

$$RT \ln \frac{(\bar{m}_N)^2 (m_M)}{(\bar{m}_M)(m_M)^2} = RT \ln K_M^N = \pi(V_M - 2V_N) + RT \ln \frac{\bar{\gamma}_M}{(\bar{\gamma}_N)^2} - RT \ln \frac{(\gamma_{\pm MX_2})^3}{(\gamma_{\pm NX})^4} \tag{4b}$$

and

$$RT \ln \frac{(\bar{m}_N)(m_M)}{(\bar{m}_M)(m_N)} = RT \ln K_M^N = \pi(V_M - V_N) + RT \ln \frac{\bar{\gamma}_M}{\bar{\gamma}_N} - 3RT \ln \frac{(\gamma_{\pm MX_2})}{(\gamma_{\pm NX_2})} \quad (4c)$$

where K is the experimentally determined selectivity coefficient, m is the molal concentration of the species, $\bar{\gamma}$ is the activity coefficient of the ion in the resin phase and $\gamma_\pm$ is the mean molal activity coefficient of the electrolyte in the external solution phase.

In these expressions only π and $\bar{\gamma}_M/\bar{\gamma}_N$ are not available; the partial molal volume of the ions at infinite dilution and the mean molal activity coefficients of the simple electrolytes can be obtained from the literature while K_M^N is experimentally accessible.

Test of Model.

The validity of this model has been demonstrated in our laboratory[15] by taking advantage of the special properties of the Linde-A zeolite, a crystalline cross-linked polymeric sodium aluminum silicate molecule of well-defined structure and ion-exchange capacity. Because of (1) the high negative charge due to the ring of O atoms in the face and corners of the unit cell and (2) the small openings available to exchanging ions in these materials, there is sizeable resistance to entry of neutral electrolyte into the zeolite. Its rigidity due to matrix inflexibility results in (1) constancy of solvent uptake over a large activity range of the solvent (see Fig. 1), and (2) a fixed geometry of the charged matrix. Because of these unique features of the zeolite the $\bar{\gamma}_M/\bar{\gamma}_N$ term of eq. 4 is constant and independent of the ionic strength of the external phase either when M^+ and N^+ are present at a fixed ratio in the zeolite or when one of the exchanging ions is present at tracer-level concentrations. The change in the value of π is calculable as a function of the solvent activity with eq. 2b since $\bar{a}_{H_2O}$ is a constant of the system.

In Fig. 1 the water adsorption isotherm (25°C) is given for the sodium A-zeolite[15]. There is little change in the amount of water taken up per gram of zeolite as the value of p/p_o (a_{H_2O}) varies from unity to a value of ~ 0.16. A sharp decrease in

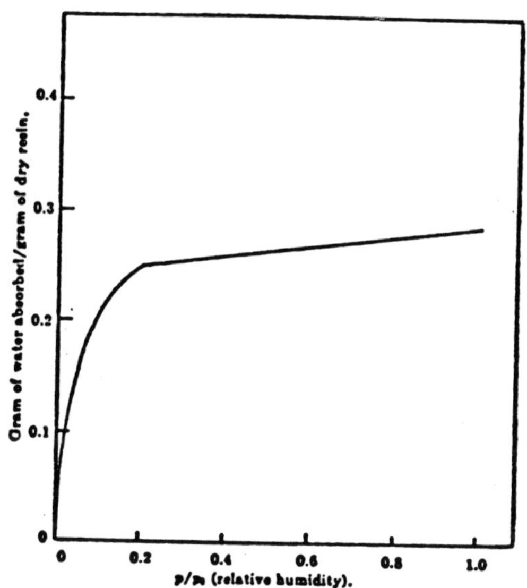

Figure 1. Water sorption isotherm of sodium A-zeolite.

water content of the zeolite occurs below this value. Similar behavior is observed for other zeolites. It is at and below this point of decreasing water content that the pressure exerted on the solvent in the zeolite reduces to zero and $\bar{a}_{H_2O} = a_{H_2O}$.

With this interpretation of the Gibbs-Donnan model a single selectivity measurement of the MX, NX system, where one of the cations $\bar{N}^{T}$ is present in trace quantity, yields the value of the RT ln $\bar{\gamma}_M/\bar{\gamma}_N \pm \pi (V_M - V_N)$ term of eq. 4 without resort to assignment of the $\bar{a}_{H_2O}$ value. By varying the concentration of the bulk electrolyte the change of the activity of water in the external phase then permits evaluation of the change in π (as long as the activity of the water in the external phase is greater than $\bar{a}_{H_2O}$) and its effect on the value of the RT ln $\frac{\bar{\gamma}_M}{\bar{\gamma}_N} + \pi (V_M - V_N)$ term. Since NX is present in trace quantity in these experiments, the mean molal activity coefficients of MX is identical with those of the pure MX solution and can be obtained from the literature[16]. The $\gamma\pm$ for trace NX in the presence of MX was calculated by use of the Harned-Cooke equation[15,16] in the form

$$\log \gamma_{o_{(NX)}} = \log \gamma_{NX_{(o)}} + \alpha_m + \beta_m 2 \qquad (5)$$

where $\gamma_{o_{(NX)}}$ is the activity coefficient of a trace of NX in the presence of MX at molality m, $\gamma_{NX_{(o)}}$ is the activity coefficient of pure NX at molality m and α and β are experimentally determined parameters.

The experimentally observed value of K_M^N was found to agree with its computation by this approach at every experimental molality to confirm unambiguously that the loss in thermodynamic rigor that is introduced by the simplifying assumptions employed in this application of the Gibbs-Donnan model to the interpretation of ion-exchange phenomenon is indeed negligible. Some representative results that were obtained in this experimental study[15] are presented in Table 1 to emphasize the above conclusion. In this table K_M^N calculated with and without consideration of the $\pi(V_M - V_N)$ term is compared with the experimentally determined value of K_M^N.

It is interesting to point out that the excellent agreement between experiment and prediction for the KA-KCl-Na^TCl system was only achieved after the β term needed for accurate evaluation of $\gamma_{o(NaCl)}$ with eq. 5 became available[17]. This occurred a short time after completion of these experiments to dramatize the success of the program.

Application of Model for Selectivity Predictions

With this successful application of the Gibbs-Donnan model an attempt was next made to extend its utility to the interpretation of ion-exchange reactions in organic exchangers. With these flexible materials, however, the $\overline{\gamma}_M/\overline{\gamma}_N$ term is no longer unique and a method for the accurate anticipation of its variation with experimental conditions had to be found to supplement the measurements of π. The most successful approach to the computation of this term has used osmotic coefficient, $\emptyset$, measurements of $<$ 0.5% cross-linked resins in the Gibbs-Duhem equation[6,7,10-12,18,19]. However, because of the unavailability of osmotic data for the polyion forms at low concentration values assumptions that needed to be made with respect to osmotic behavior in the dilute concentration range negated the utility of this approach.

The water sorption properties of polystyrene sulfonate (PSS) exchangers with different degress of crosslinking have been observed to coincide in the lower water activity region where the πV_{H_2O} term is absent[6]. This result has been further substantiated by the coincidence between our recent osmotic coefficient data[20] for the linear PSS and those obtained earlier by Soldano for the $\leq$ 0.5% cross-linked PSS gel as is shown graphically in Fig. 2; only at the highest water activity where a πV_{H_2O} term is operative in the gel as a consequence of solvent restraint is a lower amount of water taken up by the gel. These experimental results show that the cross-linking agent, divinyl benzene, does not affect the physical chemical properties of the PSS and provide fundamental justification for utilization in the Gibbs-Duhem equation of osmotic coefficient data for the linear polyelectrolyte analogue in any concentration region to compute the value of $\overline{\gamma}_M/\overline{\gamma}_N$ in the exchanger at its experimental concentration. By this approach the inherent deficiency of the earlier attempts to evaluate this term were overcome.

The osmotic coefficient was measured as a function of the linear PSS analogue concentration, for a number of ion forms (H^+, Na^+, Sr^{++}, Ca^{++}, Cd^{++}, Co^{++}, Ni^{++} and Zn^{++}) in our laboratory[20,21]. The concentration interval examined extended to the highest concentration levels encountered experimentally in the most highly cross-linked exchangers. The lowest concentration at which the osmotic coefficient was measured was defined by the sensitivity limits of the vapor pressure osmometer employed in this concentration range. The data are summarized in Fig. 3.

The trend, with dilution beyond the lowest concentrations examined in these studies, of the osmotic coefficient for fully dissociated polyelectrolyte cannot be deduced as it can

TABLE 1

SELECTIVITY DATA

Ext. Molality	K_{Calc}(no πV term)	K_{Calc}(πV term)	K_{exp}
System: NaA-$NaCl$-Cs^TCl			
0.053	2.86	2.81	2.77
0.106	2.86	2.83	2.78
0.537	2.78	2.55	2.56
1.085	2.58	2.31	2.22
2.255	2.46	1.85	1.85
3.383	2.44	1.52	1.61
4.510	2.31	1.32	1.37
6.068	2.32	1.09	1.18
System: KA-KCl-Na^TCl			
0.109	3.40	3.42	3.41
0.439	3.45	3.52	3.48
0.891	3.39	3.49	3.51
1.831	3.42	3.59	3.59
2.829	3.49	3.75	3.69
3.885	3.46	3.80	3.77

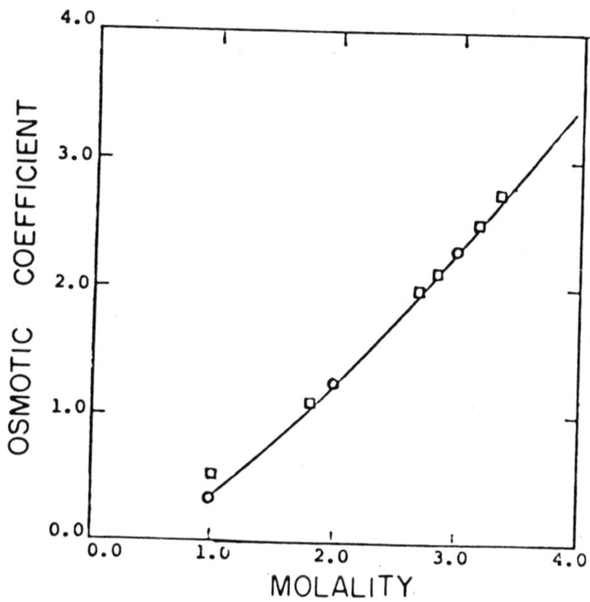

Figure 2. Concentration dependence of the molal osmotic coefficient, Ø for zinc polystyrenesulfonate (o ref. 20; □ ref. 18).

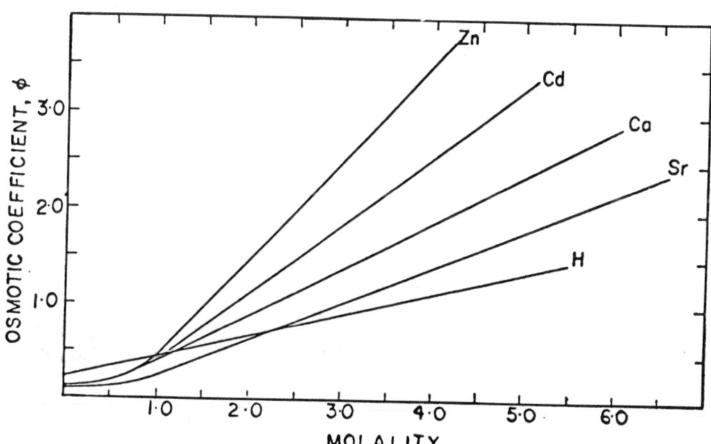

Figure 3. Concentration dependence of the molal osmotic coefficient, ϕ, for polystyrenesulfonic acid and several of its divalent metal salts in aqueous solution.

with simple electrolytes, where the Debye-Hückel limiting law applies, and the computation of mean molal activity coefficients meaningfully related to a value of unity for the polyelectrolyte at infinite dilution is impossible. As a consequence the integrated form of the Gibbs-Duhem equation given below can only be used to compute mean molal activity coefficient values, $\gamma_{\pm m}$, as a function of counterion concentration, m, relative to an indeterminate mean molal activity coefficient, $\gamma_{\pm m_r}$, at the low reference concentration, m_r. The results of this analysis of the osmotic coefficent

$$\ln \gamma_{\pm m}/\gamma_{\pm_r} = \emptyset_m - \emptyset_{m_r} + \int_{m_r}^{m} (\emptyset_m - 1)d \ln m \qquad (6)$$

data that were obtained with HPSS are presented in Fig. 4. The observed behavior of $\ln \gamma_{\pm_m}/\gamma_{\pm_{m_r}}$ as a function of ln m is characteristic. Its value is always linearly related to ln m in the lowest concentration interval. At higher counterion molalities increasingly positive deviation from the initial straight line of negative slope occurs with increasing concentration until a minimum is reached. Beyond this molality the value of $\ln \gamma_{\pm m}/\gamma_{\pm m_r}$ increases exponentially.

The value of π in a given ion-exchange gel is calculable for use in eq. 4 from the osmotic coefficient of the polyelectrolyte analogue at the same counterion molality:

$$RT\emptyset_m m W_{H_2O}/1000 = -RT \ln a_{H_2O} = \pi V_{H_2O} \qquad (7)$$

However, since the mean molal activity coefficient value of the linear analogue at every counterion concentration in the gel is always relative to an undetermined $\gamma_{\pm m_r}$ value the absolute value of the term, $\bar{\gamma}_M/\bar{\gamma}_N$, of eq. 4 is not readily available for unambiguous examination of the interpretive quality of the Gibbs-Donnan model when applied to the analysis of the ion-exchange phenomenon in the flexible, cross-linked ion-exchange resins. As a consequence, the only unambiguous test of this model was thought to be one in which the mean molal activity co efficients of the reference states are cancelled. This situation can be realized experimentally only when there is exchange of pairs of ions between a cross-linked gel and its linear polyelectrolyte analogue[22]. With such systems eq. 4 takes the form

$$RT\ln K_M^N = \pi(V_M - V_N) + RT \ln \frac{\bar{\gamma}_M \gamma_N}{\bar{\gamma}_N \gamma_M} \qquad (8)$$

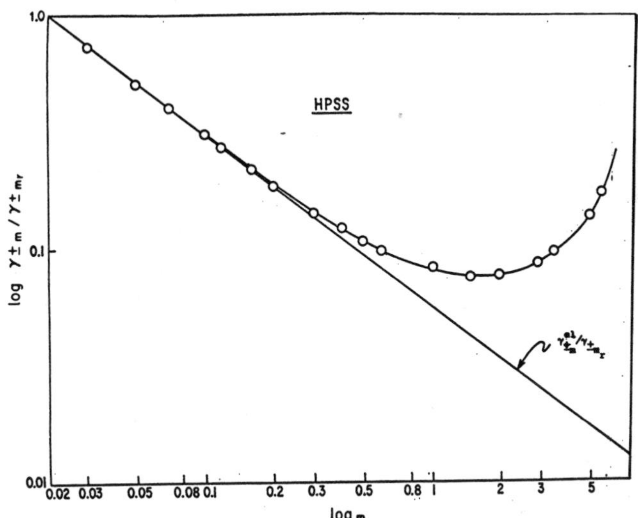

Figure 4. Plot of log $\gamma_{\pm m}/\gamma_{\pm m_r}$ versus log m.

If the model is correct the experimental K_M^N should be directly calculable.

This definitive test of the model was effected by measuring the distribution of pairs of ions, one present in macroscopic quantity and the other in trace, between PSS gels and their linear PSS analogue for comparison with the distribution predictions of equation 8. The molality of the exchanger at equilibrium was determined by the standard centrifugation method[13]. The value of π was computed with eq. (7) and $\ln \gamma_{\pm m}/\gamma_{m_r}$ in each phase was obtained for the macro ion from integration of the Gibbs-Duhem equation to the experimental concentration of the macro counter-ion component. The most appropriate method for computation of this term for the trace ion component was not immediately obvious, however. For example, it could be assumed that the trace ion behaves as if it is at the stoichiometrically equivalent concentration of the macroion[22,23]. However, at equivalent concentrations the water activity of the respective counter ion forms of PSS are different ($\emptyset_M m_M \neq \emptyset_N m_N$ when $m_M = m_N$) and this is a complicating factor; whether the thermodynamic property of the trace ion is best related to the stoichiometrically equivalent concentration of the macro component in the equilibrium mixture or to the concentration the trace component would exhibit at the water activity of the equilibrated system begs the question since one or the other approach neglects an important aspect of the problem. For example, the effect of electrostatic interaction between trace ion and the highly charged polyion on $\log \gamma_{\pm m}/\gamma_{\pm m_r}$ needs to be considered at the stoichiometrically equivalent concentration. However, when this is done the effect of ion solvation on the value of $\gamma_{\pm m}/\gamma_{\pm m_r}$ for the trace ion is either under - or overestimated at the different water activity that is associated with the trace ion at this concentration. Ideally, one would prefer to be able to account for these factors, $\gamma_\pm^{el}$ and γ^h, separately.

The shape of the curve obtained for the $\log \gamma_{\pm m}/\gamma_{m_r}$ versus log m plot (see Fig. 4) that is used to characterize these polyelectrolytes is similar to the shape of the curve that is obtained from a plot of $\log \gamma_{\pm m}$ versus $m^{1/2}$ for simple electrolyte and suggests a possible path for resolution of this problem. In simple electrolytes this property of $\log \gamma_{\pm m}$ is attributed to the product of the competing ion-ion and ion-solvent interactions[16]. At low concentrations, $\gamma_{\pm m}^{el}$ is the dominant factor and $\ln \gamma_{\pm m}$ is inversely proportional to $m^{1/2}$; as the molality is raised γ_m^h becomes increasingly important and eventually $\ln \gamma_{\pm m}$ passes through a minimum and then increases exponentially with $m^{1/2}$. By presuming these factors to be similarly operative in the polyelectorlyte as well, $\ln \gamma_{\pm m}^{el}/\gamma_{\pm m_r}$ is assumed, a priori, to be represented by the initial.

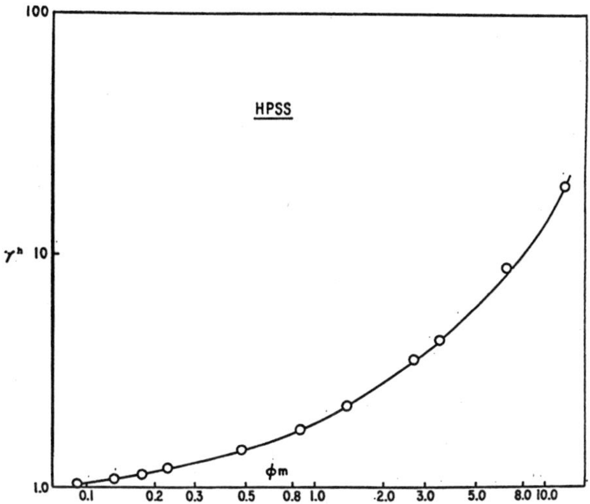

Figure 5. Plot of γ^h versus $\emptyset$ m.

straight line portion of the curve at low values of m^{24}_{el}. Extension of this line is believed to provide the value of ln $\gamma_{\pm m}/\gamma_{\pm m_r}$ over the complete concentration range. Division of $\gamma_{\pm m}/\gamma_{\pm m_r}$ by the extrapolated value of $\gamma^{el}_{\pm m}/\gamma_{\pm m_r}$ over this concentration range yields the corresponding value of γ^{h}_{m} which in turn can be analyzed as a function of the water activity associated with the polyelectrolyte of these concentrations from the type of plot that is presented for HPSS in Fig. 5. Thus appropriate estimate of the ln $\gamma_{\pm m}/\gamma_{\pm m_r}$ term for the trace ion in eq. 8 is provided by the product of $\gamma^{el}_{\pm m}/\gamma_{\pm m_r}$, obtained by extrapolation of the linear portion of the ln $\gamma_{\pm m}/\gamma_{\pm m_r}$ versus ln m curve for the trace-ion form of the polyelectrolyte to the stoichiometrically equivalent concentration of the macro-ion component in the distribution study, and $\gamma^{h}_{a_{H_2O}}$, obtained at the water activity of the equilibrium mixture from the γ versus a_{H_2O} curve analyzed as described above, for the trace ion system.

The experimentally obtained K^N_M values are compared in Table 2 with computations of this ion distribution parameter using eq. 8. Two different evaluations of the ln $\gamma_{\pm m}/\gamma_{\pm m_r}$ term for the trace ion were used in eq. 8. In the first the ln $\gamma_{\pm m}/\gamma_{\pm m_r}$ term is based on the Gibbs-Duhem integration to the stoichiometrically equivalent concentration of the macroion component[21,22]. In the second its value is based on the separate evaluation of $\gamma^{el}_{\pm m}/\gamma_{\pm m_r}$ and $\gamma^{h}_{a_{H_2O}}$ as described above[24]. In both sets of computations neglect of the effect of the macroion on the colligative properties of the trace ion is a source of potential error in this application of the Gibbs-Donnan model.

The excellent agreement between observation and prediction when computation is based on the presumption that the stoichiometry determines $\gamma^{el}_{\pm m}/\gamma_{\pm m_r}$ and the water activity defines $\gamma^{h}_{a_{H_2O}}$ is strongly supportive of this approach to the problem. The accurate prediction of ion-exchange selectivity by this application of the Gibbs-Donnan model provides strong evidence for its validity.

As was pointed out earlier the absolute prediction of the ion distribution patterns of pairs of ions between ion-exchange gels and simple electrolyte was thought not possible because the mean molal activity coefficient of polyelectrolytes can only be measured relative to its indeterminate activity coefficient at a fixed reference concentration. It was believed that in these systems only the pattern of ion-exchange selectivity as a function of polyelectrolyte gel concentration could be determined. However, with this pattern

Table 2

Selectivity Coefficient Predictions

System: HPSS gel (8% DVB), HPSS, $^{60}CO^{++}$

Molality HPSS	$K_{H(exp)}^{Co}$	$K_{H(calc\ 1)}^{Co}$	$K_{H(calc\ 2)}^{Co}$
2.41×10^{-2}	9.74×10^{-2}	2.60×10^{-1}	1.14×10^{-1}
2.42×10^{-2}	8.97×10^{-2}	2.60×10^{-1}	1.14×10^{-1}
4.61×10^{-2}	1.61×10^{-1}	3.88×10^{-1}	1.61×10^{-1}
5.82×10^{-2}	1.88×10^{-1}	4.42×10^{-1}	1.98×10^{-1}
1.06×10^{-1}	3.60×10^{-1}	6.21×10^{-1}	2.90×10^{-1}
0.91×10^{-1}	2.86×10^{-1}	6.00×10^{-1}	2.51×10^{-1}
1.628×10^{-1}	4.14×10^{-1}	8.65×10^{-1}	3.74×10^{-1}
1.666×10^{-1}	4.12×10^{-1}	8.65×10^{-1}	3.72×10^{-1}

System: HPSS gel (1-12% DVB), 0.04 m HPSS, $^{45}Ca^{++}$

%DVB	$K_{H(exp)}^{Ca}$	$K_{H(calc\ 1)}^{Ca}$	$K_{H(calc\ 2)}^{Ca}$
1	0.21	0.39	0.26
2	0.22	0.42	0.23
4	0.21	0.57	0.25
8	0.29	0.94	0.35
12	0.43	1.36	0.42

accessible evaluation of the reference state activity coefficient ratio of the exchanging counterions by use in eq. 4 of the experimental K_M^N value measured at one polyelectrolyte gel concentration provided a capability for prediction of selectivity at the other gel concentrations[22,23]. The utility of this approach for this purpose is seen in Fig. 6 from the correlation of representative ion-exchange data with prediction that is obtained once this normalization procedure has been performed. Once again, the eq. 4 based predictions obtained from the two methods of computing the ln $\gamma_{\pm m}/\gamma_{\pm m_r}$ term for the trace ion are compared to demonstrate the superiority of its synthesis from $\gamma^{el}_{\pm m}/\gamma_{\pm m}$ and $\gamma^{h}_{a_{H_2O}}$ estimates made as described above. The improvement in the correlation of ion-exchange selectivity that is obtained by this approach to the use of eq. 4, however, in some instances overcompensates the tendency to exaggerate ion-distribution trends with increasing gel concentration that results with the first approach to the computation ln $\gamma_{\pm m}/\gamma_{\pm m_r}$ for the trace ion. This result was thought to be a consequence of the neglect of the effect of the macrocounter ion on the colligative properties of the trace ion.

Most of the research that has been carried out in this laboratory to demonstrate the applicability of the Gibbs-Donnan model to the interpretation of ion-exchange phenomena has been confined to the study of the distribution of pairs of ions between gel and solution phases when one of these ions is presented at the tracer-level concentration. Recently these studies were extended to examination of the exchange of two counterions over their complete composition range[25]. The distribution of Na^+ and Zn^{++}, between polystyrene sulfonate resin crosslinked with divinylbenzene (1%, 2%, 4%, 8%, 12% and 16% by weight) and the linear polystyrene sulfonate analogue of the resin at three different concentration levels (0.01, 0.06 and 0.12 normal) was measured. The equivalent fraction of Zn and Na were varied from 0 to 1 and 1 to 0 to examine the distribution pattern of these exchanging counterions over the complete composition range. The polyelectrolyte analogue was used in these studies to permit direct assessment of selectivity, with eq 4b, the reference state mean molal activity coefficient ratio of the sodium and zinc being expected to cancel.

Over most of the composition range examined there was good agreement between the computed and measured selectivity values. Only when the fraction of Zn^{2+} approached unity was there sizeable discrepancy between the two values, the predicted value deviating by as much as a factor of five or six. A representative example of this result is provided by the data listed in Table 3. The discrepancy between prediction and experiment in the Zn-rich range was rationalized by recourse to the theoretical treatment of Manning[26]

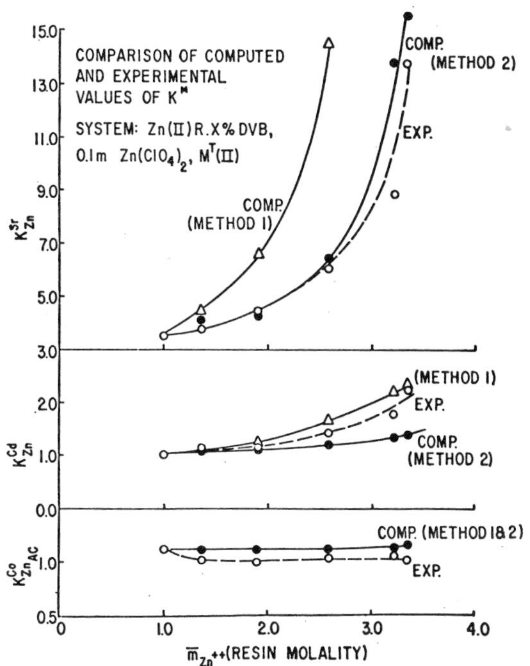

Figure 6. Comparison of Computed and Experimental Values of K_N^M.

Table 3

Calculated and Experimental Selectivity Coefficients for Na-Zn Ion Exchange Reactions Between Various Concentrations of Polystyrene Sulfonate Solutions and a 1% Cross-linked Polystrnee Sulfonate Resin.

	X_{Na}	K_{exp}	K_{cal}
A. $0.02\frac{meq}{g}$	PSS		
	0.000	0.531	0.127
	0.056	0.286	0.119
	0.548	0.088	0.134
	0.684	0.094	0.129
	0.921	0.086	0.121
	1.000	0.074	0.128
B. $0.05\frac{meq}{g}$	PSS		
	0.000	0.652	0.184
	0.118	0.508	0.187
	0.675	0.197	0.193
	0.789	0.187	0.198
	0.943	0.165	0.202
	1.000	0.163	0.186
C. $0.1202\frac{meq}{g}$	PSS		
	0.000	0.893	0.282
	0.042	0.781	0.244
	0.889	0.309	0.327
	0.944	0.305	0.347
	0.949	0.294	0.343
	1.000	0.314	0.337

who has shown that there is theoretical justification for expecting "condensation" of counter-ions on a polyion until its charge density is reduced below a certain critical value. This charge density parameter is unity for univalent ions and 0.5 for divalent ions. In the Manning model, the behavior of the uncondensed mobile ions are presumed to be described exclusively by the Debye-Hückel limiting law; the mean molal activity coefficient predicted by his approach reflects the combined effect of condensation and long-range electrostatic forces exerted on the "uncondensed" counterions. In his treatment, the prescription for condensation is idealized to infinite dilution. One would suspect then that predictions based on his model would be most nearly approached in the most dilute experimental systems employed. With the Zn-trace Na ion system, the zinc ion is expected to be condensed on the dilute polyion surface until its charge density reaches 0.5. At this charge density, the sodium ions will be mobile and free in the solution. By using Manning's "limiting law" expression[26] we have evaluated the osmotic coefficient of the trace univalent ion when the charge density, $e^2/kTb(\xi)$ (where e is the charge of the proton, ξ is the macroscopic dielectric constant water, b is the distance between neighboring charged groups of the polyion, and kT has its characteristics meaning), is equal to 0.5 due to saturation "condensation" by divalent ions in the system. With this analysis, $\phi_p = 1-1/2\xi = 0.75$. By presuming that the "uncondensed" univalent ions will behave in the same way as the fraction of "uncondensed" ion in the pure univalent ion form of the polyelectrolyte, the osmotic coefficient vs. molality plot for the sodium polystyrene sulfonate has been modified by increasing each experimental value by the difference between the "limiting law" $\phi_p = 0.75$ and the ϕ_p value of 0.25 that is obtained when the experimental ϕ_p values are extrapolated to zero concentration to obtain $\gamma_{\pm_{el}}$. On this basis one would expect the activity coefficient of the Na ion to be three times greater than in the earlier estimate. With this estimate of γ_{el} a nine-fold increase in selectivity coefficient over the expected values is predicted. We have seen that in our most dilute system the measured value is approximately 4 to 7 times greater than anticipated by our approach as $X_{Zn} \rightarrow 1$. This result is in fairly reasonable accord with predictions based on our interpretation of the Manning theory. As the concentration of polyelectrolyte increases the discrepancy between our model predictions, though sizeable, is less. In these more concentrated systems the Manning treatment, based on infinite dilution, becomes less applicable. At the higher concentrations screening of the potential field at the surface of the polyion by the univalent ion (macro-concentration) tends to lower the effective charge density on the polyion so that condensation of the Zn^{++} is less than predicted; in the resin "condensation" effects are minimal.

With sodium polyelectrolyte solution, the behavior of zinc ions at trace level will not be as seriously affected since the charge density is still sufficiently high to condense Zn ions and the model continues to be applicable.

An anternative explanation for the discrepancy between the predicted and observed distribution of Na and Zn, when the concentration ratio of Zn to Na is high was suggested by the interpretation given by Boyd and Bunzl[27] to volume changes observed to accompany the selective binding of ions by PSS gels. They concluded on the basis of the volume changes observed, that complexation of multiply-charged cations was extensive in this ion exchanger. Examination of the complexation of Zn(II), Ca(II) and Co(II) by the linear analogue of the polystyrene sulfonate gels showed, however, that such complexation by the polystyrene sulfonate polyion is very small indeed.[28] The formation constant of 0.1 measured is much too small to account for the discrepancy between prediction and observation.

Prediction of Ion-Exchange Equilibria Without System Calibration.

The direct prediction of the selectivity pattern in resin equilibrated with simple mixed electrolyte solution is not accessible with the model as developed. In the model, the mean molal activity coefficient of a particular ion form of the resin at different concentrations has had to be related to the mean molal activity coefficient of the polyelectrolye analogue at a reference concentration. A value of unity is arbitrarily assigned to the mean molal activity coefficient of each ion form at this reference concentration (0.02 meq/ml). As a consequence at least one selectivity measurement has been required to permit evaluation of the ratio of the mean molal activity coefficients of each pair of exchanging ions in their reference state. This deficiency of our approach to the anticipation of ion-exchange equilibria is removed in the following way.

i. The Effective Concentration of Mobile Counterions in Charged Polymers: The first fundamental insight to the analysis of nonideality in charged polymeric systems such as proteins was due to Linderstrøm-Lang[29]. His complicated treatment was restated much later in a more convenient form by Scatchard[30] to facilitate the study of ion-binding in complicated protein systems. The more rigorous derivations of Scatchard have been essentially duplicated with Tanford's simplified version of the Linderstrøm-Lang treatment[31].

In Tanford's development the free energy of ionization of a protein molecule is given by eq (9).

$$\Delta F^\circ_{ion} = -RT \ln K_0 = 2.303\ RT\ pK_0 \qquad (9)$$

K_o representing the intrinsic constant for the ionization reaction. In the course of the ionization additional electrostatic work must be done to increase the charge of the conjugate base from 0 to Z so that the relationship between K_o and the apparent dissociation constant of the molecule, K_z, is

$$\frac{[\Delta F^\circ_{ion}]_z - [\Delta F^\circ_{ion}]_o}{2303\ RT} = pK_z - pK_o \tag{10}$$

By considering the surface of the protein molecule to be represented by a sphere Tanford has shown that

$$[\Delta F^\circ{}_{ion}]_z - [\Delta F^\circ_{ion}] = \frac{2ZN\varepsilon^2}{2D}\left(\frac{1}{b} - \frac{\kappa}{1 + \kappa_a}\right) \tag{11}$$

when b is the radius of the protein sphere, a is its radius of exclusion, κ has the meaning customary in the Debye-Hückel theory D is the dielectric constant of the medium and ε is the protonic charge, Then

$$pK_z - pK_o = \frac{2ZN\varepsilon^2}{2.303\ RT2D}\left(\frac{1}{b} - \frac{1}{1 + \kappa_a}\right) \tag{12}$$

It is customary to equate $-\frac{N\varepsilon^2}{2DRT}\left(\frac{1}{b} - \frac{\kappa}{1+\kappa a}\right)$ to w

so that at any charge Z

$$pK_z = pK_o - \frac{2Zw}{2.303} \tag{13}$$

and

$$pH - \log\frac{\alpha}{1-\alpha} = pK_o - \frac{2Zw}{2.303} = pk_o - .8686Zw \tag{14}$$

The potential difference between the surface of a polyion and the region in which the potential is measured during pH measurements in the course of neutralization of a weak polyacid with standard base provides an experimental evaluation of this deviation term (0.8686 Zw) in such polymeric systems. The deviations from ideality of mobile H ions at the site of the neutralization reaction is obtained from the well-known equation given below:

$$pH - \log\frac{[A^-]}{[HA]} - pK_{HA} = pK_{HA_{app}} - pK_{HA} = -0.4343\frac{\varepsilon\Psi(a)}{kT} =$$

$$-0.8686\ Zw \tag{15}$$

In this equation $\Psi_{(a)}$ corresponds to the potential at the surface of the polyion, K_{HA} is the intrinsic dissociation constant (K_o) of the acid, $K_{HA_{app}}$ is its apparent dissociation constant (K_z) and $[A^-]$ and [HA] are molar concentrations of the dissociated and undissociated polyacid expressed on a monomer basis.

Arnold and Overbeek[32] in their pioneer demonstration of this plotted potentiometric titration data obtained with polymethacrylic acid as pH - log $\frac{\alpha}{1-\alpha}$ versus α. Ideally pH - log $\frac{\alpha}{1-\alpha}$ = pK_{HA} and any deviation from a straight line of zero slope in such a plot is presumed to be a quantiative measure of the deviation from ideal behavior of the system as the polyacid is progressively dissociated. A representative plot of such data obtained with PMA is presented in Figure 7. The extrapolation of the upper curve of the figure is by a straight line extension of the data as it should be in this region of low charge density where the Debye-Huckel approximation is valid. The extrapolation of the lower line must be slightly curved to reach the same intercept value[33]. Such curvature is to be expected in the higher charge density region. The K_{HA} of 1.48 x 10^{-5} that is determined from this analysis of the potentiometric data is in good agreement with the dissociation constant reported for isobutyric acid, the repeating monomer unit of this polymer. Since there is no ionization of the polymer at $\alpha = 0$ this number should correspond to the negative logarithm of the intrinsic dissociation constant of the repeating acid group of PMA, as it does, if the source of deviation is exclusively electrostatic in nature. Thus the change in the value of pK (ΔpK) with α, is attributed to the change in the electrostatic free energy of the molecule as a consequence of group-group interactions accompanying the ionization process.

ii. Equivalence in the Deviation from Ideality of Counter-Ions: We have shown that the fundamental deviation term, $\exp-\varepsilon\Psi_{(a)}/kT$, derived from the study of H^+ ion in weak polyacid systems, also provides an accurate basis for estimate of the deviation from ideality of other mobile $M^{+\nu}$ counterions present concurrently in these complicated systems when ion "condensation" perturbations are removed by screening of the charged surface with simple neutral electrolyte. The deviation term then is $\exp-\nu\varepsilon\Psi_{(a)}/kT$ where ν is the charge of the metal, $M^{+\nu}$, counter-ion. To demonstrate that this is indeed the case, a study of the complexation of trace-level concentration of Ca^{+2}, Co^{+2} and Zn^{+2} by polymethacrylic and polyacrylic acid as a function of $\underline{A}$ was made in our laboratory[34]. The distribution of trace-level concentrations of these respective metal ions between a cation-exchange resin (Na-ion form) and solution (M $NaClO_4$) in the absence and presence of various concentrations of the respective polyacids was measured at different degrees of neutralization to facilitate this objective. The partition coefficient D_o (absence of polyacid) and D (presence of ligand)

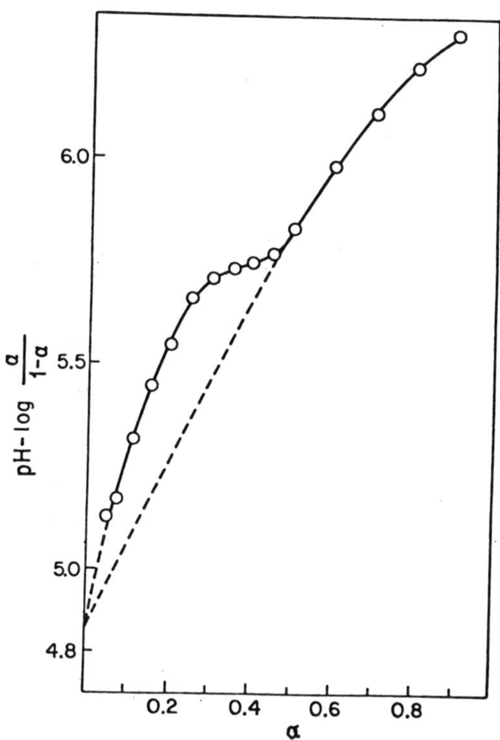

Figure 7. A representative plot of potentiometric titration data obtained with PMA.

bear the following relationship to β and $\underline{A}$ for M^{+2}-polyligand systems.

$$\frac{D_o - D}{DA} = \beta_1 \exp{-2\varepsilon\Psi_{(a)}/kT} + \beta_2 A(\exp{-2\varepsilon\Psi_{(a)}/kT}) \tag{16}$$

Analysis of the distribution results, with $\exp{-\varepsilon\Psi_{(a)}/kT}$ and $\underline{A}$ values directly available from pH measurements (binding of trace metal ion does not affect the stoichiometry of the system) made concurrently with the distribution measurements provided unambiguous verification of the general applicability of the non-ideality term so obtained in these systems. The experimental values of log $\frac{D_o - D}{DA}$ were plotted versus A. The data were extrapolated and the intercept of the ordinate defined the values of β_1. A plot of log $\frac{D_o - D}{DA\beta_1}$ versus pH-log $\frac{\alpha}{1-\alpha}$ -pK_{HA} then yielded a line with a slope of 2 which intersected the orgin. The ordinate, log D_o- D/DAβ_1 is equal to $\frac{2\varepsilon\Psi_{(a)}/kT}{2.3}$ + log $(1 + \frac{\beta_2}{\beta_1})$ while the abscissa (pH-log $\frac{\alpha}{1-\alpha}$ - pK_{HA}) is equal to $\varepsilon\Psi_{(a)}/2.3kT$; the observed result demonstrates that (1) essentially only the MA^+ species exist in these systems (i.e., $\beta_2 < 0.1\beta_1$) and (2) that the non-ideality of the divalent ion is defined by $\nu\varepsilon\Psi_{(a)}/kT$. These results are presented graphically in Figures 8 and 9.

iii. Evaluation of the Molal Activity Coefficient Ratio of Pairs of Mobile Counter Ions in the Charged Polymer Phase: We have seen that in dilute charged polymeric systems containing simple electrolyte deviation from ideality of the counterions present is identical. In describing the mass action expression for their distribution between polymer and electrolyte then, these deviation terms cancel. When the charged polymeric systems are as concentrated as they are in cross-linked resins, however, the excess free energy due to interaction of the exchanging ions with solvent becomes the important contributing factor to observed differences in counterion nonideality. Even if electrolyte imbibement by the gel is negligible the screening of polymer surface charge in the crowded three-dimensional array is sufficient to remove any selective ion "condensation" effects. Earlier we suggested that the initial linear portion of the plot of log $\gamma\pm/\gamma\pm_r$ versus log m, defined by $(\emptyset_r - 1)$ log $\frac{m}{m_r}$ provides an estimate of the electrostatic contribution to the mean molal activity coefficient $(\gamma\pm^{el}/\gamma\pm_r)$ over the complete concentration range of the polyelectrolyte. Its subtraction from the experimentally-based curve starting at values of log m near zero was then presumed to yield log $\gamma\pm^h/\gamma\pm_r^h$ as a function of the charged polymer concentration at the water activity of the equilibrated system. If we now consider the log $\overline{\gamma}_M/\overline{\gamma}_N$ term in eq. 4 derived with the Gibbs-Duhem model to be composed of electrostatic

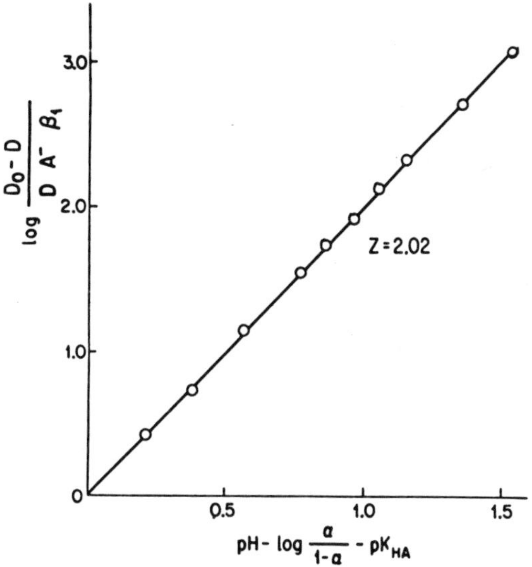

Figure 8. Demonstration of cancellation of Non-Ideality Terms (PAA).

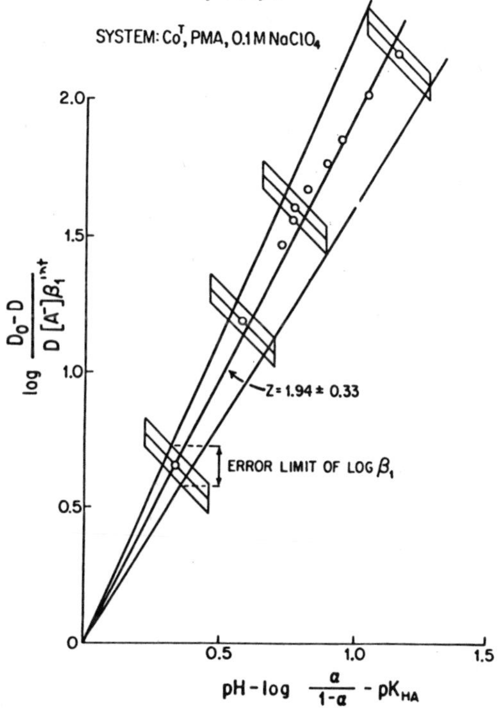

Figure 9. Demonstration of cancellation of Non-Ideality Terms (PMA).

and hydration terms as shown below

$$\log \frac{\bar{\gamma}_M}{\bar{\gamma}_N} = \log \frac{(\gamma_\pm^{el})_M}{(\gamma_\pm^{el})_N} \frac{(\gamma_\pm^{h})_M}{(\gamma_\pm^{h})_N} \frac{(\gamma_{\pm r}^{el})_N(\gamma_{\pm r}^{h})_N}{(\gamma_{\pm r}^{el})_M(\gamma_{\pm r}^{h})_M} \quad (17)$$

In the polymer

$$\gamma_\pm = (\gamma_+)^{\frac{Z}{Z+1}}(\gamma_-)^{\frac{1}{Z+1}} \quad (18)$$

where Z is the degree of polymerization. If $Z > 200$

$$(\gamma_+)^1(\gamma_-)^0 = \gamma_\pm = \gamma_+$$

and

$$\log \frac{\bar{\gamma}_M}{\bar{\gamma}_N} = \log \frac{(\gamma_\pm^{el})_M}{(\gamma_\pm^{el})_N} \frac{(\gamma_\pm^{h})_M}{(\gamma_\pm^{h})_N} \frac{(\gamma_{\pm r}^{el})_N}{(\gamma_{\pm r}^{el})_M} \frac{(\gamma_{\pm r}^{h})_N}{(\gamma_{\pm r}^{h})_M} \quad (18)$$

Since in the systems specified $(\gamma_\pm^{el})_M/(\gamma_{\pm r}^{el})_M$ is equal to $(\gamma_\pm^{el})_N/(\gamma_{\pm r}^{el})_N$

$$\log \frac{\bar{\gamma}_M}{\bar{\gamma}_N} = \log \frac{(\gamma_\pm^{h})_M}{(\gamma_\pm^{h})_N} \frac{(\gamma_{\pm r}^{h})_N}{(\gamma_{\pm r}^{h})_M}$$

At the reference concentration of 0.02 meq/ml, a_w, the activity of water, is very nearly unity and $\gamma_{\pm r}^h$ has to be unity as well so that $(\gamma_\pm^h)(\gamma_{\pm r}^h) = (\gamma_\pm^h)(1)$ and

$$\log \frac{\bar{\gamma}_M}{\bar{\gamma}_N} = \log \frac{(\gamma_\pm^{h})_M}{(\gamma_\pm^{h})_N} \quad (20)$$

at the experimental water activity; the value of $\bar{\gamma}_M/\bar{\gamma}_N$ is thus directly calculable in the ion-exchange resin at every experimental condition with $(\gamma_\pm^h)$ values resolved from $(\frac{\gamma_\pm}{\gamma_{\pm el}})$.

This same equation is obtained for $\log \bar{\gamma}_M/\bar{\gamma}_N$ even if we consider each repeating charged unit of the polymer to be osmotically active. With this estimate of the situation $\gamma_+^{el} = \gamma_\pm^{el} = \gamma_-^{el}$ so that upon cancellation of $\gamma_{\pm}^{el}$ $(\gamma_\pm^{el})$ in eq. 18

$$\log \frac{\bar{\gamma}_M}{\bar{\gamma}_N} = \log \frac{(\gamma_\pm^{h})_M}{(\gamma_\pm^{h})_N} \frac{(\gamma_{\pm r}^{h})_N}{(\gamma_{\pm r}^{h})_M} \quad (19a)$$

Marinsky and Högfeldt[35] have shown that at the higher concentration of polystyrene sulfonates the polyanion is essentially unhydrated so that $\gamma_{\pm}^{h} \simeq \gamma_{+}^{h}$; γ_{-}^{h} is very close to unity to make the $\bar{\gamma}_M/\bar{\gamma}_N$ term once again essentially identifiable with $(\gamma_{\pm}^{h})_M/(\gamma_{\pm}^{h})_N$ at the equilibrium a_w. At lower concentrations, however, $(\gamma^{h})_{-}$ may have a finite but small value somewhat greater than unity and the prediction of ion-distribution patterns in the less crosslinked resins would then be affected by neglect of γ_{-}^{h}.

iv. Test of Direct Approach to Ion-Selectivity Predictions[36,37]: The utility of eqn. 20 for predicting ion-exchange selectivity has been fully tested by demonstrating the anticipation of $K_N Ex^M$ without resort to a calibration step. For example, Boyd et al.[38] recently conducted an extensive study program designed to provide a rigorous thermodynamic basis for the prediction of the selectivity pattern of Zn^{+2} and Na^{+} during equilibration in $Zn(NO_3)_2$, $NaNO_3$ mixtures (0.1 N) of polystyrene sulfonate exchanger cross-linked to various degrees (0.5 to 24%) with divinylbenzene. The complete composition range of the resin (X_{Zn} from zero to 1) was covered in these studies. In order to facilitate the prediction of ion-exchange selectivity in this dilute electrolyte mixture they measured the water content of the pure ion forms and the mixed ion forms of the resins as a function of degree of cross-linking. These data were used in the Gibbs-Duhem equation for ternary mixtures through application of the cross differential identities which apply for exact differentials. In this way, they computed the activity coefficient ratio and the partial molar volume difference in the resin of the exchanging ions to predict ion-exchange selectivity coefficients for comparison with the experimental coefficients. In order to employ these data, however, it was necessary to calibrate first the activity estimates through experimental observations made over the complete composition range of the exchanger at one fixed cross-linking value (0.5% divinylbenzene).

We have compared in Table 4 the selectivity measurements made by Boyd et al.[38] of the NaPSS, $Zn(PSS)_2$, $Zn(NO_3)_2$, $NaNO_3$ system with predictions based upon (1) their rigorous thermodynamic analysis of the system and (2) our direct assessment of the ratio of $(\gamma_{Na}^{h})^2/(\gamma_{Zn}^{h})$ (as described earlier in this section) where $\log K_{Na}Ex^{Zn} = (\log(\gamma_{Na}^{h})^2(\gamma_{\pm Zn(NO_3)_2})^3/(\gamma_{Zn}^{h})(\gamma_{\pm NaNO_3})^4$, neglecting the $\pi\Delta V/2.3\ RT$ term which is relatively unimportant.

We see at once from Table 4 that the predictive quality of the procedure that we have developed here is fully as good as that provided by the Boyd et al.[38] approach. Their predictions are based upon calibration of the system through measurements of selectivity made with 0.5% crosslinked resin, whereas our predictions depend only upon resolution of $\bar{\gamma}_h$ from the log $\gamma_{\pm m}/\gamma_{\pm m}$, versus log

TABLE 4 (ref. 38)

COMPARISON OF LOG K_{Ex} PREDICTIONS WITH EXPERIMENTAL MEASUREMENTS

Divinyl-benzene (%)	$\log K_{Na}Ex^{Zn}$					
	$X_{Na} = 1$			$X_{Na} = 1$		
	Exp.	Predicted		Exp.	Predicted	
		Boyd et al.[38]	This lab		Boyd et al.[38]	This lab
2	0.096	0.153	0.04	0.10	0.19	-0.01
4	0.00	0.057	0.00	0.09	0.19	0.00
8	-0.22	-0.18	-0.13	-0.11	0.01	-0.20
12	-0.30	-0.33	-0.23	-0.21	-0.11	-0.36
16	-0.35	-0.47	-0.34	-0.42	-0.18	-0.52

m curves derived from our analysis of osmotic coefficient data obtained for the polymer analogues of the crosslinked resins.

In Table 5 we have presented selectivity predictions as well for pairs of ions made by the method we have introduced here. In this table we also include the experimentally measured values. Upon comparison of prediction with experiment one observes that when the macro-ion component is univalent (H^+) and the trace-ion component is divalent (Sr^{2+}, Cd^{2+}, Co^{2+}, Ca^{2+} and Zn^{2+}) agreement is quite good at the higher degrees of cross-linking (>4%). At the lower degrees of cross-linking (1 and 2%), the predicted value of $K_N Ex^{M^T}$ is too small. When the opposite experimental condition is tested, i.e. the macro-ion component is divalent (Zn^{2+}, Cd^{2+}, Sr^{2+} and Ca^{2+}) and the trace-ion component is univalent (Na^+), agreement is good for the Zn^{2+}, Na^+ system; with Cd^{2+} and Ca^{2+} the predicted $K_M Ex^{Na^T}$, while a factor of two too large, parallels the trend in selectivity with cross-linking. The computed value of $K_{Sr} Ex^{Na^T}$ is too large by a factor of almost 2 at low degrees of cross-linking but converges with the experimentally determined value at the highest degree of cross-linking (>8%). In the divalent-divalent systems the agreement between predicition and experiment is quite good for almost every pair of ions examined. Only for the Zn-Sr^T pair is their considerable discrepancy between prediction and experiment. However, the trend in selectivity is predictable with a ratio of approximately three existing between computation and experiment over the complete cross-linking range (1 to 16%).

We have found in many instances that the serious discrepancy between prediction and experiment for the various systems at the lower degrees of cross-linking can be removed by evaluation of γ_m^h at the molality of the macron-ion component rather than at the molality that the macro-ion form of the trace element corresponds to at the water activity of the equilibrium system. This observation, summarized in Table 6, may only indicate that the model we have employed becomes more fully applicable only after the concentration of PSS reaches a sufficiently large value. Indeed, neglect of ion-condensation at the lowest gel concentrations may be the responsible factor.

The remarkable effect of ion-condensation behavior on selectivity may be seen from a comparison of the exchange with PSS resins at different degress of crosslinking of Zn^{+2} and Na^+ in 0.1 M NO_3 and in 0.02, 0.05 and 0.12 M PSS[25] and of Co^{2+} and H^+ in varying concentration of $HClO_4$ and HPSS[22] (See Fig. 10).

It must be pointed out that use of eqn. 20 really needs to be restricted to the sulfonate concentration levels encountered in ion-exchange resins (> 0.5 m); in the more dilute linear polyelectrolyte

TABLE 5 (ref. 37)

SELECTIVITY PREDICTIONS FOR PAIRS OF IONS IN DILUTE ELECTROLYTE[22,23,24]

Divinyl-benzene (%)	0.168 M $HClO_4$		0.1 M $Ca(ClO_4)_2$		0.1 M $Zn(ClO_4)_2$		0.1 M $Cd(ClO_4)_2$		0.1 M Sr $(ClO)_4)_2$	
	$K_N Ex^{M^T}$ (Pred.)	$K_N Ex^{M^T}$ (Exp.)	$K_N Ex^{M^T}$ (Pred.)	$K_N Ex^{M^T}$ (Exp.)	$K_N Ex^{M^T}$ (Pred.)	$K_N Ex^{M^T}$ (Exp.)	$K_N Ex^{M^T}$ (Pred.)	$K_N Ex^{M^T}$ (Exp.)	$K_N Ex^{M^T}$ (Pred.)	$K_N Ex^{M^T}$ (Exp.)
Trace ion = Sr^{2+}										
1	1.41	3.21	0.90	1.56	0.88	2.83	1.30	2.14		
2	1.81	3.51	0.94	1.78	0.99	3.01	1.49	2.41		
4	3.06	4.30	1.05	1.72	1.34	3.59	1.80	2.91		
8	5.81	6.52	1.13	1.84	2.20	4.85	2.40	3.15		
12	9.49	10.88	1.27	1.87	3.30	7.13	2.88	3.99		
16	16.28	13.13	1.27	1.83	3.59	11.1	3.20	3.35		
Trace ion = Cd^{2+}										
1	1.22	1.82	1.01	0.76	1.04	1.03			0.99	0.60
2	1.39	1.89	0.99	0.73	0.99	1.14			1.02	0.55
4	2.02	1.97	0.91	0.62	1.12	1.16			0.90	0.53
8	2.81	2.19	0.80	0.56	1.21	1.42			0.74	0.44
12	3.31	2.54	0.72	0.56	1.50	1.80				
16	3.72	2.65	0.71	0.55	1.41	2.24				
Trace ion = Co^{2+}										
1	1.27	1.44	1.12	0.75	1.00	1.20			0.89	0.57
2	1.50	1.87	1.04	0.72	0.88	1.08			0.96	0.49
4	1.84	1.82	0.81	0.63	1.00	1.11			0.75	0.47
8	2.56	1.78	0.67	0.52	0.98	1.11			0.57	0.20

TABLE 5 (ref. 37) (Continued)

SELECTIVITY PREDICTIONS FOR PAIRS OF IONS IN DILUTE ELECTROLYTE[22,23,24]

Divinyl-benzene (%)	0.168 M $HClO_4$		0.1 M $Ca(ClO_4)_2$		0.1 M $Zn(ClO_4)_2$		0.1 M $Cd(ClO_4)_2$		0.1 M Sr $(ClO)_4)_2$	
	$K_N Ex^{M^T}$ (Pred.)	$K_N Ex^{M^T}$ (Exp.)	$K_N Ex^{M^T}$ (Pred.)	$K_N Ex^{M^T}$ (Exp.)	$K_N Ex^{M^T}$ (Pred.)	$K_N Ex^{M^T}$ (Exp.)	$K_N Ex^{M^T}$ (Pred.)	$K_N Ex^{M^T}$ (Exp.)	$K_N Ex^{M^T}$ (Pred.)	$K_N Ex^{M^T}$ (Exp.)
Trace ion = Ca^{2+}										
1	1.26	2.09	1.03*	1.18*			1.03	0.93	0.93	0.70
2	1.60	2.59	1.14*	1.26*			1.04	0.92	1.05	0.65
4	2.51	2.97	1.22*	1.37*			1.18	0.89	1.01	0.69
8	4.10	3.92	1.51*	1.49*			0.87	0.74	0.89	0.61
12	5.48	6.01	1.61*	1.88*					0.79	0.58
16	7.42	7.26	1.65*	1.40*					0.74	0.60
Trace ion = Zn^{2+}							Trace ion = Na^+			
1	1.17	1.71	1.01	0.49	0.96	0.81	0.98	0.49	0.96	0.54
2	1.35	1.75	0.87	0.48	0.88	0.71	0.86	0.38	0.93	0.50
4	1.84	1.68	0.66	0.42	1.03	0.68	0.91	0.51	0.79	0.61
8	2.15	1.72	0.81	0.42	1.29	1.24	1.11	0.50	0.68	0.57
12	2.35	1.66	0.85	0.48	2.22	1.81	1.35	0.65	0.71	0.64
16	2.64	1.50	0.86	0.49	2.56	2.53	1.51	0.90	0.71	0.71

*In 0.1 M $Co(ClO_4)_2$.

TABLE 6 (ref. 37).

SELECTIVITY PREDICTIONS AT LOW DEGREES OF CROSS-LINKING

System	Divinyl-benzene (%)	$K_N Ex^{M^T}$ (Pred.)	$K_N Ex^{M^T}$ (Exp.)
Zn^T, $HClO_4$	1	2.00	1.71
Cd^T, $HClO_4$	1	2.03	1.82
Ca^T, $HClO_4$	1	1.90	2.09
Na^T, $Cd(ClO_4)_2$	1	0.49	0.49
Na^T, $Ca(ClO_4)_2$	1	0.48	0.49
Na^T, $Sr(ClO_4)_2$	1	0.39	0.54
Na^T, $Sr(ClO_4)_2$	2	0.42	0.50
Sr^T, $Cd(ClO_4)_2$	1	1.52	2.14
Sr^T, $Cd(ClO_4)_2$	2	1.89	2.41
Sr^T, $Cd(ClO_4)_2$	4	2.47	2.91
Cd^T, $Sr(ClO_4)_2$	1	0.67	0.60
Cd^T, $Sr(ClO_4)_2$	2	0.66	0.55
Cd^T, $Sr(ClO_4)_2$	4	0.52	0.53

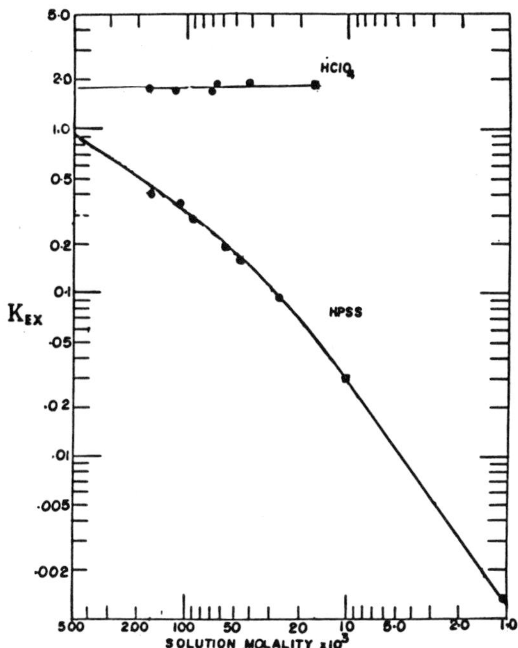

Figure 10. Variation of the selectivity coefficient, K_{EX}, with solution phase concentration for trace cobalt (o) and zinc (□) ion, hydrogen ion ion-exchanges (using polystyrenesulfonic acid ion-exchanger cross-linked with 8% divinylbenzene equilibrated in perchloric or polystyrenesulfonic acid solutions).

analogue systems the presence of simple electrolyte is required for eqn. 20 to be applicable. Only the presence either of salt or high concentrations of the fully ionized repeating functional unit with its mobile counter-ion provides sufficient screening of the highly charged polyion to reduce the remarkably selective "condensation" properties of the polyion. In fact the deviation of selectivity behavior from prediction in the lowest cross-linked resin is probably a consequence of such a residual ion-condensation effect.

There can be no question however, that the predictive quality of the new approach we have developed for the assessment of counterion distribution patterns in ion-exchange resins compares favorably with the less direct methods employed rather successfully in our earlier work.

We believe that fundamental assessment of the factors important in defining counterion distribution in charged polymers (cross-linked and linear) has been resolved with our interpretation of osmotic coefficient data. By our analysis, ion-solvent interactions are believed to contribute most importantly in the PSS-based resins to their ion-exchange selectivity patterns. The agreement obtained between the prediction of these patterns and their observed distribution without need to resort to a single measurement for calibration of the activity coefficient terms in the polymer provide strong support for the validity of the interpretation made.

v. The Elucidation of Ion Distribution Patterns in Weakly Acidic Cation-Exchange Resins: Our discussion of the prediction of ion-distribution patterns in cation-exchange resins has so far been limited to resins characterized by repeating functional units that are fully dissociated. When a weakly acidic cation exchanger is encountered, however, the acid dissociation equilbrium as well as the metal ion complexation potential of the repeating weakly acidic functional unit assume primary importance. These factors must be quantitatively analyzable to facilitate accurate assessment of the concentration of free,mobile ions in these gels. Without such an analysis the potential for correct evaluation of the mass transport properties of these resin systems is minimal.

We have already examined in some detail the non-ideality of charged polymeric systems (eqs 14 and 15; Fig 7) in order to facilitate our earlier analysis of ion-distribution patterns in fully dissociated cation-exchange gels. To achieve the additional insight required to provide the capability for accurate estimate of the free ion distribution in weakly acidic gels that is sought the following improved assessment of their complicated nature has been developed: The deviation from ideality that is accounted for in eq 15 ($2Zw$ or $\frac{\varepsilon\psi(a)}{kT}$) is not only a function of α, the degree of dissociation of the weak acid but is also a function of the ionic strength of the aqueous medium. For example, in a linear polyelectrolyte like poly (methacrylic acid) plots of pH - log $\frac{\alpha}{1-\alpha}$ versus α that are based on data obtained at different ionic strengths[32] result in a series of curves that converge as α approaches zero (Fig 11). At the highest α values were non ideality is greatest the divergence of the curves from the eventual intercept value at α = 0 is larger the lower the ionic strength. Eventually the difference in displacement of each of the curves from the intercept value becomes less as α becomes smaller until the curves merge or nearly merge when extrapolated to α =0. The point of convergence ($\pm$ 0.05 pK units) corresponds to the intrinsic pK_{HA} (pK_o) of the repeating functional unit: as $\alpha \rightarrow 0$, $\Psi_a \rightarrow 0$ and $\frac{\varepsilon\Psi(a)}{kT} \rightarrow 0$. The observed effect of ionic strength is due, of course, to its influence on the value of Ψ_a at each α value. The shielding of the charged surface by counterions is greater the higher the ionic strength to make the non-ideality term at a fixed α value smaller. At α = 0 where $\Psi_{(a)}$ is zero the value of the extrapolated pK value will be only moderately effected by the ionic strength of the medium leading to the $\pm$ 0.05 pK units discrepancy one can expect in the various extrapolations.

With crosslinked polyelectrolytes (gels) the pH of the gel is inaccessible to direct measurement. As a consequence, study of the potentiometric properties of weakly acidic gels has had to be based upon the pH of the external solution phase in equilibrium with the gel phase. The net result has been that these properties are observed to be a sensitive function of (1) the water content of the gel (which depends on the flexibility of its matrix) and (2) the ionic strength of the aqueous medium.

At a fixed ionic strength and α value, the $pK^{app}_{(HA)}$ of the repeating monomer unit appears to increase with the matrix rigidity which is in turn controlled by the quantity of crosslinking agent used in the gel manufacture. Indeed, it becomes possible to estimate the percent divinyl benzene employed in the crosslinking of a polymethacrylic acid gel from the effect of the resultant matrix flexibility on the value of pK. The effect of increasing ionic strength, on the other hand, is to decrease the measured pK.

In both instances, plots of $pK_{HA_{app}}$ (pK_z; pH-log $\frac{\alpha}{1-\alpha}$) versus α parallel each other, the value of pK^{app} as a function of α being higher the lower the ionic strength and the higher the rigidity (crosslinking) of the gel. Unlike the linear polyelectrolyte, no convergence of such curves is observed as α approaches zero (Fig. 12)[39] and a unique microscopic pK^{int}_{HA} (pK_o) cannot be resolved for the repeating functional unit of the macromolecule.

To understand this result we must return to our earlier analysis of the protolytic behavior of the weakly acidic linear polyelectro-

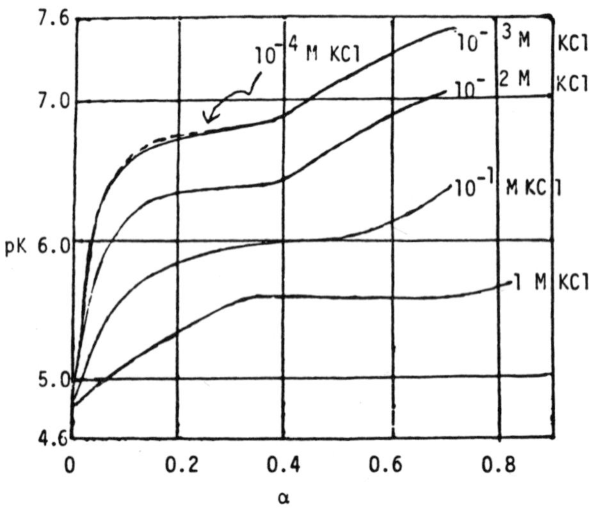

Figure 11. Potentiometric Titrations of PMA at different Ionic Strengths.

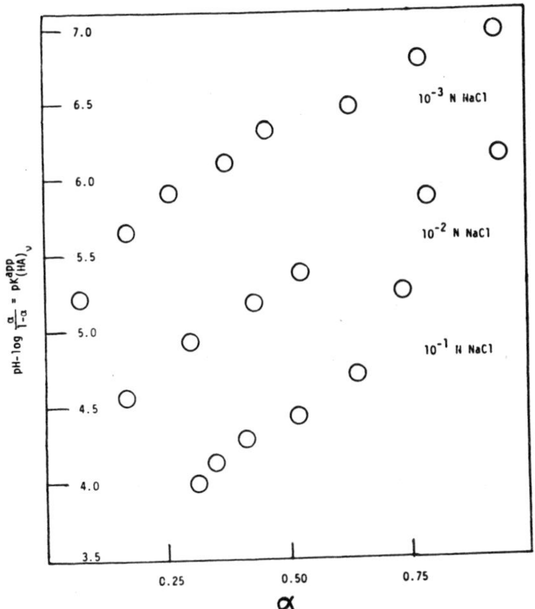

Figure 12. Potentiometric Data with Gel at Different Ionic Strengths.

lyte. Again let us define the intrinsic microscopic acid dissociation constant of the repeating functional group, HA, with eq. (9). However, at this point, it is important to realize that this system is most closely comparable in its equilibrium properties to a solution of the linear polyelectrolyte separated from simple salt solution at various concentrations by a membrane impermeable only to the macromolecule. By restricting partially the osmotic entry of water to the polyelectrolyte solution by the tension of a flexible spring attached to the membrane, the gel-simple salt system is approximated. In this situation, (H_a^+) must be related to (H_g^+), the activity of H^+ ion in the polyelectrolyte containing compartment rather than to H^+ in the aqueous medium. By assuming Boltzmann statistics apply as before (eq 15)

$$(H_a^+) = (H_g^+)e^{-\varepsilon\psi_a/kT} \tag{21}$$

with ψ_a now the potential difference between the charged polyelectrolyte framework and the bulk solution of the gel. Since (H_g^+) is not directly accessible experimentally, it must be calculated. Marinsky and Slota[40] have recently shown that in a system consisting of a solution of strong acid, HX, and its salt, MX, in equilibrium with a gel, the distribution of M^+ and H^+ between the solution and gel phases is given by[15]:

$$^*\frac{(H^+)}{(M^+)} = \frac{(H_g^+)}{(M_g^+)} \tag{22}$$

From eqs. (15) and (22) one then obtains:

$$K_{HA} = (H^+)\frac{(M_g^+)}{(M^+)} \times \frac{\alpha}{1-\alpha} \times e\frac{-\varepsilon\psi(a)}{kT}, \text{ or}$$

$$pK_{HA} = pH-\log\frac{\alpha}{1-\alpha} + p(M_g^+)-p(M^+)+\varepsilon\psi_{(a)}/kT \ln 10. \tag{23}$$

The value of (M_g^+), like the value of (H_g^+), is not measurable and must be calculated. For this purpcse, $[M_g^+]$, the concentration of M^+ in the gel phase is obtained first in the following way:[40] The volume of the gel phase is measured and a charge balance, based on (1) the stoichiometry of the neutralization reaction and (2)

---

*In this equation the osmotic pressure term $\exp\frac{(\pi(V_H - V_M)}{RT}$ is has has been omitted. In the case of H^+ and $Na^+(V_H - V_{Na}) = 0.0012$ ℓ/mole[41] and there is little error introduced from its neglect even when π is sizeable.

the requirement that the condition of electroneutrality be satisfied, is made to define the quantity of M^+ entering the gel during neutralization of HA. Imbibement of the gel phase by MX, presumed accountable by a simple Donnan equilibrium, corrects for the introduction of additional M^+. By introducing the value of $[M_g^+]$, so computed into eq. 23

$$pK_{HA} = pH - \log \frac{\alpha}{1-\alpha} + p[M_g^+] - \log \gamma_{M_g^+} - p(M^+) + \varepsilon\psi_a/kT \ln (10) \tag{24}$$

and $$pK_{HA_{app}} = pH - \log \frac{\alpha}{1-\alpha} + p[M_g^+] - \log \gamma_{M_g^+} - p(M^+) \tag{25}$$

then

$$pK_{HA} - \log \gamma_{M_g^+} - \varepsilon\psi_{(a)}/kT \ln (10) = pK_{HA_{app}} \tag{26}$$

Plots of $pK_{HA_{app}}$ versus α obtained from experiments at low ionic strength (and thus negligible imbibement of the gel by MX), when extrapolated to intercept the ordinate axis at α=0, will yield the pK_{HA} value of the repeating monomer unit, deviations from ideality being expected to vanish when the gel surface charge is zero.

The procedure, as outlined, has been applied to a crosslinked polymethacrylic acid gel (Amberlite/RD-50, Rohm and Haas Company) using dilute sodium polystyrenesulfonate solution in place of simple dilute electrolyte (MX).[40] The pK_{HA} values of 4.83 and 3.25, respectively, that were resolved are in excellent agreement with the pK_{HA} values obtained for their linear analogs[32,34,42] demonstrating the validity, as well as the utility, of this approach.

To demonstrate how great the discrepancy between $pK_{HA_{app}}$ values computed directly from the measured solution pH (eq 15) and the model-deduced $pH_{(g)}$ (eq. 25) can be these two sets of values obtained for the polymethacrylic acid gel system are plotted in Fig. 13 as a function of α.[40]

Finally, it is of great interest to observe the potentiometric behavior of the very flexible Sephadex gel when correction for the Donnan potential term that is incurred during neutralization with standard base of the gel suspended, respectively, in 0.10, 0.010 and 0.0010 M NaPSS has been carried out. The volume of the gel as a function of α in each of these media is presented in Fig. 14.[43] The $pK_{HA_{app}}$ versus α curves[43] obtained using eq 25 is presented in Fig 15 to demonstrate response typical of a linear polyelectrolyte (Fig 11). The effect of increased screening by counterions as the concentration of the gel (at each fixed α) increases resembles the effect of increased screening produced in a linear polyelectro-

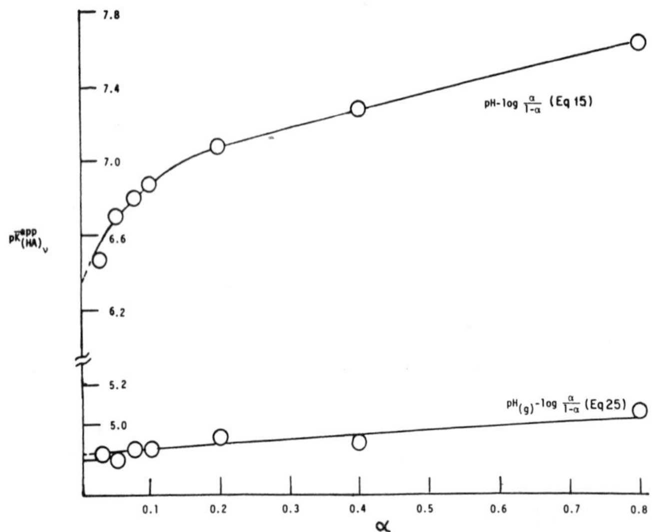

Figure 13. A Comparison of Potentiometric Titration Data Examined with Eq. 15 and Eq. 25.

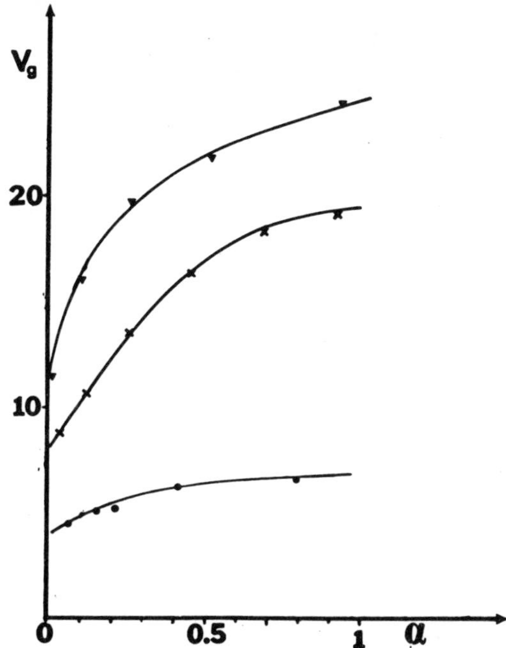

Figure 14. Variation of the volume, V_g, (in ml. meq. $^{-1}$) of Sephadex Cm-50-120 as a function of the degree of ionization α at different concentrations, C_p, of sodium polystyrene sulfonate (NaPSS); C_p = 0.1 N (o), C_p = 0.01 n (x), C_p = 0.001 N (▼).

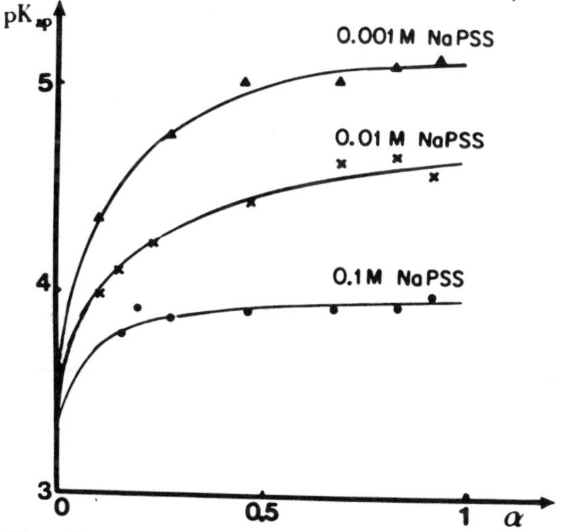

Figure 15. Apparent pK of Sephadex CM-50-120: a plot of pK_{app} versus the degree of ionization α; the pK_{app} is calculated with Eq. 25.

lyte by increasing the ionic strength of the medium while keeping the polymer concentration constant.

In the weakly acidic cation exchange gels where the distribution pattern of univalent and multivalent metal ion is during neutralization with standard base, additionally affected by complexation of the multivalent ion by, for example, carboxylate ion, which may be the repeating functional unit of the exchanger, the above approach can be used as well to computed the concentration of free M^{++} in the gel. Starting with eq 4b

$$\frac{(M^{\nu+})}{(Na^+)^{\nu}} \simeq \frac{(M_g^{\nu+})}{(Na_g^+)^{\nu}} \tag{22a}$$

Since activity coefficient ratios in the resin are accessible as described earlier the accurate assessment of free ion distribution patterns essential to the proper examination of mass transport parameters in the gel becomes practicable in these complicated systems.

iv. A Test of the Predictive Quality of Free, Mobile Ion Distribution Computations in Weakly Acidic Ion-Exchange Resins A number of years ago this author examined the complexation of Co(II), Ni(II) and Cu(II) by the weakly acidic IRC-50 resin, a copolymer of polymethacrylic acid and divinyl benzene marketed by Rohm & Haas[44,45]. In the examination of these systems 0.100 g of the resin (hydrogen form) was equilibrated in 100.0 ml of 0.05 M Na_2SO_4 solution containing MSO_4 at a total millimol content respectively 1/2, 1/4 and 1/6 that of the measured hydrogen capacity (0.100 ± 0.0005) of the resin sample. In each set of solutions the system was allowed to come to equlibrium after the resin sample was brought to different degrees of neutralization (0.1 to 0.9) with standard sodium hydroxide. The concentration of M(II) was measured polarographically and the activity of H^+ was measured potentiometrically in each equilibrated sample.

Analysis of the data was affected by comparing the exchange of H^+, rather than Na^+, with M(II). With this approach

$$[\overline{M}] \simeq \sum_{n=1} \overline{MA}_n \text{ and } [\overline{H}] \simeq [\overline{HA}] \tag{27}$$

$$\text{since } [\overline{M}(II)] << \sum \overline{MA}_n \text{ and } \overline{H^+} << [\overline{HA}] \tag{28}$$

and the treatment of the data was straightforward. By describing the equilibrium as follows:

$$K_H Ex^{M(II)} = \sum_{n=1} \overline{MA}_n (H^+)^2 (A)/(\overline{HA})^2 M(II) \tag{29}$$

the presence of complex species was resolved through a plot of $K_H Ex^{M(II)}$ versus (A) (refs. 44, 45, 46). Such a plot should yield a line intercepting the ordinate at a value equal to $(\beta_{MA^+})/(\beta_{HA})^2$ If only the MA^+ species is present the line will be straight and its slope will be zero. If a second species is present concurrently the slope of the line will correspond to $\beta_{MA_2}/(\beta_{HA})^2(\overline{V}_g)$ where V_g is the volume of the exchanger phase.

It was found that Co(II) and Ni(II) formed only one complexed species, the $M(II)A^+$ ion pair, whereas the $M(II)A_2$ species was formed as well with Cu(II).[47] To test the capability claimed for expressing the distribution of univalent and multivalent ions in these kinds of systems we have compared the value of β_{NiA^+} determined as described above with the value computed for this complex by comparing the exchange of Ni^{+2} and Na^+, lower, ${}^{exp}K_{Na}Ex^M$, in this system.

With this approach the experimentally observed gross concentration of Ni(II) in the exchanger corresponds to $(Ni_g^{+2} + NiA_g^+)$ so that

$$ {}^{Exp}K_{Ex}{}^{Ni}_{Na} = \frac{(Ni_g^{+2} + NiA)(Na^+)^2}{(Ni^{+2})\qquad(Na_g^+)^2} \tag{30}$$

Since $NiA^+ = \beta_{NiA^+}(A^-)(Ni_g^{+2})\exp(-2\varepsilon\psi_a/kT)$ (31)

$$ {}^{exp}K_{Na}Ex^{Ni} = \frac{(Ni_g^{+2})(Na^+)^2}{(Ni^{+2})(Na_g^+)^2}(1+\beta_{NiA^+}(A)\exp(-2\varepsilon\psi_a/kT)) \tag{32}$$

The quotient defined by $\frac{(Ni_g^{+2})(Na^+)^2}{(Ni^{+2})(Na_g^+)^2}$, ${}^{Th}K_{Na}Ex^{Ni}$, if $(Ni_g^{+2})/(Na_g^+)^2$ is predictable with eq 20 as claimed, should yield the value of β_{NiA^+} when used in eq 33 as shown

$$ \frac{{}^{Exp}K_{Ex}{}^{Ni}_{Na}}{{}^{Th}K_{Ex}{}^{Ni}_{Na}} - 1/(A^-)\exp(-2\varepsilon\psi_a/kT) = \beta_{NiA^+} \tag{33}$$

In expressing ${}^{Exp}K_{Na}Ex^{Ni}$ with eq 33 the concentration of Ni and Na^+ in the gel was based on the volume of the gel, V_g. To evaluate the total millimols of Na entering the gel the quantity of Na was presumed to equal to A by preserving electroneutrality in the gel phase; any invasion of the gel by Na_2SO_4 was neglected.

In the earlier studies[44,45] V_g was not measured. However, in a recent examination of the potentiometric property of IRC-50 volume measurements were made over its complete neutralization range.[40] We have presumed that in this rather inflexible resin these volume measurements are applicable in the presence of Ni(II) as well.

The results of these computations are summarized in Table 7 for the $NiSO_4$, IRC-50, 0.05 M Na_2SO_4 system investigated previously[44]. The earlier computations of β_{NiA^+} are also included in the table for comparison. To assure an equivalent basis for comparison correction for the deviation of $NiSO_4$ from ideal behavior has been made as before. The values of β_{NiA^+} deduced with eqn. 33 are in excellent agreement with the earlier computation of this parameter to confirm the capability claimed for expressing quantiatively the distribution of univalent and multivalent ions in weakly acidic cation exchangers whose repeating functional unit is capable of complexing the multivalent ion.

CONCLUSION

In order to facilitate the consideration of mass transport properties of ion-exchange resins in particle or membrane form the factors affecting the accessibility of mobile counter ions in the gel phase must be well understood. Such understanding is demonstrated in the text of this manuscript by the demonstration of a capability for (1) predicting ion-exchange selectivity and for (2) analyzing metal complexation reactions in resins. In the first instance the important role of ion solvent interaction in the selective behavior of a cation-exchange resin (crosslinked polystyrene sulfonate) has been quantitatively established. This information can be transcended to the knowledgeable consideration of ion-exchange rates. In the second case the concentration of counterious in a weakly acidic resin (crosslinked polymethacrylic acid) has been shown to be resolvable as well even during competitive complexation of H^+ and $M^{\nu+}$ to facilitate the interpretation of mass transport properties in these gels. The capability for quantiative estimate of metal complexation by the repeating functional unit that may constitute a resin as well as the concentration of the free mobile ions in the exchanger is essential to analysis of ion-exchange rates in these materials.

References

1. Gregor, H. P. J. Amer. Chem. Soc., 70, (1948) 1293.
2. Gregor, H. P., ibid., 73, (1951) 642.
3. Ekedahl, E., Hogfeldt, E. and Sillen, L. G., Acta Chim. Scand., 4, (1950) 556.
4. Hogfeldt, E., Ekedahl, E. and Sillen, L. G., ibid., 4, (1950) 828.
5. Bonner, O. D., Argersinger, W. J., Jr. and Davidson, A. W., J. Amer. Chem. Soc., 74, (1952) 1044.
6. Glueckauf, E., Proc. Roy. Soc., Ser. A, 214, (1952) 207.
7. Duncan, J. F., Aust. J. Chem., 8, (1955) 293.
8. Soldano, B. and Chestnut, D., J. Amer. Chem. Soc., 77, (1955) 1334.
9. Myers, G. E. and Boyd, G. E., J. Phys. Chem., 60, (1956) 521.
10. Boyd, G. E., Lindenbaum, S. and Myers, G. E., ibid., 65, (1961) 577.
11. Feitelson, J., ibid., 66, (1962) 1295.
12. Marinsky, J. A., ibid., 71, (1967) 1572.
13. Helfferich, F., "Ion Exchange", McGraw-Hill, New York (1962).
14. (a) Donnan, F. G. and Guggenheim, E. A., Z. Physik. Chem. (Leipzig) 162A, (1932) 356; (b) Donnan, F. G., ibid., 168A, (1934) 369.
15. Bukata, S. and Marinsky, J. A., J. Phys. Chem., 68, (1964) 994.
16. Robinson, R. A. and Stokes, R. H., "Electrolyte Solutions", 2nd ed., Butterworth Scientific Publications, London (1959).
17. Robinson, R. A., J. Phys. Chem., 65, (1961) 662.
18. Soldano, B. and Larson, Q. V., J. Amer. Chem. Soc., 77, (1955) 1331.
19. Soldano, B., Larson, Q. V. and Myers, G. S., ibid., 77, (1955) 1339.
20. Reddy, M. M., Marinsky, J. A. and Sarkar, A., J. Phys. Chem., 74, (1970) 3891.
21. Reddy, M. M. and Marinsky, J. A., ibid., 74, (1970) 3884.
22. Reddy, M. M. and Marinsky, J. A., J. Macromol. Sci-Phys., B5(1), (1971) 135.
23. Reddy, M. M., Amdur, S. and Marinsky, J. A., J. Amer. Chem. Soc., 94, (1972) 4087.
24. Marinsky, J. A., Reddy, M. M. and Amdur, S., J. Phys. Chem., 77, (1973) 2128.
25. Yang, R. and Marinsky, J. A., J. Phys. Chem., 83, (1979) 2737.
26. Manning, G. S., J. Chem. Phys., 51, (1969) 924, 3249.
27. Boyd, G. E. and Bunzl, K., J. Amer. Chem. Soc., 96 (1974) 2054.
28. Baldwin, R., Ph.D. Thesis, State University of New York at Buffalo, Buffalo, NY, 1978.
29. Linderstrøm-Lang, Compt. Rend. Trav. Lab, Carlsberg, 15, (1924) No. 7.

30. Scatchard, G., Ann. N.Y. Acad. Sci., 51, (1949) 660.
31. Tanford, C., J. Amer. Chem. Soc., 72, (1950) 441.
32. Arnold, R. and Overbeek, J. Th. G., Rec. Trav. Chim. Phys.-Bas, 69, (1950) 592.
33. Nagasawa, M. and Holtzer, A., J. Am. Chem. Soc., 86, (1964) 538.
34. Travers, L. and Marinsky, J. A., J. Polymer. Sci., Symposium No. 47, (1974) 285.
35. Marinsky, J. A. and Hogfeldt, E., Chemica Scripta, 9, (1976) 233.
36. Marinsky, J. A., J. Chromatography, 201, (1980) 5.
37. Marinsky, J. A., Reddy, M. M. and Baldwin, R. S., "Water in Polymers", ACS Symposium Series No. 127, Amer. Chem. Soc., edited by S. P. Rowland, (1980) 387.
38. Boyd, G. E., Myers, G. E. and Lindenbaum, S., J. Phys. Chem., 78, (1974) 110.
39. The data for Fig. 12 are from Marinsky, J. A., Bunzl, K. and Wolf, A., Talanta, 27, (1980) 461-8.
40. Slota, P. and Marinsky, J. A., "Ions in Polymers", ACS Symposium Series No. 187, Amer. Chem. Soc., edited by A. Eisenberg, (1980) 313.
41. Mukherjee, P., J. Phys. Chem., 65, (1961) 740.
42. Gekko, K. and Nozuchi, H., Biopolymers, 14, (1975) 2555.
43. Merle, Y. and Marinsky, J. A., presented for publication in Talanta (1982).
44. Marinsky, J. A. and Anspach, W. M., J. Phys. Chem., 79, (1975) 433.
45. Marinsky, J. A. and Anspach, W. M., J. Phys. Chem., 79, (1975) 439.
46. Marinsky, J. A., Coord. Chem. Rev., 19, (1976) 125.
47. Marinsky, J. A., J. Phys. Chem., 86, (1982) 0000.

THE MOLECULAR BASIS OF IONIC SELECTIVITY IN MACROSCOPIC SYSTEMS*

George Eisenman

Dept. of Physiology
UCLA Medical School
Los Angeles, Ca. 90024

1 INTRODUCTION

The purpose of this paper is to review the fundamental principles underlying ionic selectivity in "macroscopic" systems such as ion exchangers and ion exchange membranes, whose counterion concentrations are constrained by electroneutrality. I will bring up to date an earlier theory of equilibrium selectivity (1,2) so as to include the considerable progress that has occurred since 1961, particularly for species other than the Group Ia cations. However, I will not deal in detail with kinetic aspects of selectivity since these are more appropriately covered elsewhere in the context of biological membranes (3) which, because of their extreme thinness, can deviate from electroneutrality locally. Indeed, much of the progress in selectivity has occurred in the field of "microscopic" (e.g., biological) membranes, whose thickness (<100 Angstroms) is small relative to the Debye length so that the usual electroneutrality constraint does not apply locally. This has the consequence that, whereas the sites of macroscopic systems are always effectively "saturated" (i.e., completely occupied), those in biological membranes are often empty, so that they are not necessarily constrained to strict ion exchange behavior and are said to obey "independence" (3). It should also be appreciated that equilibrium selectivity considerations can, in principle, be generalized so that they can be applied to such non-equilibrium phenomena as membrane potential selectivity. This is done by applying the same energetic principles as for equilibrium selectivity to the energetics of the "transition states" of

*Supported by the USPHS (GM 24749) and the NSF (PCM 76-20605).

Eyring's theory of absolute reaction rates (4). The activation energy of these transition states underly the observed mobility, and this procedure was first proposed by Hille (5) for the analysis of cell membrane potentials and currents.

2 QUALITATIVE FORMULATION OF SELECTIVITY PATTERNS FOR GROUP Ia CATIONS

The existence in chemistry and biology of specific ionic effects following neither the simple sequence associated with Hofmeister series of hydrated radii nor that of the naked atomic radii but in apparently quixotic "irregular" sequences was recognized a long time ago (6), and a number of suggestions were made quite early to account for these. The first was Jenny's (7,8) proposal that such sequences could be interpreted as various stages in a transition between hydrated and dehydrated sequences where "the most hydrated ion will be the first affected by the dehydration process." The factors governing dehydration were unspecified until Bungenberg de Jong (9) suggested an atomic parameter, the "polarizability" of the interacting anion, as a possible cause for a series of "polarizability sequences" which he explicitly enumerated. These sequences, which are the ones implicitly required by the Jenny postulate, never have ions other than Li or Cs as the most preferred species and therefore do not correspond to the selectivities observed in ion exchangers, glass electrodes, or biological phenomena, in which the K or Na preferences are noteworthy. Gregor and Bregman (10) interpreted the selectivity of organic cation exchangers in terms of a theory similar to Jenny's but, not having data for Rb and Cs, did not notice its inadequacy.

Eisenman et al (11), as quoted in (12), starting from a large data base of glass electrode selectivity in which preferences for K or Na were also prominent features (13), and recognizing the existence of a parallel selectivity in ion exchanger minerals, postulated that the glass was an ion exchanger membrane and proposed that a completely different (in fact, opposite) set of selectivity sequences would be expected to occur from variation of the "electrostatic field strength" of an interacting anionic site. The "Eisenman sequences" constituted a pattern of eleven particular orders of selectivity, listed below, which were expected to occur with increasing negative electrostatic field strength of the interacting site (the quantitative basis for these sequences will be given shortly).

Li > Na > K > Rb > Cs	Sequence XI	(at highest field strength)
Na > Li > K > Rb > Cs	" X	
Na > K > Li > Rb > Cs	" IX	
Na > K > Rb > Li > Cs	" VIII	
Na > K > Rb > Cs > Li	" VII	
K > Na > Rb > Cs > Li	" VI	
K > Rb > Na > Cs > Li	" V	
K > Rb > Cs > Na > Li	" IV	
Rb > K > Cs > Na > Li	" III	
Rb > Cs > K > Na > Li	" II	
Cs > Rb > K > Na > Li	" I	(at lowest field strength)

Notice in these sequences that each alkali cation, in turn, is favored and moreover that a particular "pattern" of selectivity inversions occurs. These sequences were shown to have a wide natural occurence and were subsequently generated (1) on a quantitative basis from simple models for the relevant Gibbs´ free energies of binding vs. hydration (also see Fig. 4.9 of (14)). Interestingly, the "Eisenman sequences" are the reverse of the "polarizability" sequences of Bungenberg de Jong and Jenny, which can be obtained simply by reading these sequences from back to front (i.e., by replacing the symbol > by <). The reasons for this have been discussed elsewhere (3) and will be briefly noted under TOPOLOGY where it is pointed out that the particular "Eisenman sequences" correspond to the situation where a selectivity optimum for each alkali cation occurs as a monotonic function of cationic size; whereas the "polarizability" sequences correspond to a selectivity minimum for each cation.

Underlying the *qualitative* selectivity rule proposed by Eisenman et al (11) was the assumption that *with increasing negative electrostatic field strength the least strongly hyrated cation would be the first to be desolvated* (see Fig. 1 of (3). The atomistic asymmetry underlying this assumption was made explicit in Eisenman´s further analyses (1,2) of the atomic interactions underlying simple models for cation-site vs. cation water interaction energies. Consider Fig. 1 which plots the hypothetical selectivity (discused below) which would be observed solely as a consequence of the electrostatic interactions between a cation and a monopolar anionic site vs. those between a cation and a multipolar water molecule. In this and all subsequent figures the cation most strongly selected from water by an anionic "site" is the lowest on the chart. Above the graph are tabulated the cationic sequences (increasing specificity downwards), corresponding to the eleven rank order designations described above. The model in Fig. 1 is purely heuristic and is not meant to apply literally to any realistic ion exchange system (more realistic models will be described below). It does illustrate how simply the elementary asymmetry between ion-site vs. ion-water interactions can lead for a monopolar site to the set of "Eisenman sequences" for the free energy of ion exchange.

It is of interest that a simple multiplication of the cation-water energies of Fig. 1 by a factor of 4.8 yields a fair prediction of the experimentally observed differences in cationic hydration energies (cf. Fig. 7). This is physically reasonable since the total hydration energy is primarily that of the four-fold coordinated first hydration shell plus a smaller additional energy from orientation of more distant water molecules.

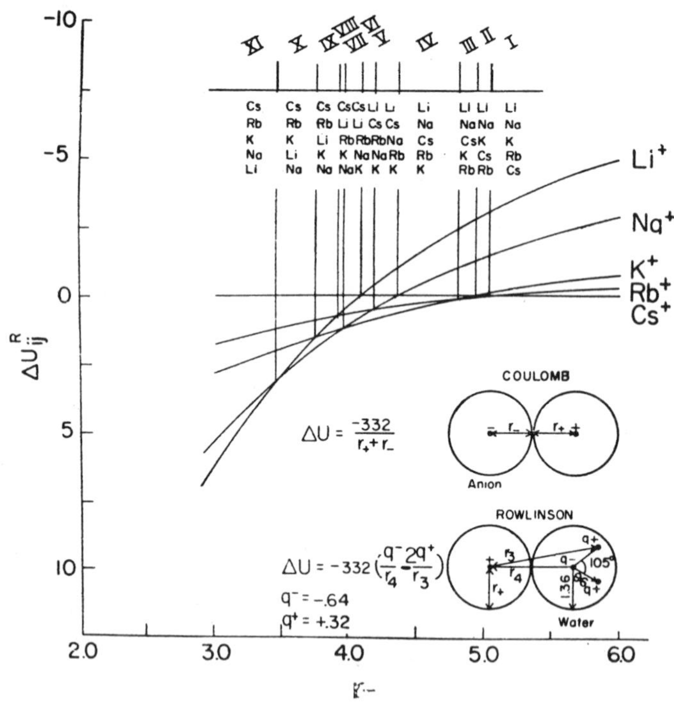

Fig. 1
The atomistic asymmetry underlying the cationic selectivity rule as illustrated by the hypothetical cation exchange between monopolar anion and a single multipolar water molecule (after Rowlinson). Units in kcal./mole and Å in all Figs.

3 QUANTITATIVE FORMULATION OF GROUP Ia CATION SELECTIVITY

A major advance in selectivity theory was the recognition (1) that equilibrium selectivity among the group Ia cations could be rigorously formulated as a balance between the energies of ion-water vs. ion-site interactions which were calculated for a variety of hypothetical situations in a number of ways, all of which yielded the aforementioned set of 11 sequences with only minor anomalies. The calculations included: (A) a purely heuristic electrostatic model for the competition between a single multipolar water molecule and a single monopolar site, (B) more physically realistic coulombic models for ion-site interactions in 1-fold and 6-fold coordination states using experimentally known hydration energies, (C) halide models for sites using thermochemical data for the energies of ion-site interaction in various coordination states (ion pairs or crystal lattices) and for hydration energies, and (D) models using free-solution data for the effects of varying the degree of hydration, where the energies of a given hydration state were assessed from the known free energies of dilution for the halide salts. Because the conclusions are sometimes mistakenly considered to be restricted to electrostatic models, it should be pointed out that the latter two classes of calculations contain all the energy terms which contribute to the thermochemical data and are therefore not restricted to coulomb energy considerations. The rationale for their use is the simple one of using experimental data for the halides, whenever available, as prototypes for anionic groups more generally (a useful relationship between the field strengths of of halides and oxyanions, via the pK_a, has been given in Fig. 9). Although these models were explicit, quantitative, and unambiguous in their predictions, it should be emphasized that they did not purport to be *ab initio* calculations for any real molecular structure except for crude calculations (2) for the silicate and aluminosilicate binding sites of glass.

It will probably be helpful in following the detailed arguments to summarize at the outset the principal conclusions we will reach regarding the factors governing selectivity among the alkali metal cations and H^+. Three important factors are identified. (I) The principal factor is the electrostatic field strength of the anionic groups with which alkali metal cations and H^+ interact. (II) The number of water molecules present in the vicinity of the interacting cations and anionic group influences the magnitude, but not the sequence, of specificity. (III) The spatial distribution of the individual anionic groups also influences selectivity through the overlap of the electrostatic forces of the groups. Decreasing spacing between sites leads to an increase in "effective field strength" without changing the overall selectivity pattern among ions having the same charge; however it affects the pattern of selectivity between cations of different charge such as K^+ and Ca^{++}, as will be shown below.

3.1 Free Energy and Equilibrium Selectivity

The selectivity of a typical cation exchange reaction is given by its the Gibbs' free energy change, F_{ij}:

$$IX\ (\text{state } s) + J^+\ (\text{aqueous, dilute}) \rightarrow JX\ (\text{state } s) + \\ + I^+\ (\text{aqueous, ditute}) + \Delta F_{ij} \qquad (1)$$

IX and JX represent an ion exchanger in state s (where s can be any one of the states of Fig. 2 which can be used to model appropriate degrees of ligand coordination state, hydration, etc., as will be described below). F_{ij} is related to the equilibrium constant of the exchanger by Eq. 2 and to its membrane potential, E, by Eq. 3.

$$\Delta F_{ij} = -RT\ln K_{ij} \qquad \text{where } K_{ij} = \frac{(I^+)\,JX}{(J^+)\,IX} \tag{2}$$

$$E = \text{const.} + \frac{RT}{F}\ln\,[(I^+) + K_{ij}\,(J^+)] \tag{3}$$

The permeability ratio, P_j/P_i, in Eq. 3 determines the selectivity and has been shown elsewhere (cf. 15) to be the product of the ion exchange equilibrium constant and the mobility ratio, u_j/u_i.

$$P_j/P_i = K_{ij} \times (u_j/u_i) \tag{4}$$

3.1.1 Membrane Potentials. A word on membrane potentials is in order here. Since the specific ion permeability of any membrane is determined jointly by the mobility of ions within the membrane together with the equilibria at the membrane-solution interfaces (recall Eq. 4) it is meaningful to examine the physical bases for equilibrium specificity and mobility specificity separately even though both depend, at least partially, on the same atomic forces (this is discussed in more detail in (3)). The present paper summarizes the principal atomic factors controlling equilibrium ionic specificity. A theoretical analysis of the origin of mobility specificity, which would be desirable for completeness, is yet to be accomplished; but see (15) for discusion of these factors in the glass electrode membrane potential.

3.1.2 The Energies Underlying Equilibrium Selectivity. What do we know of the energies underlying the equilibrium selectivity of Eq. 1? Those representing the interactions of the cations with water are constants characteristic of each cation and are known experimentally with great accuracy. The differences of their values referred to Cs^+ in Kg-cal per mole, as extracted from Rossini et al. (16) are: H^+,-192.7; Li^+, -54.3; Na^+,-30.4; K^+,-12.7; Rb^+,-7.7; Cs^+, 0. On the other hand, the free energies of interaction of these ions with the exchanger depend upon details of the chemical composition of the exchanger. The problem of analyzing the origin of equilibrum specificity is thus reduced to the problem of characterizing the affinities of the various cations for an ion exchanger as a function of its pertinent chemical composition.

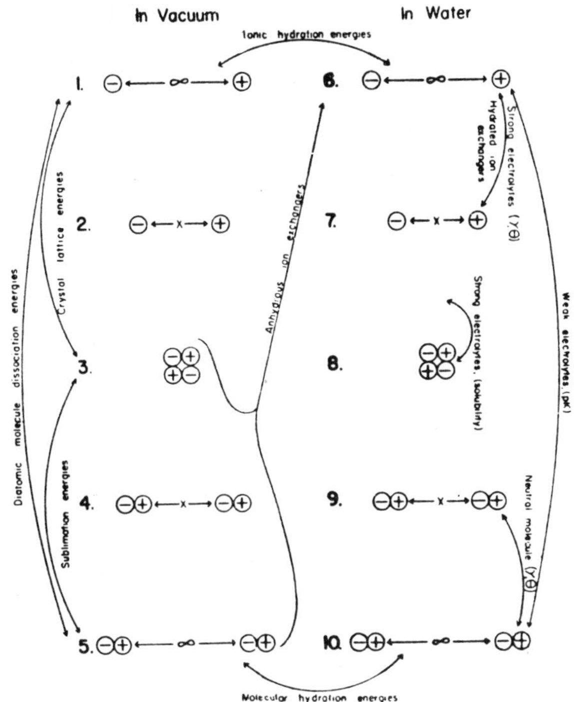

Fig. 2
The states of ions and their interconnecting processes. States 1–5 have no matter intervening between the ions. States 6–10 have water as represented by the shading. The degree of fixation of the ions in the states is unspecified here, since it does not enter into the internal energy calculations. The arrows indicate common phenomena representable by transitions between these states.

The rigorous calculation of the energies of cation interaction with such molecular anionic groups as might be found in a typical ion exchanger is yet to be accomplished (but see the very recent ab initio calculations by A. Pullman (53)). Nevertheless, an approximation is feasible. The most obvious simplification is to represent the principal features of the interaction between cations and sulphonate or carboxylate groups by a monopolar (e.g. halide) model. Under this approximation, there are two, mutually complementary, methods of evaluating the free energy. In one, referred to as the "thermo-chemical" method, the empirical methods of thermochemistry are used to evaluate the free energy change. This method provides precise values for the free energy, independent of a knowledge of the nature of forces involved or of the details of molecular structure. The other, the "theoretical" method, involves calculation of the free energy change from considerations of the elementary forces involved. Examples of both methods will be given below.

3.2 The States of Ions and Their Interconnecting Processes

The considerations of Fig. 1 can be extended to theoretical calculations of the specificity required by the macroscopic free energy changes (resulting from the differences in microscopic elementary atomic interactions) which occur when ions are taken between appropriate states as summarized diagrammatically in Fig. 2. Fig. 2 Schematizes the states in which cations and sites (represented by anions) can exist in nature. States 1-5 have no matter intervening between the species; while states 6-10 are supposed to have water intervening. The arrows indicate common phenomena representable by transitions between these states. For example, the hydration energy of a cation is represented by the arrow labelled "ionic hydration energies" which represents the process (Process 1-6) of taking an ion from vacuum (state 1) and transferring it into water (state 6). The corresponding free energy change for this process can be symbolized as F^{16}, where superscripts indicate the initial and final states.

The energies of these states can be calculated from simple theoretical models for them (as will be exemplified in Fig. 3) or, alternatively, can be assessed from thermochemical or other experimental data (as will be exemplified in Fig. 4). They can also be modelled by a combination of such procedures, for example by theoretical calculations for ion-site interaction energies while using experimentally known values for the hydration energies. The figure is worthy of close study since many equilibrium specificity phenomena (indicated by the labels beside the arrows) can be represented by transitions among the states of this figure, and the nomenclature used in this paper corresponds to the states. For example, the process of hydration of a cation, I^+, is represented by its transition between states 1 and 6 (Process 1-6); and the corresponding free energy change is symbolized by ΔF_I^{16}.

3.3 Typical Examples of Simple Calculations Based on Coulomb and Born Energies in Anhydrous systems.

A particularly simple case for a prototype ion exchange is one in which the anionic "sites" (X^-) are contained at various separations in a three-dimensional phase which totally excludes water molecules. Theoretical selectivity isotherms for this situation are illustrated in Fig. 3.

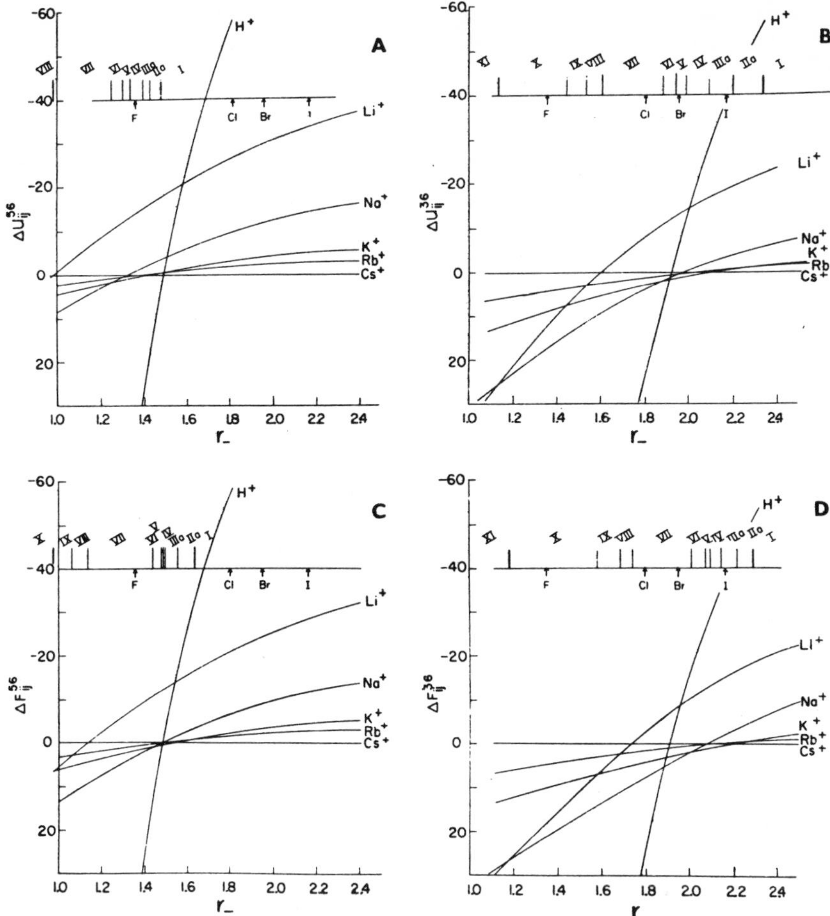

Fig. 3
Theoretical specificity isotherms for H, Li, Na, K, Rb, and Cs as a function of r_-. Internal energies and free energies of ion exchange between anhydrous exchange phases and dilute aqueous solutions, are presented as calculated theoretically from equations 5, 6.

A and C: internal energy and free energy for infinitely separated sites.

B and D: internal energy and free energy for six-fold coordinated sites.

In B and D the factor 1.56 is also used for H^+. The Roman numerals refer to the regions of field strength in which the indicated selectivity rank order prevails. The vertical demarcation lines between the numerals correspond to the intersections of the isotherms in this and all succeeding Figs. F, Cl, Br, I refer to the radii of these ions.

The energy change underlying the selectivity of this situation is given by the difference of the energies of hydration of the ionic species in the aqueous phase (process 1-6 of Fig. 2) vs. the difference of their energies of interaction in the ion exchange phase (process 1-3 for a crystal lattice model, or process 1-5 for a diatomic ion pair model). Both internal energies (U) and Gibbs' free energies (F) are plotted, with appropriate subscripts and superscripts.

Since the properties of I^+, J^+; and H_2O are constant, the only variables upon which these energies can depend are the state (S) of the species IX and JX and the force field of the anion (represented by its radius, r_-). Moreover, since we are interested in the theoretical effects of varying the exchanger phase, experimental values have been used for the hydration energies, rather than assessing them theoretically (but cf. Fig. 7 for an indication of the satisfactory agreement between several theoretical calculations of differences of hydration energies and experimentally determined values). References (1) and (2) should be consulted for further details.

The state of the species IX and JX depends only upon (a) the separation between the anions X^-, (b) their degree of mobility or fixation, and (c) the extent to which particles other than I^+ and J^+ (e.g. water molecules) are admitted into their vicinity. For the present example we are considering water to be totally excluded from the ion exchanger phase, a restraint which will be relaxed later. In this situation there are two limiting states of importance. These correspond to ion pairs with widely separated anionic sites (i.e. greater than 5 A apart) which is state 5, and binding to anionic sites in their closest possible energetic packing which occurs in state 3 (the crystal lattice).

In the former case, the internal energies are given approximately by Coulomb's law for rigid cations in kcal./mole as:

$$\Delta U_{JX}^{15} = -\frac{332}{r_{J+} + r^-}; \qquad \Delta U_{IX}^{15} = -\frac{332}{r_{I+} + r^-} \tag{5}$$

In the crystal lattice case they are given by the Born-Lande (17) equation for the crystal lattice energy.

$$\Delta U_{JX}^{13} = 1.56\left(-\frac{332}{r_{J+} + r^-}\right); \qquad \Delta U_{IX}^{13} = 1.56\left(-\frac{332}{r_{I+} + r^-}\right) \tag{6}$$

The factor 1.56 by which Eq. 6 differs from Eq. 5 consists of the product A x (1 - 1/n), where A is the Madelung constant (1.75 for 6 fold coordination, 1.76 for 8 fold (CsCl) coordination) and n is the Born repulsion exponent (n = 9 for the group Ia halides).

The internal energy of the ion exchange can be seen to depend only upon the radius of the anion, r_-; so that a set of selectivity isotherms for the various cations can be plotted solely as a function of r_-, as has been done for the internal energies in Figs. 3A and 3B. As in Fig. 1, the lower the position of the ion, the more it is preferred by the exchanger. Similarly, Figs. 3C and 3D were constructed for the free energies by using the internal energies of A and B together with experimentally observed values for the entropies of the crystal state and calculated estimates for the diatomic case (the energy correction due to entropy in this situation was less than 1 Kcal/mole for all ions).

In all cases it is apparant that virtually the same selectivity pattern is observed, as indicated by the roman numerals above the figures, the only exceptions being the small "anomalies" indicated by the sequences IIa and IIIa which are discussed elsewhere (2,3).

3.4 Thermochemical Models

It is possible to examine selectivity patterns in a totally different way, involving no theoretical calculations whatsoever, by formulating the problem as follows: "what can we learn about selectivity from a prototype system where we take the halide anions as a model for our anionic sites?" In this formulation we utilize the same concepts of states and interconnecting processes of Fig. 2 but, instead of trying to calculate the Gibbs' free energy theoretically, we use tabulated thermochemical data for the heats and free energies of formation of the alkali halides in their diatomic gaseous states (state 5), in their crystal lattice (state 3), and in water (state 6). An illustration of the results of such a calculation is given in Fig. 4. This figure plots thermochemical specificity isotherms for H^+, Li^+, Na^+, K^+, Rb^+, and Cs^+ for comparison with Fig. 3. The free energies of D are experimental values (18,16). The free energies of C are experimental in the case of H but for the other cations they were calculated (1) by combining the experimentally observed heat contents of A with experimental values for the entropies of hydration and calculated entropies of diatomic dissociation (1,2).

Clearly, the observed selectivity pattern is very similar to that calculated in Fig. 3. Therefore, since the experimental models encompass energy terms over and above purely electrostatic, it is apparant that the observed selectivity pattern applies to anionic groups which may contain as large non-Coulomb contributions as are present in the energies of the diatomic molecules or the crystal lattice of the alkali halides. It does seem as if one can learn a considerable amount about group Ia selectivity just from taking the halides as prototype anions. It is also clear that the "Eisenman pattern" is not restricted solely to Coulombic interactions with the binding sites.

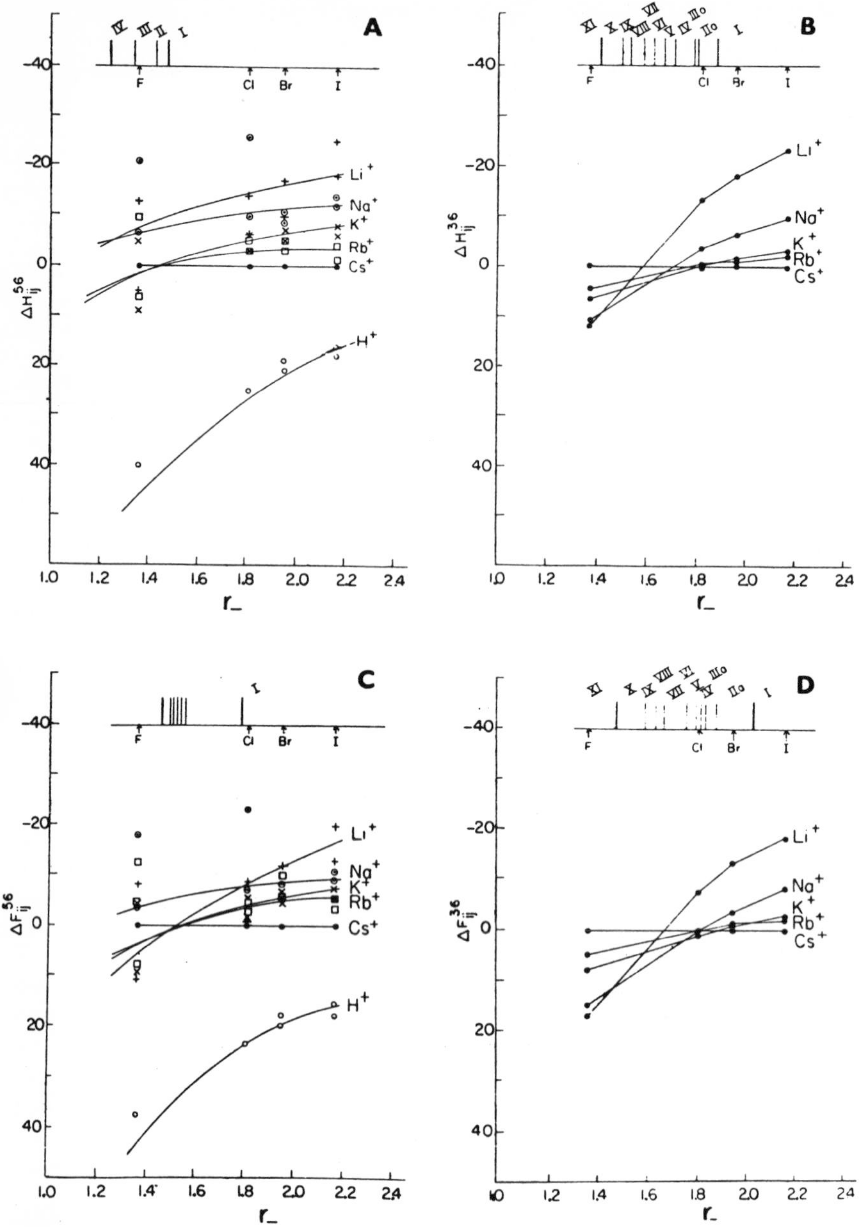

Fig. 4
Experimentally observed specificity isotherms for H^+, Li^+, Na^+, K^+, Rb^+, and Cs^+ for comparison with Fig. 3.

Some insight into the reasons why this is found can be gained from examining Figs. 5 and 6, which compare theoretically calculated energies for the alkali halides with the values observed experimentally, both for the crystal (Fig. 5) and for the gaseous diatomic molecule (Fig. 6). Since the 1918 Born-Lande theory (17) represented a major breakthrough in expressing the essence of the crystal lattice energy theoretically, it is hardly surprising that the experimental and theoretical values agree so closely. Indeed, among the group Ia cations, the only serious deviations are seen for the fluoride anion (which may really represent an inadequacy in the radius taken for the Li ion and a consequent difficulty with Pauling's radius ratio (19)). Of course, the theory fails to represent the highly polarizable (and polarizing) Ag ion satisfactorily, with increasing discrepancies occuring with increasing size (and polarizability) of the halide anion. Since the crystal, through its symmetry, minimizes polarizability energies, it may not be too unreasonable that a theory which is predominantly coulombic should be so satisfactory for the alkali halides in this situation.

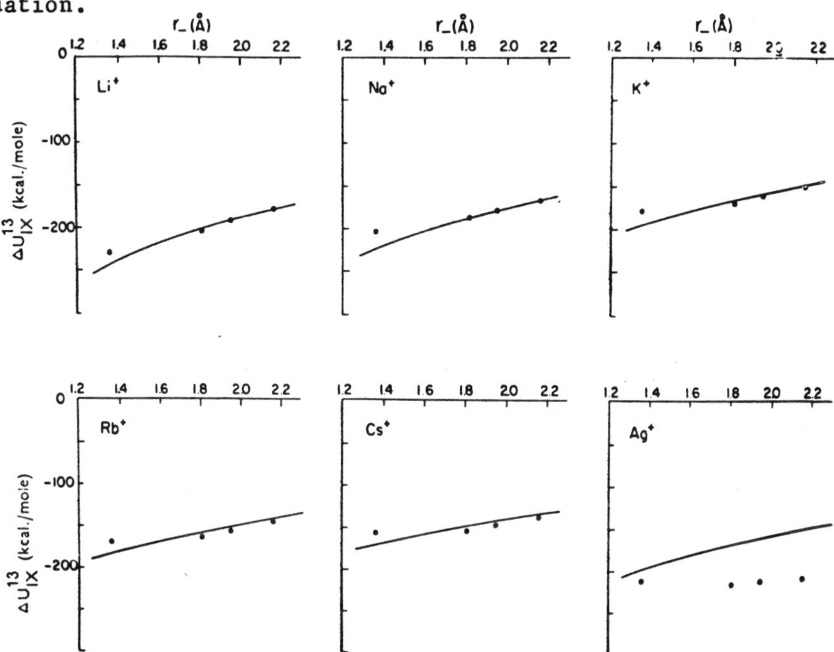

Fig. 5. Comparison of theoretical energies of crystal lattice (17) with experimental data (16).

It is therefore quite interesting to examine the situation of the gaseous diatomic ion pair where polarizability contributions should be at their maximum. The extent to which the observed experimental data (20), except in the case of the proton, are accounted for by a a coulombic calculation using Eq. 5 is more than a little surprising. It seems to indicate that the coulomb energy really comprises the principle part of the interaction energy of group Ia cations with anions as polarizable even as Iodide, even in the diatomic ion pair situation in vacuum where non-coulomb forces should be most pronounced.

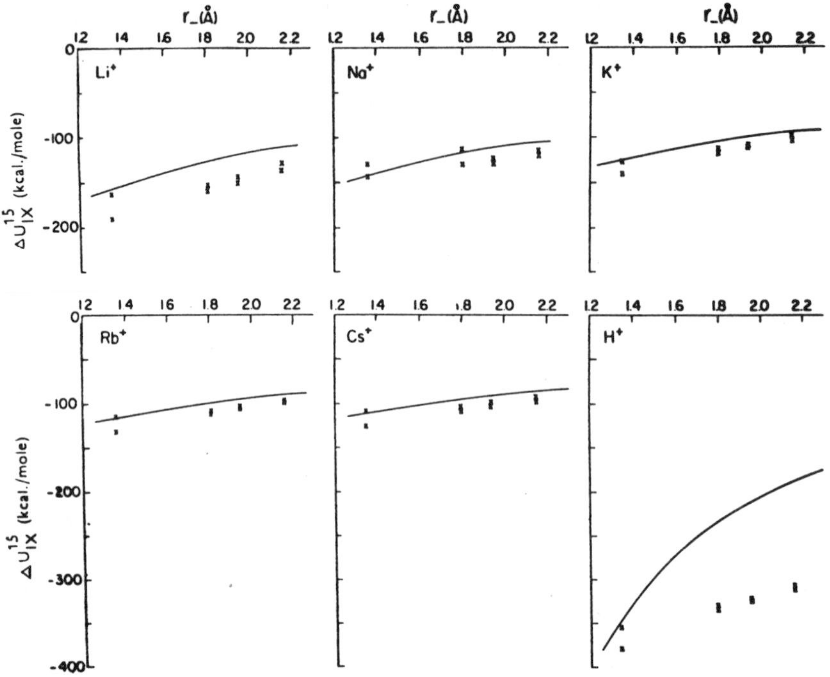

Fig. 6. Comparison of theoretical electrostatic energy of gaseous diatomic alkali halide molecules with experimental data (20). Theoretical energy indicated by solid line; experimental data by points.

Electrostatic energies and selectivity. This seems an appropriate place to comment on the assertion, sometimes incorrectly made that there are no electrostatic energy differences in the interaction between an anion and a series of like-charged cations, and the conclusion is reached that special "chemical" forces are therefore needed for specificity. This misconception is based upon the failure to realize that, while the electrostatic fields at a given (large) distance from Na^+ and K^+ are identical, the anion-cation separation in the configuration of closest possible approach differs sufficiently to produce large energy differences as was recognized by Bronsted (21, p. 782) among others. A simple example for the NaCl and KCl gaseous molecules will emphasize this. The observed interatomic separations are 2.51 and 2.79 A respectively (22).The corresponding electrostatic energies are

-132 and -119 kcal./mole, or 13 kcal./mole preference by chloride for Na^+ over K^+. The extent to which electrostic forces alone account for the interaction energies of our analysis has been illustrated in Figs. 5-7. Fig. 5 compares experimental values for the crystal lattice energies with those calculated and from electrostatic and Born repulsion forces, using the Born-Lande equation; while Fig. 6 compares experimentally observed diatomic dissociation energies with those calculated (curves) from electrostatic forces alone. Fig. 7 compares the differences of experimental heats of hydration among the cations with "electrostatic" values calculated by taking 4.8 times the electrostatic energy in Fig. 1 for the interaction of each cation with a single multipolar H_2O. Notice the excellent accounting of the energies for the 5 alkali metal cations and also that, when non-coulomb forces are expected to be important as in the case of Ag^+ in Fig. 5 and H^+ in Fig. 6, the electrostatic energy is insufficient to account for the total energy. Nevertheless, even for H^+ and Ag^+ the deviation between electrostatic and experimental energies is relatively small for the least polarizable anion (F^-), which has a polarizability only slightly smaller than the average oxyanion.

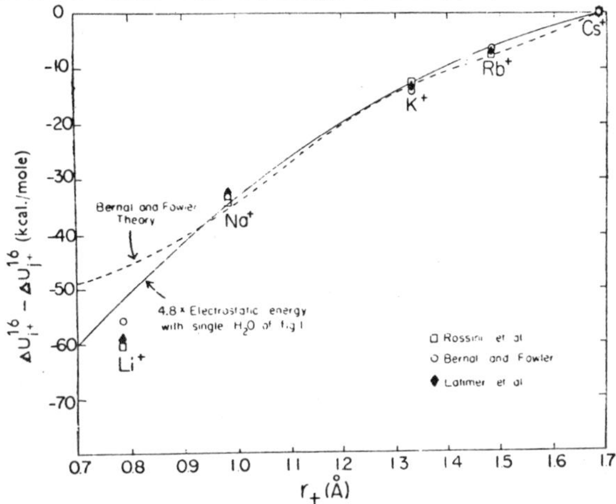

Fig. 7. Comparison of "Theoretical electrostatic hydration energy" (4.8 x energy of Fig. 1) with experimental data (16). Dashed curve plots the theoretical energies calculated from the theory of Bernal and Fowler (51).

3.5 Conclusions for Anhydrous Systems

We therefore conclude (subject only to the restriction to be removed below that water be excluded from the immediate vicinity of the interacting ions) that the primary physical variable controlling equilibrium cationic specificity is the field strength of the anion as measured inversely by r_-.

A subsidiary parameter related to the field strength of the individual detecting ions, the spatial distribution of their ensembles, also controls selectivity through the overlap of the electrostatic forces of the individual ions (cf. equations 6 and 7). Decreasing the intersite spacing leads to an increase in "effective field strength" of the anionic sites without changing the overall selectivity pattern (cf. A with B and C with D in Fig. 3 and 4). Thus, an exchanger having infinitely separated sites of "fluoride" type has approximately the same field strength as an exchanger having closely spaced sites of "iodide" type. Site spacing is especially important in controlling the differential inter-group selectivity as between the univalent and divalent groups of ions, closer spacing favouring the divalents.

3.6 Effects of Admitting Water Into the Exchanger

Our analysis next extends the applicability of the above conclusions to systems admitting water into the vicinity of the sites by relaxing the restriction of exclusion of water molecules using the properties of aqueous electrolyte solutions as a model system. Here we use a purely "thermochemical" procedure to obtain these energies, by modelling the complicated situation of a water swollen cation exchanger using a "free solution model" in which the energies of the states of IX and JX are represented by those of aqueous solutions of halides at an appropriate molality m (state 7). In this situation the Gibbs´ free energy change for the appropriate hypothetical ion exchange process between states 6 and 7 is given by the experimentally known differences of free energy of dilution of IX and concentration of JX, which are calculatable from the osmotic coefficients and activity coefficients through Eq. 7 (23).

$$\Delta F_{ij}^{67} = 2RT(\Theta_{IX} - \ln\gamma_{IX}m) - 2RT(\Theta_{JX} - \ln\gamma_{JX}m) \tag{7}$$

Eq. 7 represents the free energy of exchanging I^+ at molality m in state 7 for J^+ at infinite dilution in state 6. and are the osmotic coefficient and activity coefficients for the appropriate halide salts of I and J at molality m.

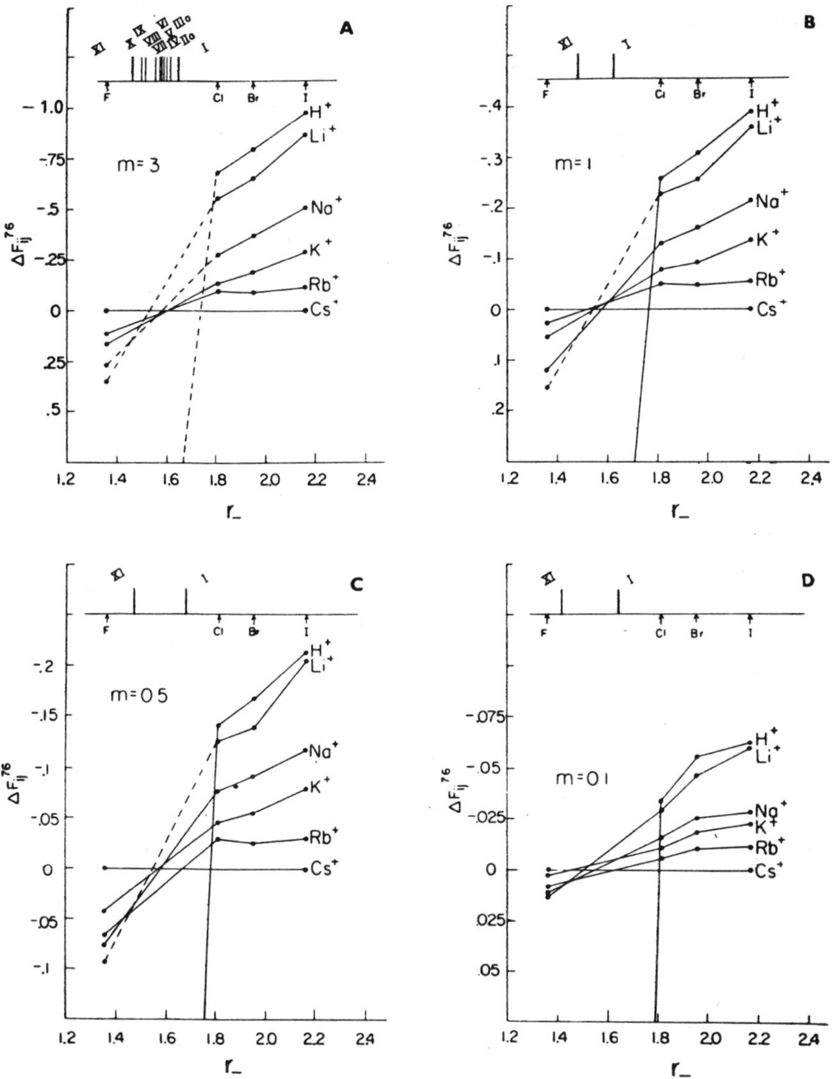

Fig. 8. The effect of the admission of water into an ion exchange phase, using the free energies of hypothetical cation exchange in aqueous solution as a model. A: m = 3, B: m = 1, C: m = 0.5, D: m = 0.1.

The results of such a calculation (see (2) for additional details), using tabulated experimental data (24), are illustrated in Fig. 8 which plots the selectivity as a function of r_- for the alkali halides calculated by Eq. 6 at four decreasing molalities to represent the effect of increasing the ratio of water molecules to "sites". Here the selectivity properties of an ion exchanger at a given internal molality are assumed to be identical to those of a free solution at the same molality, which has moreover an anion of r_- corresponding to the "field strength" of the exchanger site (see later discussion of oxyanions for how one might choose an appropriate halide model for a sulphonate group.

The pattern of selectivity is again found to be governed predominantly by r_-; and the specific selectivity pattern is even preserved in detail for the admission of quite a large number of H_2O's per site (note that at m = 3.0 there are 55.5 H_2O's present for every 3 sites). These same specificity results are also contained in and extractable from Scatchard's (25) extension of the Debye-Huckel theory of concentrated solutions of alkali halides. Analysis of Glueckauf's and Kitt's (26) study of the energies of polystyrene sulphonate cation exchangers at high resin molalities supports this conclusion, which is also in agreement with the proposition that increasing polymerization increases the magnitude of selectivity for comparable functional groups (27).

These results suggest that the primary factor controlling the magnitude of selectivity among ions at a given field strength (hence in a particular rank order) is the amount of water admitted into the vicinity of the site.

3.7 Molecular Anions

The analysis also applies the above enunciated general principles to molecular anions, of which the most important are the oxyanions, by extending the work of Kossiokoff and Harker (28), Ricci (29) and Pauling (19) on the relationship between oxyanion field strength and the acid dissociation constant (pK_a) through the realization that the pK_a should depend upon r_- if one recognizes that the energy underlying the pK_a of an acid is that of process 10-6 (recall Fig. 2) which differs from the free energy change of process 5-6 (whose selectivity we already calculated in Fig. 3C) only by the (relatively small) free energy of hydration of the undissociated acid molecule. Specifically, we conclude that the pK_a is related to the free energy change of process 10-6 by Eq. 8

$$pK_a \simeq -20\, r_- + 30 \tag{8}$$

To the extent that the free energy of hydration of the undissociated acid molecule is constant from acid to acid (or relatively small) r_-, the "anionic field strength" of an energetically equivalent singly charged anion can be assigned as a monotonic function of the pK_a, as has been done in Fig. 9 by assuming a linear relation between r_- and pK_a for HF and HCl. In this way it becomes possible to infer an approximate r_- value for any oxyanion whose pK_a is known. This has been done in the lowest portion of Fig. 9 for a variety of important anions of biology and chemistry. The selectivity properties to be expected for these oxyanions are then obtained by reading directly upward to the isotherms of the upper portion of Fig. 9 (which replot those from Fig. 3C).

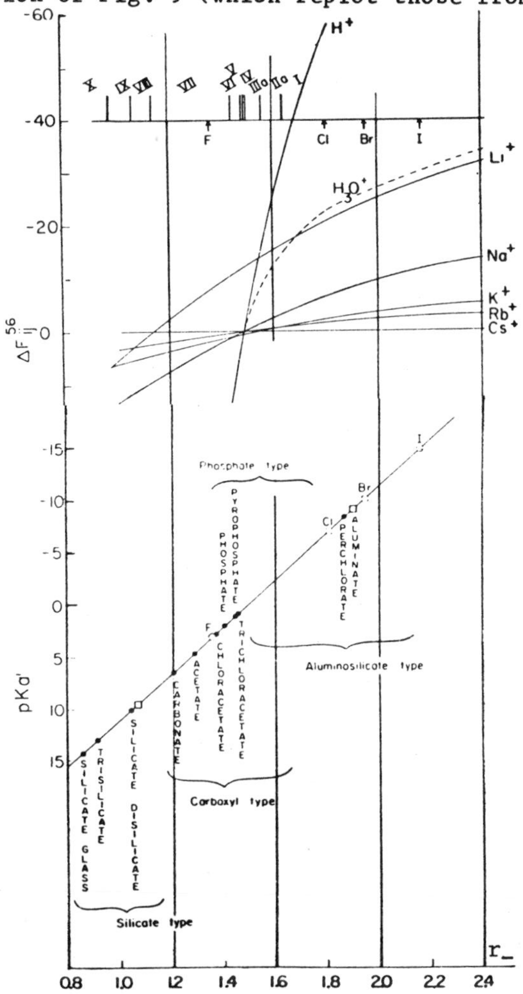

Fig. 9. The relationship between pK_a, r_-, and cation specificity for oxyanions whether free solution or polymerized The possible existence of the species H_3O^+ in the otherwise anhydrous exchanger is indicated by the dotted line (1).

The most striking feature of experimentally observed selectivity is its unique dependence on ionic size, which can show a pronounced optimum for an intermediate sized species (e.g., K or Na) while simultaneously disfavoring larger and smaller sized ions. This should be apparent from Figure 1 since each cation, in turn, is preferred as r_- is varied, and can be seen by scrutinizing the list of "Eisenman sequences". This is very clearly shown by plotting selectivity as a function of cation size, as is examined in considerable detail elsewhere (3). It is a characteristic of the "Eisenman sequences" that on such a plot selectivity is described by an upwardly convex, essentially monotonic, uninflected curve having a single maximum, which can occur at a radius corresponding to any one of the group Ia cations. Fig. 10 presents typical plots illustrating "Eisenman sequences" for the cation selectivity of a variety of complexones (52).

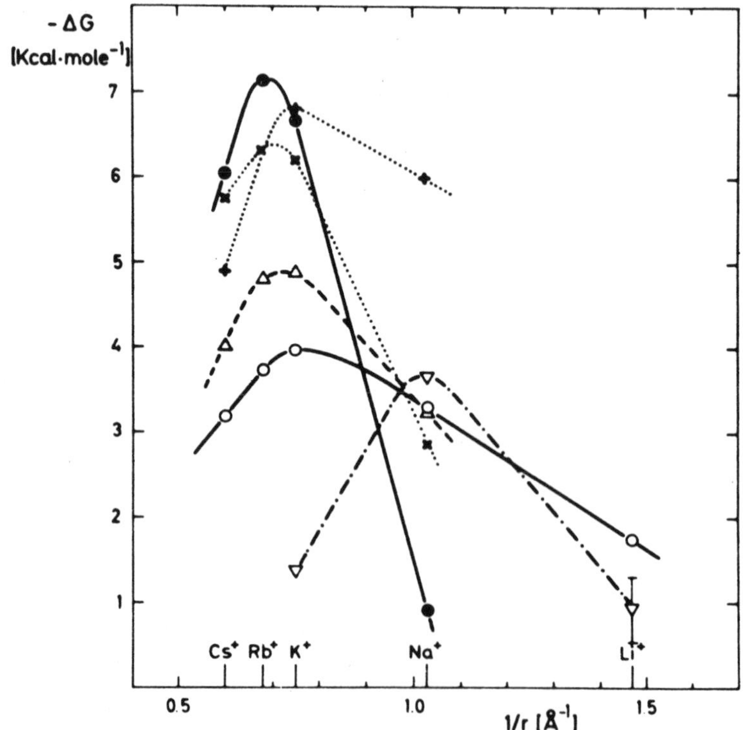

Fig. 10. (after Fig. 40 of (52)). Specificity of alkali complexing agents. Free enthalpies of complex formation in methanol at 25°C vs. reciprocal cation radii. ● valinomycin; O enniatin B; Δ nonactin; ∇ antaminide; x dibenzo-30-crown-10; + dibenzo-18-crown-6.

In contrast, on such a plot the "polarizability" sequences would be described by upwardly concave curves having a single minimum, so that only Li or Cs can be the most preferred cation. This form of plot is useful because it makes it possible, without having to remember the particular sequences, to diagnose whether one is dealing with an "Eisenman sequence, a "polarizability" sequence, or a deviant from one of these. On such a plot an "anomaly" in an Eisenman sequence appears as an additional optimum or, when less severe, as an inflection (see (3) for further details.

5 EXTENSION TO OTHER SPECIES

5.1 Monovalent cations other than group Ia.

Although we will not discuss the subject further here since it is fully covered elsewhere (31), an examination of Selectivity would be incomplete if it did not mention that monoatomic monovalent cations other than those of group Ia, such as Tl, Ag, and H, can provide information as to ligand type and orientation of "selectivity filters" and "binding sites," as can polyatomic ions such as NH_4 and its derivatives. Hille (32,33) used a variety of polyatomic cations to obtain structural information about the "selectivity filter" of the Na Channel and Eisenman and Krasne (31) have continued this approach by developing a series of "selectivity fingerprints" for such species for a variety of ion binding molecules of known structure.

5.2 Halide Anions.

Detailed studies on selectivity coefficients, free energies and enthalpies for the chloride-sulphate system for different types of anionic resins have been reported by Liberti and his colleagues (38), who provide good references to the anion exchange literature.

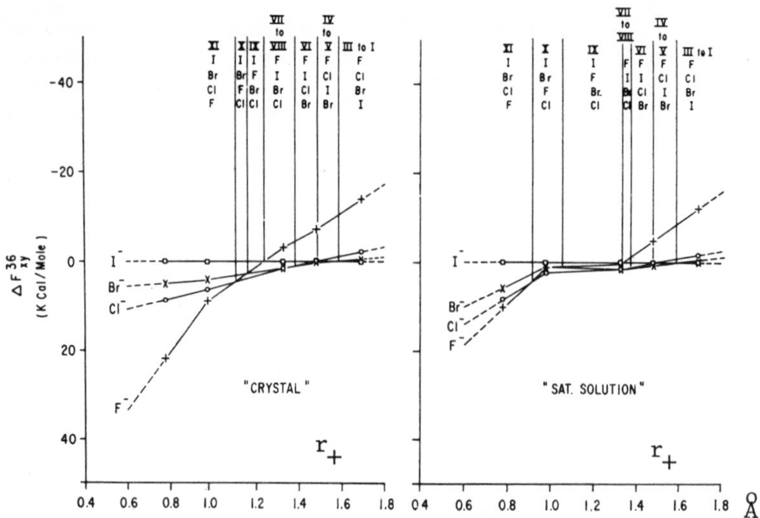

Fig. 11. Anionic selectivity pattern. The abscissa in Angstrom units. The ordinate is free energy. The left side corresponds to the anhydrous crystal state, the right side to the saturated aqueous solution state. Calculations were based on the "thermochemical" method and were carried out in the same manner as in constructing Fig. 20 of (2).

The general principles for anions are, of course, the same as for cations; and anionic selectivity sequences and patterns of these are available in the literature (15) and (34). The selectivity pattern for the halide anions is illustrated in Fig. 11 as calculated (15) from the thermochemical data for the "crystal" and "saturated solution" states. The sequences of decreasing selectivity among the halide anions are indicated above the figures by Roman numerals as defined below, including the unstable At^- species, for completeness.

AI	At	>	I	>	Br	>	Cl	>	F
AII	I	>	At	>	Br	>	Cl	>	F
AIII	I	>	Br	>	At	>	Cl	>	F
AIV	Br	>	I	>	At	>	Cl	>	F
AV	Br	>	I	>	Cl	>	At	>	F
AVI	Br	>	Cl	>	I	>	At	>	F
AVII	Cl	>	Br	>	I	>	At	>	F
AVIII	Cl	>	Br	>	I	>	F	>	At
AIX	Cl	>	Br	>	F	>	I	>	At
AX	Cl	>	F	>	Br	>	I	>	At
AXI	F	>	Cl	>	Br	>	I	>	At

5.3 Divalent Cations.

Truesdell (35) extended Eisenman's electrostatic models to the divalent cations of group IIa by assuming the substrate to contain either isolated pairs of sites of charge -1 or sites of charge -2 and showed that divalent cations were preferred over monovalents by -1 charged sites separated by less than 5 Angstroms. Eisenman (15) examined monovalent vs divalent selectivity further using a thermochemical alkali halide crystal model, and also carried out calculations on simple coulomb models for the effects of allowing local departures from electroneutrality, so as to include the possibility of competition on a molar as well as an equivalent basis. Subsequently, Sherry (36) examined more refined models and achieved considerable success in accounting for the ion exchange properties for monovalent and divalent cations of zeolites.

The general question of monovalent vs. divalent selectivity has been examined by Simon and Morf (37) in the course of treating the ion selectivity of cyclic carriers. They examined theoretically a number of important factors involved in divalent vs. monovalent discrimination. Their identification of the role played by the overall dimensions of the complex is of particular interest.

5.4 Monovalent vs. divalent cation binding

The selectivity of an ion exchange reaction between monovalent cations vs. divalent cations, corresponding to:

$$\tfrac{1}{2}B^{++}_{aq.} + AX \leftrightarrow \tfrac{1}{2}\,BX_2 + A^{+}_{aq.} + \Delta F^{\circ} \qquad (9)$$

is illustrated in Fig. 12. This figure illustrates that divalent cation selectivities invert with field strength in the same manner as do those of monovalent to divalent cations but, with increasing field strength, divalent cations become increasingly preferred.

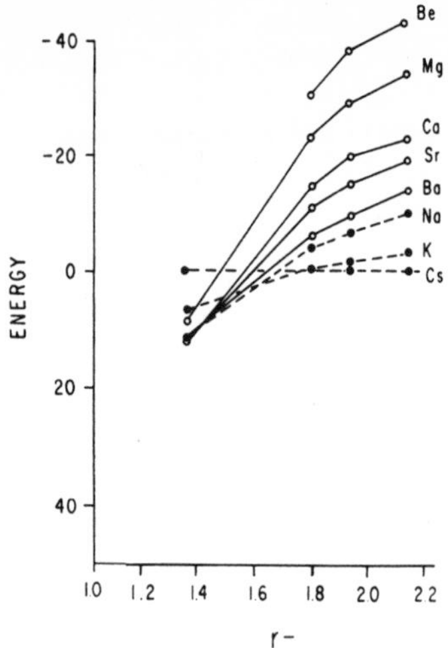

Fig. 12. Equilibrium selectivities calculated using internal energies as approximations to free energies by the "thermo-chemical" method for a completely anhydrous system having highly coordinated sites and counterions (the alkali halide and alkaline earth halide crystals). The abscissa represents the anionic radius in Angstrom units. The ordinate is energy in KCal/gram equiv.

As mentioned above, variation in site spacing leads to no important changes in selectivity pattern within a series of ions bearing the same charge; but among species bearing unequal charges, such as K^+ vs. Ca^{++}, site spacing becomes a very important factor in selectivity. This is illustrated in Fig. 13 which compares the selectivities of monovalent vs. divalent cations for pairs of widely separated sites at the left and the selectivities for closely spaced sites at the right. Fig. 13 llustrates that while widely separated sites greatly prefer monovalent cations, more closely packed sites have an increased affinity for divalent cations. The abscissa represents the anionic radius in Angstrom units. The energies for the site and counterion spacings diagrammed on the figures were calculated by Coulomb's law and experimental values were used for the differences of free energies of hydration.

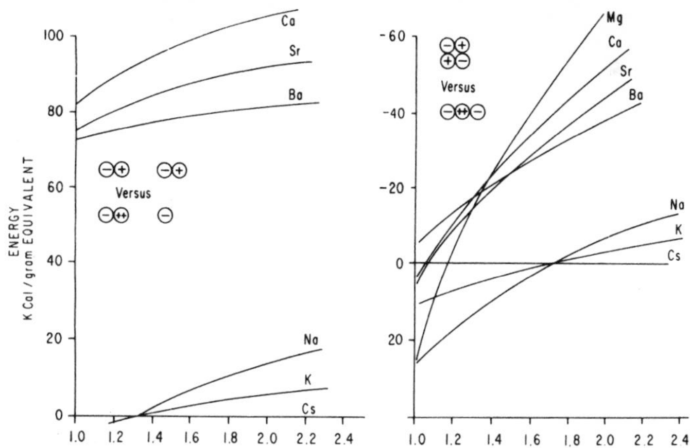

Fig. 13. Effects of site spacing on group IA vs. group IIA selectivity.

The effect of the number of sites available to interact with each counterion becomes more pronounced if one ion exchanges for another on a molar basis, regardless of charge, instead of the above equivalent exchange where the product of number times charge is constant. Thus, if local electroneutrality can be preserved by a co-ion species, a typical molar exchange reaction is:

$$B^{++}_{aq.} + [AX_2]^- \leftrightarrow BX_2 + A^+_{aq.}, \tag{10}$$

which, in effect, is the chelation reaction of a bidentate ligand with singly charged chelating groups. The reaction would be:

$$B^{++}_{aq.} + [AX_3]^= \leftrightarrow [BX_3]^- + A^+_{aq.}, \tag{11}$$

for a tridentate ligand.

The group Ia vs. group IIa specificity of a molar exchange are compared with equivalent exchanges in Fig. 14, which represents, from left to right, the equivalent exchange of Eq. 9, the molar exchange with a bidentate ligand of Eq. 10, and the molar exchange with a tridentate ligand of Eq. 11. It can be seen that increasing the number of ligand sites in a molar exchange markedly favors the binding of ions of higher charge.

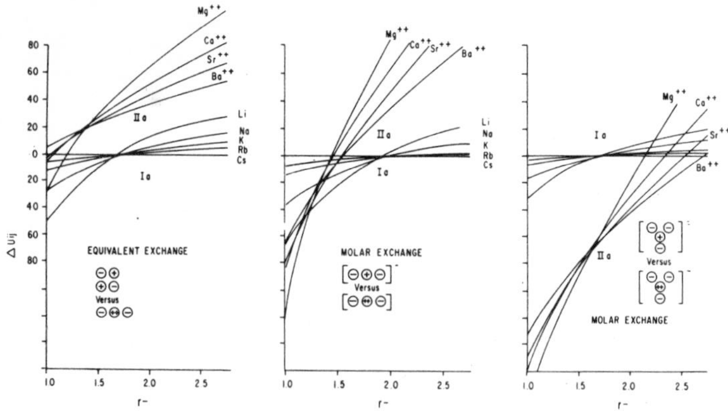

Fig. 14. Comparison of an equivalent exchange (left) with the molar exchange of a bidentate chelator (middle) and the molar exchange of a tridentate chelator (right).

The above results can be summarized as follows. Among ions bearing the same charge: the electrostatic field strength of the site determines the specificity sequence and magnitude; the coordination number (or spacing) of sites and counterions alters the "effective field strength" but leaves the pattern of selectivity unchanged; likewise the extent of hydration of the system affects only the magnitude of selectivity without altering the pattern. Among ions bearing different charges: in addition to field strength effects, the number of sites per counterion becomes an important factor in determining selectivity; and when a molar (as opposed to equivalent) exchange can occur, site coordination becomes the dominant factor.

6 NEUTRAL LIGANDS

Besides the classical charged ion exchange sites considered so far, the above selectivity principles can also be applied to neutral ligand groups which can also bind ions in competition with water molecules. Water immiscible solvents with polar groups are one example of such systems and can be important in solvent extraction processes. Related systems are the macrocyclic ion complexones whose selectivity properties have been so thoroughly discussed elsewhere (39-42) that only a few basic principles will be reviewed here in order to bring neutral ligands into the context of the ion exchange processes.

6.1 Calculations for a Simple Electrostatic Model

An understanding of the elementary factors underlying the specificity of neutral molecules which bind cations can be gained from aanalyzing a model in which the negatively charged ion exchange site is replaced by a neutral dipole, as illustrated in

Figure 15.

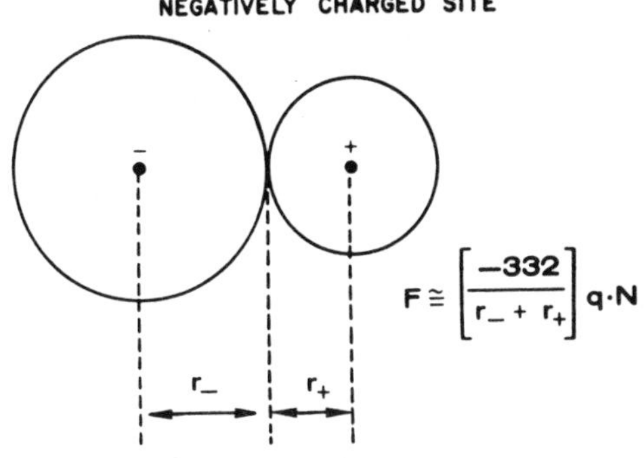

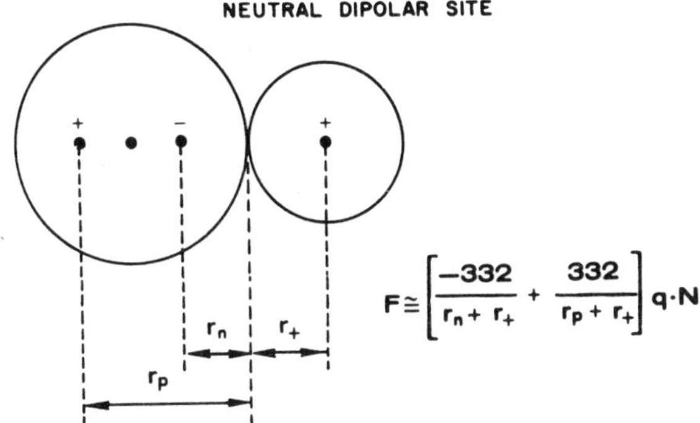

Fig. 15. Models for a negatively charged ion exchange site and a neutral dipolar ligand.

The partial molar free energy of the complexed ion is given for the case of such a neutral dipolar site by the equation on the figure, where q is the fractional value of electronic charge, N is the coordination number of the ligands, r_+ is the (Goldschmitt) cationic radius, and r_n and r_p are the distances from the surface of the dipole of the negative and positive charges, respectively. (This calculation lumps all repulsions between ligands as an effective diminution of coordination number or of effective charge).

Fig. 16 plots the calculated values for these energies, using Latimer's experimental values (18) for the free energies of hydration (see (43) for details). Each subfigure in Fig. 16 presents a set of isotherms in the manner of Figure 1. The ordinate is the free energy in kcal/mole (the lower the position of the cation, the more it is preferred). The abscissa of each subfigure represents the separation (in Angstroms) of the positive pole from the negative pole ($r_p - r_n$). This separation is zero energy of interaction with the dipole. Therefore, the values of energies along the right hand edge of each subfigure are the same for a

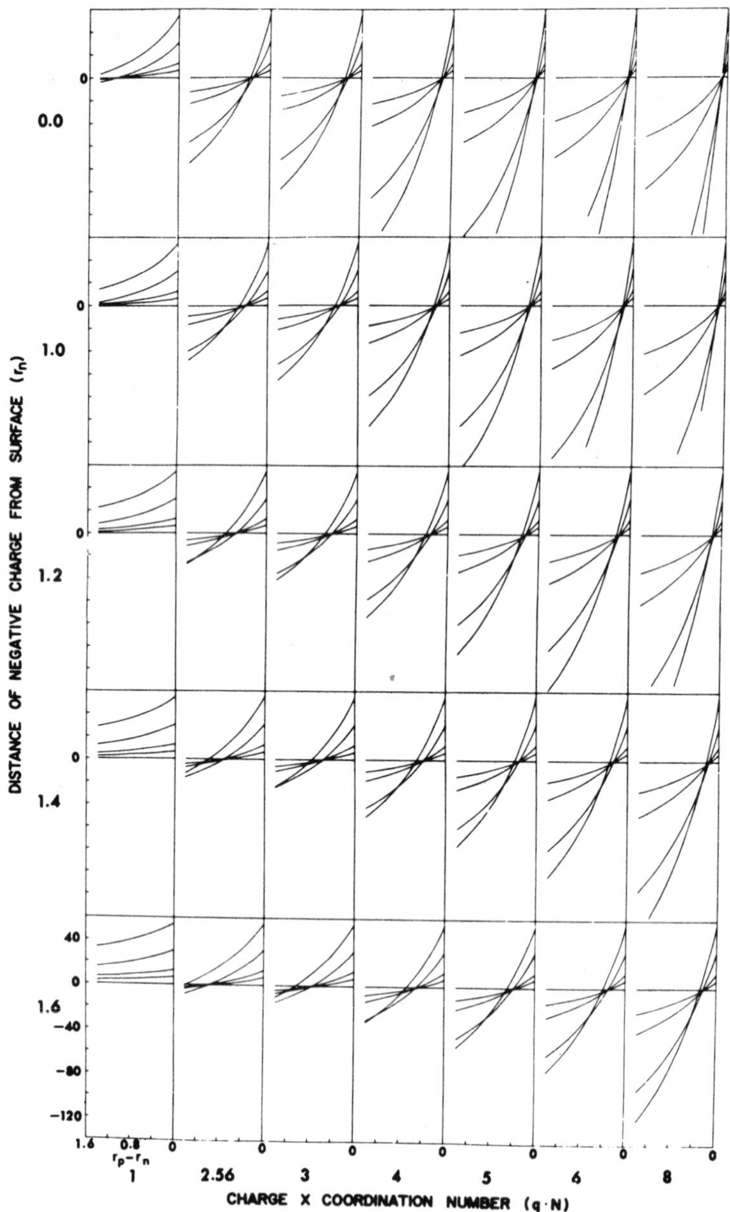

Fig. 16. Selectivity isotherms for a neutral dipolar site as a function of various charge distributions and ligand coordination numbers.

given cation on each subfigure (corresponding simply to the differences of free energies of hydration). The individual isotherms are not labelled; but in all cases the horizontal line represents the reference ions Cs^+, while the sequence of isotherms at the right hand edge of each subfigure is Li^+, Na^+, K^+, Rb^+, Cs^+ from top to bottom (corresponding to the differences of free energies of hydration).

The horizontal sets of isotherms are calculated for a given (constant) distance of the negative pole of the dipole from the surface (i.e., "0.8" means $r_n = 0.8$ A); while the vertical sets of isotherms correspond to a particular product of charge times coordination number (q x N) equal to 3 would correspond to dipoles of unit charge in three-fold coordination or to dipoles of charge 0.5 in six-fold coordination. The family of isotherms for the value (q x N) of 2.56 represents a particularly reasonable case of a tetrahedral coordination of oxygens having a net negtive charge of -0.64 corresponding to the Rowlinson-type water molecule of Fig. 1. The ranges of (q x N) and of r_n should encompass all values of dipoles likely to be encountered for carbonyl or ether oxygens in nature.

This model examines what sort of specificity would be expected if the ligand oxygens of the sequestering molecules were free to assume their distance of closest approach around each cation with negligible configurational differences of the molecule from one cation to another. Under these circumstances, a very crude approximation of the energies of binding is given by the attractive energy of interaction between cations and dipolar models for the ligand oxygens, neglecting differences in deformation energy and entropy among cations.

Not surprisingly, when the dipole separation is sufficiently large ($r_p - r_n >> r_n$), the selectivities approach those calculated for the simple monopolar model, since in this situation the energetic contribution due to the positive pole becomes negligible. Of more interest is the fact that over a wide range of dipolar charge distributions, the sequences of cation selectivity are essentially the same as those given by the monopolar model (cf. Fig. 1).

This is in accord with the experimental observation that a wide variety of selectivity data for the effects of macrocyclic molecules on artificial membrane systems (as well as on solvent extraction) are characterized by these selectivity sequences (41,42); and it appears that the energies of interaction of cations with the dipolar ligand groups, in competition with the hydration energies of the ions, can indeed account for the salient features of the selectivity for neutral ion sequestering molecules.

6.2 Thermochemical Considerations

The solvation of ions by appropriate solvents can serve as a useful prototype for the ′solvation′ of ions by neutral ligands. This has been reviewed elsewhere for cations (44) and anions (45-46). In particular, solvents exist which seem likely to solvate ions through the same ligand groups as those in ion carriers (e.g. ester carbonyls, amide carbonyls and ether oxygens); and the intrinsic selectivities of these ligand groups can serve as excellent prototypes for the selectivity of the same groups inside of a carrier molecule. This is of some interest, particularly in relation to the question of cation vs anion selectivity, for which it is possible to get some insight by comparing the energies of interactions of cations and anions with the amide (or imide) nitrogen and the carbonyl groups using Somsen′s (47) thermochemical data for the enthalpies of solvation of the alkali halide salts in amide solvents. These energies for cations and anions, separated using the Halliwell and Nyburg (48) convention, have been used to rationalize not only the observed cation selective properties of amide carbonyls (44) but also as the initially surprising ability of such ligands to bind anions (45,46).

7 CONCLUSIONS

Modern selectivity theory states the problem of equilibrium selectivity in terms of the Gibbs´ free energy change of an appropriate process and shows how the classical attractive forces of chemistry and physics lead directly to the observed selectivity sequences. An extension to kinetic phenomena is possible by analogous considerations for the energetics of the transition state.

From the relative insensitivity of the observed selectivity pattern to the details of coordination and hydration states, one can also tentatively conclude that the pattern reflects an underlying asymmetry at the molecular level in the ion-site vs. ion-water interaction energies whose essence is is that ion-site interaction energies fall off as a function of cation size as a slower power of the cation radius than do ion-water interaction energies.

Two particular conclusions are worth emphasizing, namely: that the primary physical variable controlling equilibrium cationic specificity is the field strength of the anion (representable in a coulomb model by an equivalent anionic radius, r_-) and that the primary factor controlling the magnitude of selectivity among ions at a given field strength is the amount of water admitted into the vicinity of the site.

The above concepts have been tested and found to be useful in a variety of physical systems (34,36,49), a recent example being Eberl´s (50) use of the concepts of both the "equivalent anionic radius" and the "free solution model" to predict succesfully the group Ia cation selectivity and interlayer water content in clay minerals.

REFERENCES

1. Eisenman, G. On the Elementary Atomic Origin of Equilibrium Ionic Specificity. In: Symposium on Membrane Transport and Metabolism. (New York, Academic Press, 1961),p. 163-179.
2. Eisenman, G. Cation Selective Glass Electrodes and Their Mode of Operation. Biophysical Journal 2, Part 2 (1962) 259-323.
3. Eisenman, G. and Horn, R. Ionic Selectivity Revisited: The Role of Kinetic and Equilibrium Proccesses in Ion Permeation Through Channels. Journal of Membrane Biology (1982) In Press.
4. Glasstone, S., Laidler, K.J. and Eyring, I.I. The Theory of Rate Processes. (New York, McGraw-Hill Book Company, 1941).
5. Hille, B. Ionic Selectivity of Na and K Channels of Nerve Membranes. In: Membranes--A Series of Advances, Chap. 4. (New York, Marcel Dekker, 1975), p. 255-323.
6. Hober, R. Physical Chemistry of Cells and Tissues. (Philadelphia, Blakiston, 1945).
7. Wiegner, G. and Jenny, H. Ueber Basenaustausch an Permutiten. Kolloid Zeitschrift 42 (1927) 268-272.
8. Jenny, H. Studies on the Mechanism of Ionic Exchange in Colloidal Aluminum Silicates. Journal of physiological Chemistry 36 (1932) 2217-2258.
9. Bungenberg de Jong, H.G. In: Colloid Science, (Amsterdam, Elsevier Publishing Company, 2, 1919) p. 287-.
10. Gregor, H.P., and Bregman, J.I. Studies on Ino-Exchange Resins. IV. Selectivity Coefficients of Various Cation Exchnagers Towards Univalent Cations. Journal of colloid science 6 (1951) 323-347.
11. Eisenman, G., Rudin, D.O. and Casby, J.U. Principles of Specific Ion Interactions. In: 10th Annual Conference on Electrical Techniques in Medicine and Biology of the A.I.E.E., I.S.A., and I.R.E. (1957) Boston.
12. Mattock, G. pH Measurement and Titration. (London, Heywood and Co. 1961), p. 130-134.
13. Eisenman, G., Rudin, D.O. and Casby, J.U. Glass Electrode for Measuring Sodium ion. Science 126 (1957) 831-834.
14. Ling, G.N. (1962). "A Physical Theory of the Living State." Ginn (Blaisdell), Boston, Massachusetts.
15. Eisenman, G. Some Elementary Factors Involved in Specific Ion Permeation. Proceedings of the XXIIIrd International Congress of Physiological Sciences, Tokyo 87 (1965) 489-506.
16. Rossini, F.E., Wagman, D.D., Evans, W.H., Levine, S. and Jaffe, I. Selected Values of Chemical Thermodynamic Properties. Circular of the National Bureau of Standards, No. 500. (Washington, D.C., U. S. Gov. Printing Office, 1951).
17. Born, M. and Lande, A. The Calculation of the Compressibility of Cubic Crystals from the Space Lattice Theory. Verh. dtsch. phys. Ges. 20 (1918) 210-

18. Latimer, W.M., Pitzer, K.S. and Slansky, C.M. The Free Energy of Hydration of Gaseous Ions, and the Absolute Potential of the Normal Calomel Electrode. Journal of Chemical Physics 7 (1939) 108-111.
19. Pauling, L. The Nature of the Chemical Bond. (Ithaca, Cornell University Press, 1948).
20. Gaydon, A.G. Dissociation Energies and Spectra of Diatomic Molecules. (New York, Dover, 1950).
21. Bronsted, J.N. Studies on Solubility. I. The Solubility of Salts in Salt Solutions. Journal of American Chemical Society 42 (1920) 761-786.
22. Syrkin, Y.K. and Dyatkina, M.E. Structure of Molecules and the Chemical Bond. (London, Butterworths, 1950).
23. Cruickshank, E.H. and Meares, P. The Thermodynamics of Cation Exchange. Part 2-Comparison between Resins and Concentrated Chloride Solutions. Trans. Faraday Society 53 (1957) 1299-1308.
24. Robinson, R.A. and Stokes, R.H. Electrolyte Solutions, 2nd edition (New York, Academic Press, 1959).
25. Scatchard, G. Concentrated Solutions of Strong Electrolites. Chemical Review 19 (1936) 309-327.
26. Glueckauf, E. and Kitt, G.P. A Theoretical treatment of Cation Exchangers III. The Hydration of Cations in Polystryrene Sulpjonates. Proceedings Royal Society 228A (1955) 322-341.
27. Harris, F.E. and Rice, S.A. A Model for Ion Binding and Exchange in Polyelectrolyte Solutions and Gels. Journal of physiological Chemistry 61 (1957) 1360-64.
28. Kossiokoff, A. and Harker, D. The calculation of the Ionization Constants of Inorganic Oxygen Acids From Their Structures. Journal of American Chemical Society 60 (1938) 2047-2055.
29. Ricci, J.E. The Aqueous Ionization Constants of Inorganic Oxygen Acids. Journal of American Chemical Society 70 (1948) 109-113.
30. Eisenman,G. The Molecular Basis for Ion Selectivity and its Possible Bearing on the Neurobiology of Lithium. Neurosciences Research Program Bulletin, The Neurobiology of Lithium 714 (1976) 154-161.
31. Eisenman, G. and Krasne, S. the Ion Selectivity of Carrier Molecules, Membranes and Enzymes. MTP International Review of Science, Biochemistry Series. (London, Butterworths, 1975), p. 27-59
32. Hille, B. Potassium Channels in Myelinated Nerve. Journal of General Physiology 61 (1973) 669-686.
33. Hille, B. The Permeability of the Sodium Channel to Organic Cations in Myelinated Nerve. Journal of General Physiology 58 (1971) 599-619.
34. Diamond, J.M. and Wright, E. Biological membranes: the physical basis of ion and non-electrolyte specificity. Annual Review of Physiology 31 (1969) 581-646.

35. Truesdell, A.H. Theory of Divalent-Cation Exchange Selectivity. Geological Society of America Special Paper 76 (1964) 170a.
36. Sherry, H.S. The Ion-Exchange Properties of Zeolites. In: Ion Exchange (New York, Marcel Dekker, 1969), p. 89-133.
37. Simon, W. and Morf, W.E. Alkali Cation Specificity of Carrier Antibiotics and Their Behavior in Bulk Membranes. In: Membranes, A Series of Advances. G. Eisenman, Ed., Vol. 2 (New York, Marcel Dekker, 1973), p. 329-375.
38. Boari, G., Liberti, L., Merli, C., and Passino, R. Exchange Equilibria on Anion Resins. Fourth International Symposium on Fresh Water from the Sea 3 (1973) 25-48.
39. Ovchinnikov, Yu.A., Ivanov, V.T., and Shkrob, A.M. Membrane-Active Complexones. (Amsterdam, Elsevier, 1974).
40. Pedersen, C.J. Cyclic Polyethers and Their Complexes with Metal Salts. Journal of the American Chemical Society 89 (1967) 7017-7036.
41. Eisenman, G., Szabo, G., Ciani,S., McLaughlin, S. and Krasne, S. Ion Binding and Ion Transport Produced by Neutral Lipid-Soluble Molecules. In: Progress in Surface and Membrane Science 6 (1973) 139-241.
42. Morf, W.E. The Principles of Ion-Selective Electrodes and of Membrane Transport (Amsterdam, Elsevier, 1981).
43. Eisenman, G. Theory of Membrane Electrode Potentials: An Examination of the Parameters Determining the Selectivity of Solid and Liquid Ion and of Neutral Ion-Sequestering Molecules. National Bureau of Standards Special Publication 314 (1969) 1-56.
44. Krasne, S. and Eisenman, G. The Molecular Basis of Ion Selectivity. In: Membranes, A Series of Advances, G. Eisenman, Ed., Vol. 3 (New York, Marcel Dekker, 1973), p. 277-328.
45. Eisenman, G. and Margalit, R. Amphoteric Complexes of a Neutral Ionophore Having Tertiary Amide Ligands--A Model for Anion Binding to the Polypeptide Backbone. In: Frontiers of Biological Energetics. (New York, Academic Press, 1978), p. 1-11.
46. Margalit, R. and Eisenman, G. Some Binding Properties of the Peptide Backbone Inferred from Studies of a Neutral Non-Cyclic Carrier Having Imide Ligands. Proceedings of the Sixth American Peptide Symposium, 1979), p. 665-679.
47. Somsen, G. Solution and Solvation Enthalpies of Salts in Several Solvents. (Warsaw, Proceedings 1st International Conference on Calorimetry and Thermodynamics, 1969), p. 959-965.
48. Halliwell, H.F. and Nyburg, S.C. Enthalpy of Hydration of the Proton. Trans. Faraday Society 59 (1963) 1126-1140.
49. Reichenberg, D. Ion-Exchange Selectivity, In Ion Exchange, A Series of Advances, J. A. Marinsky, Ed., Vol. 1 (New York, Marcel Dekker, 1966), p. 227-276.

50. Eberl, D.D. Alkali Cation Selectivity and Fixation by Clay Minerals. Clays and Clay Minerals 28 (1980) 161-172.

51. Bernal, J.D. and Fowler, R.H. A Theory of Water and Ionic Solution, with Particular Reference fo Hydrogen and Hydroxyl Ions. J. Chem. Phys. 1 (1933) 531-548.

52. Grell, E., Funck, T., and Eggers, F., Structure and Dynamic Properties of Ion-Specific Antibiotics. In Membranes, A Series of Advances, G. Eisenman, Ed., Vol. 3 (New York, Marcel Dekker, 1975), p. 1-126.

53. Gersh, N., Etchebest, de la Luz Rojas, O., and Pullman, A. A Theoretical Study of the Selective Alkali and Alkaline-Earth Cation Binding Properties of Valinomycin. International J. of Quantum Chemistry: Quantum Biology Symposium 8 (1981) 109-116.

54. Gersh, N. and Pullman, A. A Theoretical Study of the Interaction of Nonactin With Na, K, and NH_4. International J. of Quantum Chemistry (1982), In press.

ION EXCHANGE KINETICS--EVOLUTION OF A THEORY

Friedrich G. Helfferich

Department of Chemical Engineering
The Pennsylvania State University
University Park, PA 16802

ABSTRACT

The theory of ion exchange kinetics has come a long way from its origins over a century ago but still has much distance to cover. In the earliest speculations about "base exchange" in soils, forces other than those of inanimate nature were suspected to have their part. This view was superseded by an almost equally speculative one regarding ion exchange as a chemical reaction of a kinetic order corresponding to its stoichiometric coefficients. Only the concentrated, extensive, theoretical and practical studies in connection with the Manhattan project in World War II revealed ion exchange as essentially a statistical redistribution of ions by diffusion, with a rate limited by mass transfer resistances in either the particle or the external fluid. The next step ahead was taken in the 1950's with the realization that ions, as carriers of electric charges, are subject to the electric field their own diffusion generates. In other words, they obey the Nernst-Planck equations more closely than Fick's simpler laws. Actually, this was a rediscovery since the Nernst-Planck equations had been proposed as early as 1913 for ionic diffusion in glasses and had become stock in trade of membrane science ever since Teorell and K. H. Meyer published their pioneering studies in 1935.

Today the Nernst-Planck equations, with or without correction terms for activity coefficients, are still the dominant theory, the standard against which new propositions are measured almost as a matter of course. Significant refinements within the Nernst-Planck approach to date have been the inclusion of transport of coions and solvent (through their own differential transport equations) and of reactions such as neutralization, complex formation, etc. (through appropriate source-or-sink terms in the differential equations). Moreover, systems with three or more counterions species, requiring closer attention to the boundary condition at the particle-solution interface and to simultaneous mass transfer resistances in the particle and the fluid, have received more wide-spread interest.

Apart from special situations--e.g. involving structural rearrangements in zeolites or the greater complexities of diffusion in macroporous ion exchange resins--the Nernst-Planck equations have proved successful within reason. Among their more notable achievements have been the explanation of the differences between forward and reverse exchange rates of counterions of different mobilities and the correct prediction of the effects of reactions (e.g. neutralization of weak-acid or weak-base groups) on exchange rates and concentration profiles. Nevertheless, the days of pre-eminence of the Nernst-Planck equations may be numbered.

The shortcomings of the Nernst-Planck equations may be classified into two groups. The first includes effects that may be very important in some systems while less so in others, and tend to be evaded through appropriate choice of experiments because their treatment is inconveniently complex. Among these effects are the variation in swelling condition, and thus in particle size as well as ionic mobilities, accompanying ion exchange, and the even more obnoxious slow relaxation of resin networks giving rise to "non-Fickian" behavior, in which the diffusion coefficient becomes a function of time as well as of concentrations. The shortcomings of the second group are more fundamental in that they question the Nernst-Planck equations' basic premise of the existence of "individual" diffusion coefficients of ions, replacing them with an array of interaction coefficients as in the Stefan-Maxwell equations or the formalism of thermodynamics of irreversible processes.

Most scientists today seem to agree that the Nernst-Planck approach is a sound and useful approximation, but is inaccurate and insufficient to explain all.

* * * * *

This presentation attempts to provide background and context for the discussion of the state of the art and ongoing developments in kinetics of ion exchange, to be taken up in the subsequent contributions to the session on kinetics at the NATO Advanced Study Institute at Maratea, Italy, June, 1982. An effort is made to put the evolution of our thinking in perspective, so as to prepare the stage for the description of recent advances by their respective authors.

BEGINNINGS

The history of the kinetic and mechanistic theory of ion exchange is interesting to trace. It is typical in that it reflects the general progression of attitude toward phenomena of the world we live in--from wonder to speculation to basic theory to struggle with the complexities of reality--and is perhaps exceptional in that on occasion theory seems to have been well ahead of observation and experimental verification.

Ion exchange in nature, in sands and soil, was known to the ancient Greeks (1) and was rediscovered as "base exchange" by soil scientists of the 19th century (2,3). As that name implies, the mechanism was thought to involve chemical compounds rather than ions, whose existence was still unknown at the time. By the few who cared to seek a mechanistic explanation, the phenomenon was attributed to forces beyond ordinary inorganic chemistry and physics.

As ion exchange became established in our century as a tool in the laboratory and in industry, it was studied chiefly by the practical chemist interested in effects and performance rather than mechanism and kinetics, and what little on the latter was published was almost universally ignored. It was only in the general context of the extensive work on the Manhattan project in World War II--the development of the atomic bomb--that the first detailed and systematic experimental and theoretical studies of ion exchange including its mechanism were undertaken.

REACTION OR DIFFUSION?

In some of the first systematic kinetic studies it was taken more or less for granted that ion exchange is essentially a chemical reaction best described in terms of rate coefficients and kinetic orders corresponding to the stoichiometry of exchange (4-7).

$$d \langle \bar{C}_1 \rangle / dt = k C_1^{\nu_1} \langle \bar{C}_2 \rangle^{\nu_2} - (k/\alpha_{12}) C_2^{\nu_2} \langle \bar{C}_1 \rangle^{\nu_1} \qquad (1)$$

Soon, however, the realization that the rate of ion exchange increases with decreasing particle size of the ion exchanger led to the inescapable conclusion that mass transfer rather than an actual reaction must be the rate-controlling mechanism. A classic paper by G. E. Boyd and coworkers (8) stemming from the Manhattan project and published in 1947 after declassification, argued this point convincingly and laid the issue to rest--if only for the time being.

Boyd addressed more than the question of reaction versus diffusion. He identified the potentially rate-controlling steps as diffusion in the liquid to and from the particle surface, diffusion within the particle, and actual exchange reaction at a fixed site, the slowest of these steps being the bottleneck and therefore limiting the overall rate (see Figure 1). From the solutions of Fick's Law of diffusion

$$J_i = D \text{ grad } C_i \tag{2}$$

for film and particle, Boyd derived rate laws. Moreover, he devised an experimental criterion to distinguish between these three possibilities on the basis of the dependence of the rate on particle size (diameter or radius): no dependence for reaction control; inverse proportionality for control by diffusion in the liquid; and inverse proportionality to the square of particle size for control by diffusion within the particle. Experimental evidence pointed universally to diffusion control, and not a single instance of reaction control was found at the time.

This finding was enthusiastically received by electrochemists who had studied conductivities of ion exchangers and concluded that at least the common, gel-type resins are essentially strong electrolytes, that is, media in which the counterions are <u>not</u> chemically bonded to fixed sites (9,10). It also put on a firmer and more general basis the much older and long forgotten hypothesis that ion exchange is a mass transfer phenomenon (11).

The reaction-kinetic model of ion exchange, however, was to survive. Through a freak of mathematics, a formal description of ion exchange as a reversible second order-second order reaction proves much more tractable than diffusion rate laws when applied

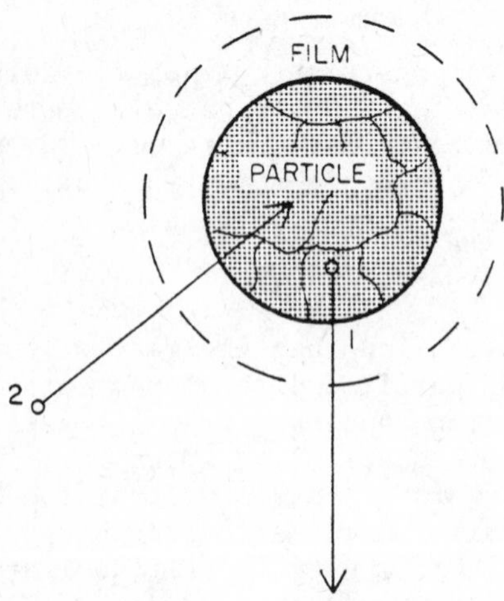

Figure 1. Mass transfer paths in ion exchange of counterions 1 and 2.

to calculations of ion exchange columns (6) and so was long retained by engineers who, rather than foregoing this convenience, related the reaction rate coefficient to mass transfer properties (12-14). More recently, reactions in ion exchange have surfaced once again, as shall shortly be seen.

When the shortcomings of the reaction kinetic model became apparent, the engineer, still disinclined to burden himself with nonsteady-state diffusion calculations for the interior of the particle, developed a mass transfer model based on linear driving force approximations for liquid and particle (15):

$$d\langle \bar{C}_1\rangle/dt = k_{liquid}(C_1 - C_1^*) = k_{solid}(\langle \bar{C}_1^*\rangle - \langle \bar{C}_1\rangle) \qquad (3)$$

(Asterisks refer to equilibrium between liquid and particle at the interface.) This approach postulates that the rate is proportional to the distance from equilibrium. It is simpler to handle than intra-particle diffusion calculations and reflects the two different mass transfer resistances better than can the reaction-kinetic model. A procedure based on linear driving force approximations is now widely used for column calculations (16).

PARTICLE AND "FILM"

Boyd and those following in his footsteps viewed ion exchange with the eye of the physical chemist, not the chemical engineer. This is reflected in their adoption of the idealization of a Nernst "film" around a particle, a fictitious, unstirred layer in which all fluid-side resistance to diffusion is localized (17). So the two possible types of mass transfer came to be called particle-diffusion control and film-diffusion control. It is possible, of course, to relate the characteristic Nernst parameter, the thickness of the fictitious film, to the mass transfer coefficient and dimensionless numbers (Sherwood, Reynolds, Schmidt) the engineer is accustomed to work with (16,18). The introduction of yet another dimensionless number, N, as yet unnamed, enabled the user to predict whether ion exchange under given conditions would be controlled by film or particle diffusion (18,19):

film-diffusion control if $N \gg 1$

particle-diffusion control if $N \ll 1$

$$\text{where}\quad N \equiv \frac{\bar{C}\bar{D}\delta}{CDr^\circ}(5 + 2\alpha_{12})$$

(derived from a comparison of half times of film- and particle-diffusion controlled exchanges).

The "effective interdiffusion coefficients" D and $\bar{D}$ in the liquid (film) and ion exchanger, respectively, can be viewed as empirical quantities to be determined experimentally. The coefficient in the liquid, D, should obviously be some average of the diffusion coefficients of the exchanging ions in that liquid (usually water). To estimate the value of the coefficient $\bar{D}$ in the ion exchanger, a relation orginally derived for ion exchange membranes was found to be reasonably successful (20):

$$\bar{D} = D[\varepsilon/(2 - \varepsilon)]^2$$

[derived from a model accounting for the tortuosity of diffusion paths and obstruction of cross-sectional area by the solid matrix of the ion exchanger (see Figure 2)].

Solutions to diffusion problems in various geometries are readily available [e.g., see (21)]. Thus, with the ability to estimate the film thickness and interdiffusion coefficients and with the other relevant parameters known or easily measurable, the user was given a workable, although approximate, theory which proved successful within reason for many simple ion exchange systems--but by no means for all.

REDISCOVERY OF THE ELECTRIC POTENTIAL GRADIENT: THE NERNST-PLANCK EQUATIONS

The pioneers who first studied ion exchange kinetics in a systematic manner had been chemists and engineers, not electrochemists. For them, the choice of Fick's Law, with an effective diffusion coefficient assumed constant until proved otherwise, was the natural choice. It soon turned out, however, that things are not as simple. Ions, as carriers of electric charges, are subject

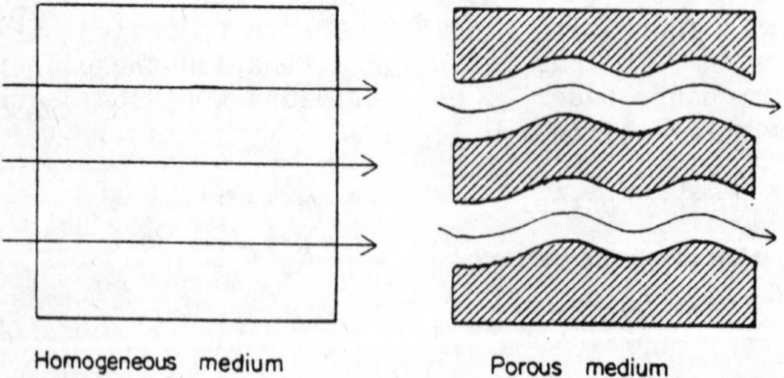

Figure 2. Comparison of diffusion paths in free solution and porous solid [from Helfferich (18)].

to electric forces. Their diffusion generates an electric potential gradient, a "diffusion potential," whose action on the ions has to be taken into account in their equation of motion. The electrochemist does this with the so-called Nernst-Planck equations (22-24):

$$J_i = -D_i \left(\text{grad } C_i + z_i C_i \frac{F}{RT} \text{ grad } \phi\right) \tag{4}$$

flux — diffusion — electric transference

(here shown for an ideal system).

Actually, the recognition of the importance of the electric potential gradient in ion exchange (19,25,26) was a <u>re</u>discovery. The Nernst-Planck equations had been used with great success in the theory of ion exchange membrances since the classic studies by Teorell (27) and K. H. Meyer (28) in the mid-1930's. Still well before that time they had been proposed by Warburg (29) for ion exchange in glasses. Their introduction into ion exchange kinetics merely required the application of principles well developed for other fields.

Physics of the Nernst-Planck Equations

The shortcomings of the Fick's Law approach and the physical significance of the Nernst-Planck equations are best seen in a simple example: particle-diffusion controlled binary ion exchange under ideal conditions (absence of coions, no activity effects, and constant mobilities of the counterions). Electroneutrality requires the sum of the counterion concentrations to equal the concentration of fixed charges and thus be constant:

$$\bar{C}_1 + \bar{C}_2 = \bar{C} = \text{const.}$$

(for ions of equal valence). The concentration gradients then must be equal to one another in magnitude (and opposite in sign):

$$\text{grad } \bar{C}_2 = -\text{grad } \bar{C}_1$$

If the diffusion coefficients are not equal, the diffusion fluxes according to Fick's Law then would be unequal:

$$\bar{J}_2 = -\bar{D}_2 \text{grad } \bar{C}_2 \neq \bar{J}_1 = -\bar{D}_1 \text{grad } \bar{C}_1 \quad \text{if } \bar{D}_1 \neq \bar{D}_2$$

But this would amount to a net transfer of electric charge and result in a violation of electroneutrality. Conservation of electroneutrality--in the absence of an electric current--requires the Fick's-Law coefficients of the two ions to be equal although their mobilities (as obtainable, for example, from conductivity or tracer diffusion measurements) differ. What value will the common coefficient have, how is it related to the individual mobilities, and what is the mechanism ensuring the equality of fluxes? The Fick's Law approach has no answers to these questions.

What actually happens, as Nernst and Planck recognized, is that a minute initial disparity of diffusion fluxes causes a minute deviation from electroneutrality, immeasurable except through observation of electric potential differences. The electric potential gradient produces electric transference of both ions, in the direction of the diffusion flux of the slower ion, in effect retarding the faster ion and accelerating the slower one:

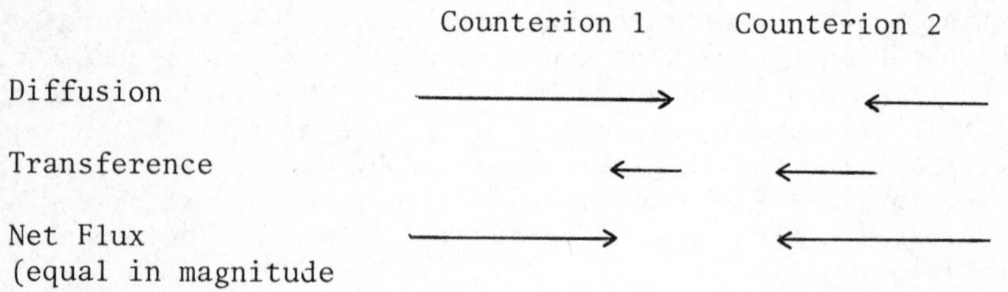

The electric potential gradient thus is the mechanism which enables the system to maintain electroneutrality--or, rather, a state so close to electroneutrality that the deviation is negligible in just about all cases of practical interest in ion exchange kinetics.

Asymmetry and "Minority Rule"

A remarkable prediction from the Nernst-Planck equations is that forward and reverse exchange of the same two ions of different mobilities should occur at different rates and with different behavior of the concentration profiles (26). With particle diffusion controlling, the rate is higher and the concentration profiles in the particle are more sharply stepped if the counterion initially in the ion exchanger is the faster of the two (see Figures 3 and 4). Surprisingly, no such direct comparisons between forward and reverse exchange had been made before this asymmetry in behavior was predicted. Its experimental verification [Figure 3, (32-36)] may thus be viewed as strong evidence for the basic soundness of the Nernst-Planck approach.

For binary ion exchange, the Nernst-Planck equations have led to a further remarkable prediction whose validity extends well beyond ion exchange to coupled diffusion processes in general: that of "minority rule" (19).

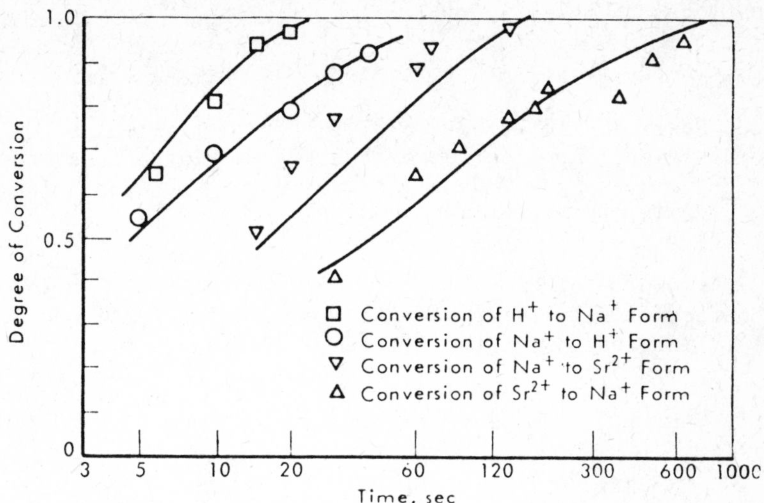

Figure 3. Comparison of forward and reverse exchange rates. Data: from experiments by Fedoseeva et al. (30,31); curves: Nernst-Planck theory [from Helfferich (32)].

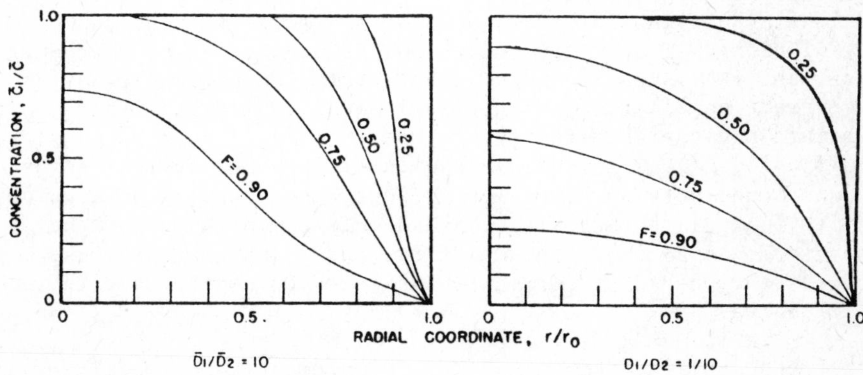

Figure 4. Radial concentration profiles at different degrees of conversion, F, in exchange of counterions of different mobilities [from Helfferich and Plesset (26)].

Mass transfer of the two counterions obeying the Nernst-Planck equations can, of course, still be described formally by Fick's Law equations

$$\bar{J}_1 = -\bar{D} \text{ grad } \bar{C}_1, \qquad \bar{J}_2 = -\bar{D} \text{ grad } \bar{C}_2$$

in which, however, the effective diffusion coefficient $\bar{D}$ now varies with composition. Application of the Nernst-Planck equations gives

$$\bar{D} = \frac{(\bar{C}_1 + \bar{C}_2)\bar{D}_1\bar{D}_2}{\bar{C}_1\bar{D}_1 + \bar{C}_2\bar{D}_2}$$

(shown here, for simplicity, for ions of equal valence and ideal systems). Interestingly, the limiting cases are

$$\bar{D} = \bar{D}_1 \qquad \text{if } \bar{C}_1 \ll \bar{C}_2$$

$$\bar{D} = \bar{D}_2 \qquad \text{if } \bar{C}_2 \ll \bar{C}_1$$

that is, the effective diffusion coefficient is that of the minority ion. Binary interdiffusion is not a democratic process but, in the parlance of the activist 1960's, is ruled by a participating minority!

The physical reasons for this behavior are not hard to see. We must keep in mind that diffusion by itself is a purely statistical phenomenon in which no physical force whatever drives a molecule in any preferential direction. (To declare the concentration gradient a "driving force" is a loose figure of speech, can be justified only if force is understood in a generalized, mathematical sense, and is misleading here.) But diffusion by itself was seen to generate an electric potential gradient, which enables the system to maintain electroneutrality by superimposing electric transference on diffusion. This potential gradient *is* a true physical force and, as such, acts on every ion present, thus causing a large corrective flux of the majority ion and only a minute one of the minority ion. So it is the majority, the taxpayers, who must shoulder the burden of correction while the minority have things their own way.

For systems with three or more counterions, no two fluxes are as strictly coupled. Here, conditions come nearer to democratic ideals: The major parties must strike a compromise while small minorities are left to do their own thing.

The prediction of the principle of "minority rule" on purely theoretical grounds, by now amply verified experimentally, can certainly be counted among the successes of the Nernst-Planck approach. However, one should keep in mind here that the principle is generally valid for binary interdiffusion under any constraint generating a corrective force--indeed, therein lies much of the utility of the concept. Therefore, the experimental evidence supports the general picture of diffusion modified by a corrective force, not necessarily the exact nature of the Nernst-Planck equations.

NERNST-PLANCK IN MORE ELABORATE SETTINGS

In their first classic applications to ion exchange the Nernst-Planck equations were integrated for ideal systems, under the constraints of electroneutrality and absence of an electric current, for the geometry of a homogeneous sphere, and under the simplest conceivable initial and boundary conditions (26). But it would be inaccurate--and less than fair to Nernst and Planck--to consider the "Nernst-Planck approach" confined to such simple cases. The Nernst-Planck equations are differential equations and, as such, are applicable under much more general and complex conditions. Indeed, Planck's early work (24) included transient behavior prior to attainment of the balance of diffusion and transference fluxes, a problem for which the constraints of electroneutrality and absence of electric current must be relaxed.

In the same or similar ways, the Nernst-Planck equations have been applied to, or proposed for, a great variety of more complex systems in ion exchange and membrane permeation. The deviation from electroneutrality at the particle-solution interface and in and around extremely small ion exchanger particles is one example, a case in which good use can be made of the theory of electric double layers and of ultra-thin membranes. This point will be discussed in one of the contributions to this session.

Another problem requiring a more elaborate treatment is that of ion exchange with macroporous resins, for which the simple model of a homogeneous sphere is obviously inadequate. Here, the flux equations must be solved for a more complex geometry [e.g., see (37)].

Along different lines, application of the Nernst-Planck equations to real systems calls for the inclusion of an additional term accounting for the effect of gradients of activity coefficients:

$$J_i = - D_i(\text{grad } C_i + C_i \text{grad } \ln f_i + z_i C_i \frac{F}{RT} \text{grad } \phi) \qquad (5)$$

as had been done in the theory of membranes long before the equations were applied to ion exchange kinetics (10,38). This correction term may be especially important in ion exchange with zeolites (39,40), as one of the contributions to this session will illustrate.

Still another complication to be accounted for in the application of the Nernst-Planck equations is that the ion exchanger contains coions--although at low concentration--and solvent in addition to counterions. These additional mobile components also undergo mass transfer under the direct or indirect effect of the electric potential gradient (32,41). Coions are transferred in the direction of the flow of the faster counterion. Any volume element of "pore liquid" in the ion exchanger contains more counterions than coions and thus has a net electric charge, which makes it subject to electroosmotic transfer in the direction of flow of the slower counterion. Thus, coion and solvent are transferred in opposite directions, and a transient accumulation of one and depletion of the other accompanies the exchange of ions of different mobilities. For the coion, a third Nernst-Planck equation can be written. The description of the transfer of solvent, however, calls for a more elaborate model that includes the interactions between ions and solvent molecules and adds a convective term to the Nernst-Planck equations. Here, theory can once again build on developments worked out for membranes [e.g., see (42)].

Also, in the application of the Nernst-Planck equations, it may be necessary to account for concentration gradients in both the film and particle (43,44). The nonlinear equilibrium at the particle surface then deserves special attention. Experience has shown this to be especially important in multi-ion systems (45).

DIFFUSION AND REACTION

As we had seen, the question whether diffusion or reaction controls the rate of ion exchange had been settled in favor of the latter possibility and the concept of ion exchange as a mere statistical redistribution of ions had been universally accepted.

There are, however, situations in which a "reaction"--that is, formation or dissolution of a chemical bond--was undoubtedly involved. Among the reactions that often accompany ion exchange are acid-base neutralization, dissociation of weak electrolytes in solution or weak ionogenic groups in ion exchangers, complex formation, or combinations of these. Some examples are listed in Table 1. Indeed, in some of these instances, abnormally low apparent interdiffusion coefficients in ion exchangers had been observed.

The distinguishing feature of systems with reactions is that the basic material balance must account for two phenomena rather than only one. The local increase or decrease in concentration

Table 1. Examples of Ion Exchange With Reaction.

$\boxed{RSO_3^- + H^+}$	$+ Na^+ + OH^-$	$\longrightarrow$ $\boxed{RSO_3^- + Na^+}$	$+ H_2O$
$\boxed{RCOO^- + Na^+}$	$+ H^+ + Cl^-$	$\longrightarrow$ $\boxed{RCOOH}$	$+ Na^+ + Cl^-$
$\boxed{RNH_2}$	$+ H^+ + Cl^-$	$\longrightarrow$ $\boxed{RNH_3^+ + Cl^-}$	
$\boxed{2\ RSO_3^- + Ni^{2+}}$	$+ 4Na^+ + EDTA^{4-}$	$\longrightarrow$ $\boxed{2\ RSO_3^- + 2Na^+}$	$+ 2Na^+ + NiEDTA^{2-}$

(Boxes symbolize ion exchanger particles.)

of a species now results from mass transfer and reaction

$$\partial C_i/\partial t = \underbrace{-\operatorname{div} J_i}_{\text{mass transfer}} + \underbrace{R_i}_{\text{reaction}}$$

rather than from mass transfer alone (Fick's Second Law, without reaction term) as in ordinary ion exchange. The effect on kinetic behavior can be dramatic; the rate may be decreased by several orders of magnitude, the dependence on variables such as solution concentration is profoundly altered, and quite distinct mechanistic features may appear (46).

Progressive Shell Mechanism

A particularly interesting type of behavior results in some of the combinations of ion exchange with reactions, namely, if a fast and practically irreversible reaction in the particle eliminates or generates one of the exchanging ions (46). An example is the neutralization of a weak-acid cation exchanger in free-acid form by a strong base:

$$\boxed{RCOOH} + Na^+ + OH^- \rightarrow \boxed{RCOO^- + Na^+} + H_2O$$

The base reacts immediately with acid groups as soon as it reaches unconverted ion exchanger, RCOOH. As a result, conversion to the Na^+ form proceeds with a very sharp "front" from the surface toward the center of the particle. At any time, an unconverted, shrinking "core" still completely in free-acid form is surrounded by a converted "shell" completely converted to the Na^+ form and growing in thickness (see Figure 5). [Incidentally, in this particular case, the OH^- ion as a coion is the minority ion in the shell and

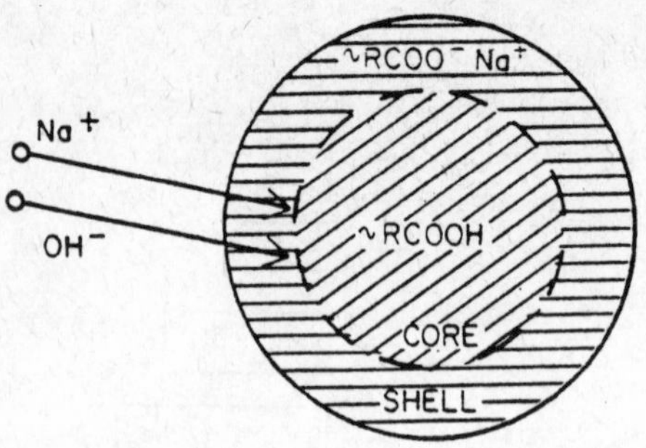

Figure 5. Progressive shell mechanism in neutralization of weak-acid ion exchanger.

thus controls the rate of mass transfer to the core; that is, ion exchange is controlled by diffusion of a coion (46)!]

Such behavior, variously called a progressive shell, shell-core, shrinking core, or moving boundary mechanism, was predicted for a number of systems on purely theoretical grounds and has by now been verified in several cases (47). Two contributions in this session will deal with this type of mechanism.

Rate Control in Systems With Reaction

There are two ways in which a chemical reaction may affect the rate of ion exchange. In the first case, the reaction is slow compared with diffusion. In the limit, diffusion is fast enough to level out any concentration gradients within the ion exchanger particle. The reaction (at the fixed site) then is the sole rate-controlling factor and, characteristically, the rate of ion exchange is entirely independent of particle size.

In the second case, the reaction may be faster than diffusion--possibly so much faster that it is in local equilibrium at any point--but is binding ions on whose diffusion ion exchange depends. Such binding of ions to fixed sites inhibits their diffusion and thereby slows the exchange rate (48,49). The rate, then, is controlled by (slow) diffusion which, in turn, is affected by the equilibrium of the (fast) reaction. Being diffusion-controlled, the exchange rate depends on particle size. This is the type of situation for which the theories mentioned in the preceding paragraphs were developed. It is also a situation which often, although not always, leads to a progressive shell mechanism.

There is a close parallel in heterogeneous catalysis, where the two limiting situations are rate control by a (slow) reaction on the one hand, and extreme diffusion limitation on the other [e.g., see (50-52)]. Here as in ion exchange, the second situation is a combination of fast reaction with slow diffusion, both affecting the overall rate. It would be in the interest of a uniform and consistent notation to reserve the term "reaction-controlled" in ion exchange for the first situation only, where a slow reaction alone controls the rate, which then is independent of particle size. The second situation, the combination of slow diffusion with fast reaction and called diffusion limitation in catalysis, is better characterized as diffusion-limited reaction or reaction-retarded diffusion.

With few exceptions such as complexing of some trivalent transition-metal ions, reactions of ions in solution are well known to be very fast, at least on the time scale of diffusion. Reaction-retarded diffusion is therefore a much more likely situation in ion exchange than is reaction control. Indeed, so far, no case of genuine reaction control has been convincingly demonstrated.

BEYOND NERNST-PLANCK

The extensions of kinetic theory of ion exchange referred to so far have all involved refinements of the conditions of integration of the basic differential flux equations (4) or (5), not of these equations themselves (except for the case of solvent transport). Indeed, the ability of the Nernst-Planck equations to describe--or even predict!--observed ion exchange rates rather well even in such complex systems lends credence to the basic soundness of the approach. Nevertheless, the Nernst-Planck equations are certainly not the last word in ion exchange theory, and efforts have been and are being made to improve upon their description of mass transfer in ion exchange. Several such contributions will be presented in this session.

Complications of the Real World

Many effects of the real world may cause the Nernst-Planck approach to be incomplete or inadequate. Often, ion exchange is accompanied by processes other than ionic mass transfer, processes not accounted for in the Nernst-Planck equations or conditions of their integration although they affect the rate of ion exchange. In many such cases, the physical situation is fairly obvious but the mathematical treatment encounters great difficulties, as so often happens in other areas of science. These are the kinds of effects which the researcher, who stands to gain little insight from much work, tends to avoid by judicious choice of systems or conditions, and which the practical chemist or engineer likes to overlook, preferring a workable, approximate theory over a more accurate but unwieldy one.

The most prominent effect of this type is the change in swelling condition of an ion exchange resin in the course of ion exchange. As the resin swells or shrinks, the mobilities of the ions increase or decrease, so that the diffusion coefficient in the Nernst-Planck equations is no longer constant. A dependence of this coefficient on the relative concentrations of the counterions could be accommodated within the Nernst-Planck equations, but this would be of little help. Since the electric potential gradient induces electro-osmotic solvent transfer, the local solvent content is not in equilibrium with the local counterion concentrations (32). Thus, the ionic mobilities depend on history as well as local ion concentrations--that is, diffusion is "non-Fickian." Moreover, swelling and shrinking not only affects the mobilities, but also alters the particle size of the ion exchanger, letting the frame-of-reference problem for mass transfer become highly complex.

Even in rigid ion exchange materials such as zeolites, which do not swell or shrink, the exchange of one counterion for another of different size usually affects the ease of motion and so leads to variations of the diffusion coefficients.

In resins of very high degree of crosslinking or if the change in swelling condition is very large, the configuration change of the polymer matrix may be a process as slow as, or slower than, counterion diffusion. This effect of slow matrix rearrangement is well known from nonionic polymers (53) and also causes diffusion to be "non-Fickian."

In some zeolites, ion exchange may be accompanied by discontinuous changes in crystal structure, a phenomenon that can profoundly affect ion exchange equilibria and rates and usually produces hysteresis (54).

Even in quite ordinary ion exchange without such complications, the Nernst-Planck equations are of only very limited use for diffusion in the "film." Although one might be tempted to integrate across the Nernst film, the result cannot be expected to bear much resemblance to reality. The film is fictitious and in reality we are faced with a combination of diffusive and convective mass transfer in proportions that vary from bulk solution to particle surface. As is apparent from Nernst's original argument (17), the outer limit of the "film" is defined only as the point where the concentration profile, if linearly extrapolated from the particle surface, reaches the concentration level of the bulk solution.

Nonlinear profiles calculated for an unstirred film from the Nernst-Planck or other nonlinear flux equations thus have little or no meaning. A better approach is to use a model that more accurately reflects the hydrodynamic situation (55), even if it simplifies ionic diffusion. Such approaches have been suggested but are not widely used; if the result is only approximate anyway, the practical engineer prefers the much simpler "linear driving force" approximations, equations (3), to the complexities of hydrodynamic models.

The list of complications could be continued almost at infinitum, but those mentioned here are generally believed to be the most serious ones.

Basic Critique of Nernst-Planck Equations

Of greater fundamental interest than complications arising from other phenomena accompanying ion exchange is the problem of adequacy in principle of the Nernst-Planck equations. The inherent premise of the Nernst-Planck approach is that diffusion and transference are additive phenomena, both characterized by a coefficient describing the mobility of the respective ion, and that the fluxes are coupled only through the electric potential gradient. [The original Nernst-Planck equations contain two coefficients, for diffusion and transference; in equations (4) and (5) the Nernst-Einstein relation for ideal systems was used to express the electrochemical mobility in terms of the diffusion coefficient.] But to what extent is the belief in a single-ion mobility justified?

The validity of the view that the "mobility" is a property of the respective molecule (or ion) cannot be taken for granted. In classical gas kinetics, for example, conservation of energy and momentum see to it that both components of a binary mixture have the same diffusion coefficient, with a value depending on the properties of both species; the molecule with greater molecular velocity does not diffuse faster because it changes its direction more drastically when colliding. For solutes in a liquid, a better case can be made that the ease of motion should be a property of the moving molecule (and the solvent), and in very dilute solutions this should be strictly true as the molecule then "sees" virtually only solvent molecules, not those of other solutes. Also, the unquestionable success of the Nernst-Planck equations in electrochemistry of dilute and moderately concentrated solutions attests to the soundness of this argument. However, ion exchangers are not dilute solutions and should be viewed more critically.

In recent years, several more fundamental approaches to mass transfer in ion exchangers have been taken and do indeed cast doubt on the validity of the basic premises of the Nernst-Planck equations. Regardless of whether statistical considerations (56,57), thermodynamics of irreversible processes (58,59), or the Stefan-Maxwell equations (59,60) are taken as the starting point, a more complex picture emerges in which fluxes are coupled not only through the electric potential gradient and more than one coefficient per species is required to characterize kinetic behavior. Several such approaches will be presented in this session and that on Mass Transfer in Ion Exchange Membranes.

STATE OF THE ART

Today, the user has a spectrum of theories of ion exchange kinetics at his disposal, ranging from simple approximations to complex, but still not entirely rigorous approaches.

The engineer designing an ion exchange column, having to cope with many other mathematical complications, will almost invariably prefer the simplest rate law and work with linear driving force approximations whose coefficients have been fitted to observed behavior or estimated on the basis of known properties (16). This is a workable approach. It has, however, one fundamental weakness. Any driving force law, no matter how refined, regards the momentary exchange rate as a function of only the concentrations in bulk solution and the _average_ concentrations in the particle, entirely glossing over the effect of concentration profiles in the particle. Nevertheless, since an entirely accurate procedure seems out of reach anyway, the driving-force approach is likely to remain the designer's method of choice for the time being.

For the research-oriented user, the Nernst-Planck approach (18) with appropriate constraints etc. still seems to provide the best "work horse." Taking into account the principal coupling effect of ionic fluxes, the Nernst-Planck equations achieve a significant improvement over Fick's Law at a relatively minor increase in complexity. However, even more important than this improvement is that the flux equations--Fick, Nernst-Planck, or other--are used under the proper conditions. For instance, the electric coupling accounted for in the Nernst-Planck equations rarely alters the exchange rate by more than a factor 2 or 3, whereas an accompanying reaction may easily do so by several orders of magnitude (46).

The researcher specifically interested in mass transfer in ion exchange materials is unlikely to be satisfied with the Nernst-Planck equations. His work may well lead to the replacement of the Nernst-Planck approach by a more refined, comprehensive new theory. The key question here, as yet unanswered, is that of the point of diminishing returns. To account for _all_ complications that may, and often do, accompany ion exchange is impracticable, so that even the finest mass transfer model may give only approximate results. To become universally accepted, the new theory or model will have to demonstrate that it yields an improvement significant enough to warrant the greater complexity.

LIST OF SYMBOLS

C concentration

D diffusion coefficient

f activity coefficient

F Faraday constant

or Degree of Conversion

J flux

k reaction rate coefficient

or mass transfer coefficient (per specific surface area)

r distance from center of particle

r^{o} particle radius

R gas constant

or reaction term (moles formed by reaction per unit volume and unit time)

T absolute temperature

z electrochemical valence (negative for anions)

α separation factor

δ film thickness

ε fractional pore volume (fractional volume occupied by solvent)

ν stoichiometric coefficient

ϕ electric potential

Indices, superscripts, etc.:

1, 2, ..., i refer to species, usually ions

over-bar refers to quantitites in ion exchanger

$\langle\ \rangle$ denotes average

REFERENCES

1. Aristotle, Works, Vol. 7, p. 933 (Clarendon Press, London, 1927).
2. Thompson, H. P., J. Roy. Soc. Agr. Engl. 11 (1850) 68.
3. Way, J. T., J. Roy. Soc. Agr. Engl. 11 (1850) 313.
4. Du Domaine, J., R. L. Swain and G. A. Hougen. Cation Exchange: Water Softening Rates. Ind. Eng. Chem. 35 (1943) 546.
5. Nachod, F. C. and W. Wood. Reaction Velocity of Ion Exchange. J. Am. Chem. Soc. 66 (1944) 1380; 67 (1945) 629.
6. Thomas, H. C. Heterogeneous Ion Exchange in a Flowing System. J. Am. Chem. Soc. 66 (1944) 1664.
7. Juda, W. and M. Carron. Equilibria and Velocity of the Na-H Exchange. J. Am. Chem. Soc. 70 (1948) 3295.
8. Boyd, G. E., A. W. Adamson and L. S. Myers, Jr. Exchange Adsorption of Ions by Organic Zeolites, II. Kinetics. J. Am. Chem. Soc. 69 (1947) 2836.
9. Heymann, E. and I. J. O'Donnell. Physicochemical Investigation of Amberlite IR-100, II. Resin Conductance. J. Colloid Sci. 4 (1949) 405.
10. Manecke, G. and K. F. Bonhoeffer. Elektrische Leitfähigkeit von Anionenaustauschermembranen. Z. Elektrochem. 55 (1951) 475.
11. Schulze, G. Die Ionendiffusion im Permutit und Natrolith. Z. physik. Chem. 89 (1915) 168.
12. Thomas, H. C. The Kinetics of Fixed-Bed Ion Exchange, in Ion Exchange, F. C. Nachod, ed., p. 29 (Academic Press, New York, 1949).
13. Selke, W. A. and H. Bliss. Application of Ion Exchange. Cu-Amberlite IR-120 in Fixed Beds. Chem. Eng. Progr. 46 (1951) 509.
14. Gilliland, E. R. and R. F. Baddour. The Rate of Ion Exchange. Ind. Eng. Chem. 45 (1953) 330.
15. Glueckauf, E. Principles of Operation of Ion-Exchange Columns, in Ion Exchange and its Applications, p. 34 (Society of Chemical Industry, London, 1955).

16. Vermeulen, T., G. Klein and N. K. Hiester. Adsorption and Ion Exchange, in Perry's Chemical Engineers' Handbook, R. H. Perry and C. H. Chilton, eds., Section 16 (McGraw-Hill, New York, 1973).
17. Nernst, W. Theorie der Reaktionsgeschwindigkeit in heterogenen Systemen. Z. physik. Chem. 47 (1904) 52.
18. Helfferich, F. Ion Exchange, Chap. 6. (McGraw-Hill, New York, 1962, reprint University Microfilms International, Ann Arbor, MI, No. 2003414).
19. Helfferich, F. Kinetik des Ionenaustauschs. Angew. Chem. 68 (1956) 693.
20. Mackie, J. S. and P. Meares. The Diffusion of Electrolytes in a Cation-Exchange Resin Membrane. Proc. Roy. Soc. (London) A 232 (1955) 498.
21. Crank, J. The Mathematics of Diffusion (Clarendon Press, Oxford, 1956).
22. Nernst, W. Zur Kinetik der in Lösung befindlichen Körper. Z. physik. Chem. 2 (1888) 613.
23. Nernst, W. Die elektromotorische Wirksamkeit der Ionen. Z. physik. Chem. 4 (1889) 129.
24. Planck. M. Über die Erregung von Electrizität und Wärme in Electrolyten. Ann. Phys. and Chem. 39 (1890) 161.
25. Schlögl, R. and F. Helfferich. Comment on the Significance of Diffusion Potentials in Ion Exchange. J. Chem. Phys. 26 (1957) 5.
26. Helfferich, F. and M. S. Plesset. Ion Exchange Kinetics. A Nonlinear Diffusion Problem. J. Chem. Phys. 28 (1958) 418.
27. Teorell, T. An Attempt to Formulate a Quantitative Theory of Membrane Permeability. Proc. Soc. Exp. Biol. 33 (1935) 282.
28. Meyer, K. H. and J. F. Sievers. La perméabilité des membranes. I. Théorie de la perméabilité ionique. Helv. chim. Acta 19 (1936) 649.
29. Warburg, E. Über die Diffusion von Metallen in Glas. Ann. Physik 40 (1913) 327.
30. Fedoseeva, O. P., E. P. Cherneva and N. N. Tunitskii. Kinetics of Ion Exchange, II. Exchange Kinetics With Hydrogen Ion Participation. Zhur. Fiz. Khim. 33 (1959) 936.
31. Fedoseeva, O. P., E. P. Cherneva and N. N. Tunitskii. Kinetics of Ion Exchange, III. Complete Exchange of Uni- and Bivalent Ions. Zhur. Fiz. Khim. 33 (1959) 1140.
32. Helfferich, F. Ion-Exchange Kinetics, III. Experimental Test of the Theory of Particle-Diffusion Controlled Ion Exchange. J. Phys. Chem. 66 (1962) 39.
33. Kuo, J. C. W. and M. M. David. Single Particle Studies of Cation-Exchange Rates in Packed Beds: Ba^{2+}-Na^{+} System. AIChE J. 9 (1963) 365.
34. Hering, B. and H. Bliss. Diffusion in Ion-Exchange Resins. AIChE J. 9 (1963) 495.

35. Morig, C. R. and M. Gopala Rao. Diffusion in Ion Exchange Resins: Na^{+}-Sr^{2+} System. Chem. Eng. Sci. 20 (1965) 889.
36. Turner, J. C. R., M. R. Church, A. S. W. Johnson and C. B. Snowdon. An Experimental Verification of the Nernst-Planck Model for Diffusion in an Ion-Exchange Resin. Chem. Eng. Sci. 21 (1966) 317.
37. Frisch, N. W. Catalysis Kinetics of a Macromolecular Ion-Exchange Resin. Chem. Eng. Sci. 17 (1962) 735.
38. Schlögl, R. and F. Helfferich. Zur Theorie des Potentials von Austauschermembranen. Z. Elektrochem. 56 (1952) 644.
39. Barrer, R. M., R. F. Bartholomew and L. C. V. Rees. Ion Exchange in Porous Crystals, II. The Relationship Between Self- and Exchange-Diffusion Coefficients. J. Phys. Chem. Solids 24 (1963) 309.
40. Brooke, N. M. and L. C. V. Rees. Kinetics of Ion Exchange. Trans. Faraday Soc. 12 (1968) 3383.
41. Spalding. G. E. Predictive Theory of Coion Transport Accompanying Particle-Diffusion Controlled Ion Exchange. J. Chem. Phys. 55 (1971) 10.
42. Schlögl, R. Stofftransport durch Membranen (Steinkopff, Darmstadt, 1964).
43. Grossman, J. J. and A. W. Adamson. Diffusion Process for Organolite Exchangers. J. Phys. Chem. 56 (1952) 97.
44. Špan, J. and M. Ribarič. Self-Diffusion of Ions Into Ion-Exchange Resins. J. Chem. Phys. 41 (1964) 2347.
45. Baypai, R. K., A. K. Gupta and M. Gopala Rao. Single Particle Studies of Binary and Ternary Cation Exchange Kinetics. AIChE J. 20 (1974) 989.
46. Helfferich, F. Ion-Exchange Kinetics, V. Ion Exchange Accompanied by Reactions. J. Phys. Chem. 69 (1965) 1178.
47. Dana, P. R. and T. D. Wheelock. Kinetics of a Moving Boundary Ion Exchange Process. IEC Fundamentals 13 (1974) 20.
48. Conway, D. E., J. H. S. Green and D. Reichenberg. The Kinetics of Na-H Exchange on a Monofunctional Cation Exchange Resin Containing Carboxyl Groups. Trans. Faraday Soc. 50 (511) 1954.
49. Schwarz, A., J. A. Marinsky and K. S. Spiegler. Self-Exchange Measurements in a Chelating Ion-Exchange Resin. J. Phys. Chem. 68 (1964) 918.
50. Carberry, J. J. Chemical and Catalytic Reaction Engineering, Chap. 9 (McGraw-Hill, New York, 1976).
51. Hill, C. G., Jr. An Introduction to Chemical Engineering Kinetics and Reactor Design, Chap. 12 (Wiley, New York, 1977).
52. Butt, J. H. Reaction Kinetics and Reactor Design, Chap. 7 (Prentice Hall, Englewood Cliffs, 1980).
53. Park, G. S. An Experimental Study of the Influence of Various Factors on the Time-Dependent Nature of Diffusion in Polymers. J. Polymer Sci. 11 (1953) 97.

54. Barrer, R. M. and L. Hinds. Ion Exchange in Crystals of Analcite and Leucite. J. Chem. Soc. (1953) 1879.
55. Van Brocklin, L. P. and M. M. David. Coupled Ionic Migration and Diffusion During Liquid-Phase Controlled Ion Exchange. IEC Fundamentals 11 (1972) 91.
56. Glasstone, S., K. J. Laidler and H. Eyring. The Theory of Rate Processes (McGraw-Hill, New York, 1941).
57. Shewmon, P. G. Diffusion in Solids (McGraw-Hill, New York, 1963).
58. De Groot, S. R. and P. Mazur. Thermodynamics of Irreversible Processes (North Holland, Amsterdam, 1962).
59. Lightfoot, E. N. and E. M. Scattergood. Suitability of the Nernst-Planck Equations for Describing Electrokinetic Phenomena. AIChE J. 11 (1965) 175.
60. Graham, E. E. and J. S. Dranoff. Applications of the Stefan-Maxwell Equations to Diffusion in Ion Exchangers. IEC Fundamentals, in press.

PLANNING AND INTERPRETING KINETIC INVESTIGATIONS

Lorenzo Liberti

Istituto di Ricerca Sulle Acque
Consiglio Nazionale delle Ricerche
5,via De Blasio,70123 Bari,Italy

1.WHY A KINETIC INVESTIGATION?

A kinetic investigation on a certain ion exchange process may be aimed at different objectives.For the pure researcher it may be finalized either to ascertain the kinetic mechanism in a new system or,more generally,to confirm experimentally the existing rate theories,still largely relying on hypotheses more than on facts.The applied researcher may desire to know,at a certain stage of development of a new application (usually after the equilibrium evaluation),how the kinetic feature of the resin investigated compares with its selectivity performance. Finally,the design engineer may need to determine experimentally mass transfer coefficients and other kinetic parameters necessary to design an ion exchange plant.

In this lecture the three points of view,i.e.,that of pure or applied researcher as well as that of the design engineer,will be taken into consideration.To this aim,the various types of apparatus to be chosen from,the available theoretical models,the adoption of proper experimental conditions and the selection of different (and often complementary) investigation techniques will be described separately and consecutively.Furthermore,the utilization of experimental kinetic data will be illustrated by reference to the $Cl^-/SO_4^=$ exchange process.

2. SELECTION OF THE APPARATUS FOR RATE STUDIES

In spite of the wide difference among the possible aims of kinetic investigations, basically the same type of apparatus may be used to answer all the above questions, i.e., a batch system or a shallow-bed column. The fundamental data gained therein may then be related to practical information through well established engineers' correlations.

In practice, however, the design engineer distrusts instinctively 'micro'-investigation and prefers the laboratory column approach, based on the analysis of the ionic concentration profiles (breakthrough curves) in the effluent, under operating conditions similar to the real ones.

2.1 Batch System

It is the more widely used system, common to kinetic investigations in pure chemistry, biology, etc. In a thermostated batch reactor, proper volumes of ion exchange resin and solution of known initial composition are mixed and vigorously stirred, while the time variation of a representative property of the system (e.g., pH, conductivity, electric potential, radioactivity, etc.) is recorded.

Rapid stirring of the solution (1000-3000 rpm) is essential to minimize mass transfer resistance related to ion diffusion through the liquid film surrounding the resin beads (see later). Use of paddle or magnetic stirrers may produce severe cracking of the resin beads. An interesting solution is offered by the Kressman-Kitchener centrifugal stirrer (1), where the ion exchanger (approx. 10 mgs) is held in the central part of the stirrer. Centrifugal action forces the inner solution to leave the stirrer through the radial holes in the casing, being instantaneously replaced by fresh solution entering the cage at the bottom (see Fig.1).

Developed in 1949 this device since then has had only minor modifications and is still the preferred choice to obtain a rapid flow of solution circulating around the resin bead (and hence to minimize polarization concentration effects). Furthermore it allows for almost instantaneous separation of ion exchanger and solution by raising

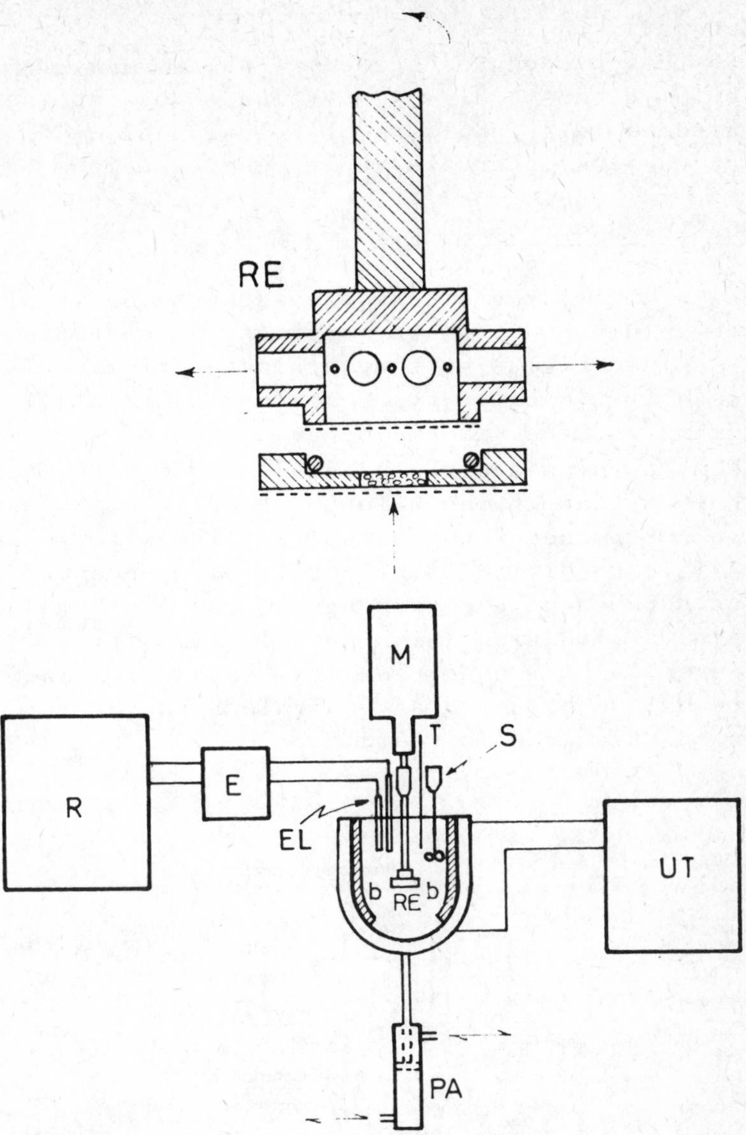

Fig.1.Schematic diagram of the apparatus employed to study the $Cl^-/SO_4^=$ exchange kinetics.
(RE,centrifugal stirrer;E,potentiometer;R,recorder;M, stirrer;T,thermometer;EL,electrodes;PA,pneumatic piston; UT,ultrathermostat;B,jacketed vessel;bb,baffles;S,supplementary stirrer)

the (still rotating) stirrer out of the solution (interrupted test).

With this procedure the fractional attainment of equilibrium, U, can be directly evaluated by fitting experimental data (e.g., analysis of solution aliquots withdrawn at various intervals) to proper equations (see later).

2.2 Shallow Bed Technique

Now a thin layer of ion exchanger beads is placed in micro-column and a known solution is forced to enter the layer (see Fig.2), so that again the time variation of a representatitve property of the effluent solution gives a description of the kinetic behavior of the system(2).

The main advantage of the shallow bed over the batch technique is that if the solution flow is high enough the exchange occurs under the so called 'infinite solution volume' (ISV) condition (i.e., negligible concentration in solution throughout the exchange of the ion released by the resin), a situation which greatly simplifies calculation.

However, the ISV condition may also be realized in the batch technique by using an 'equivalent ratio' $\geqslant$ 100 between ions in solution and in resin.

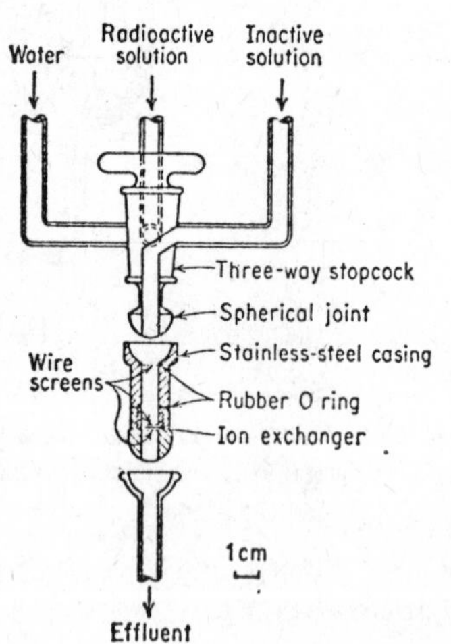

Fig.2. Shallow-bed apparatus for ion exchange kinetics. From ref.(2)

Of course, analytical problems may arise in ISV condition, where trace amounts of one ion in an excess of the other are to be monitored. Radioactive measurements are especially suitable for this purpose.

With the shallow-bed technique the exchange flux as a function of time is obtained experimentally. This may be converted into U by graphical integration of the flux function.

By both the batch and the shallow-bed techniques the time variation of the system may be measured either in discontinuous or in continuous manner.

2.3 Laboratory Column

As it will be seen later, the batch and the shallow-bed techniques permit, under favorable circumstances, some fundamental information about the system investigated to be obtained, such as interdiffusion coefficients. From this, the data of interest for the design engineer (e.g., resin exchange capacity to breakthrough, exhaustion time, etc.) may be calculated by means of complex, and often empirical, mathematical correlations. On the other hand, it has long been recognized that experimental results obtained in laboratory column experiments compare well with those obtainable in industrial columns, if experimental conditions are similar (i.e., use of real solutions, same flow rates, etc.). Crucial parameters are column bed depth (usually $\gg$ 60 cm to permit full development of the exchange zone within the column) and diameter ($\gg$ 40 times the average resin particle diameter to minimize problems with wall effects(3); with commercial resins this requires columns with $\gg$ 1.2 cm diameter).

This explains why design engineers usually prefer traditional laboratory column experiments, where a solution of known composition is fed at a strictly constant flow rate down a column containing a limited amount of resin (typical values: 0.4 l of resin in a 2 cm i.d. column for a bed depth of 120 cm).

Analysis of effluent aliquots withdrawn at known time intervals permits one to obtain the so called breakthrough curves (time concentration histories), extremely useful for many practical purposes, such as graphical calculation of over-all and dynamic resin exchange capacities, evaluation of service time, throughput, etc.

Furthermore, breakthrough curves may be used to determine, with acceptable accuracy, the type of equilibrium involved, respectively favorable or unfavorable if the pattern of the breakthrough curve remains constant or flattens when the bed depth is increased(4).

Column experiments permit also one to discriminate between film and particle diffusion kinetics(see later). To obtain this latter information, it is necessary to consider the variation of the mid-point slope of the breakthrough curve with feed flow rate. The mid-point slope is expected to be independent, or to depend on the square root of the feed spatial velocity for particle or film diffusion controlled kinetics respectively(5).

Table I resumes advantages and disadvantages of the different types of apparatus described.

Tab.I. Characteristics of apparatuses for rate studies.

	Batch system	Shallow bed	Laboratory column
Outputs	U vs. t	Flux vs. t	Breakthrough curve
Advantages	-minimum film resistance -direct D,$\bar{D}$ calculation -choice of analytical parameters -FSV and ISV	-ISV easy -direct D,$\bar{D}$ calculation	-practical information for design -choice of analytical parameters
Disadvantages	-indirect information for design -ISV difficult	-indirect information for design -radioactive measurement -high liquid resistance	-FSV rather than ISV -indirect D,$\bar{D}$ calculation -high liquid resistance

As anticipated,column experiments are especially suited when practical information on the system investigated is required.When more detailed information on rate mechanism or interdiffusion and mass transfer coefficients are required,then the more sophisticated batch or shallow bed apparatus is definitely to be preferred for kinetic studies.

3.THEORETICAL MODELS

Two fundamental questions are to be independently solved in a theoretical kinetic investigation (see Fig.3):

i) what is the rate determining step?

ii) what is the mechanism of ion diffusion in the resin?

As for the first question,the model of the 3 resistances in series for ion exchange processes is accepted almost universally.According to this model,ions leaving solution must diffuse first through the stagnant liquid film surrounding the resin bead(film diffusion),then through the resin particle (particle diffusion) up to the resin active sites where the chemical exchange reaction finally occurs (chem-reaction).Depending on which of (and whether any) these steps is the slowest,the rate of exchange is said to be controlled by film diffusion (fdc),particle diffusion (pdc) or chemical substitution reaction (chem-control).

Needless to say that,unless special precautions are taken,two or even all three resistances may be simoultaneously limiting the exchange kinetics,a fact that makes the difficulty of theoretical interpretation of rate data almost unsurmountable.

A proper application of Fick' and Nernst-Planck' theory for ion diffusion under combined chemical and electrical potentials has allowed a solution for the rate laws of ion exchange under purely fdc or pdc mechanism to be found since 1947 with the pioneering work of Boyd and coworkers(2).

Even in these limiting situations,however,a further distinction between the simpler 'infinite'(ISV) and the

A) Rate determining step? i) film diffusion

ii) particle diffusion

iii) chemical reaction

B) Diffusion mechanism ?

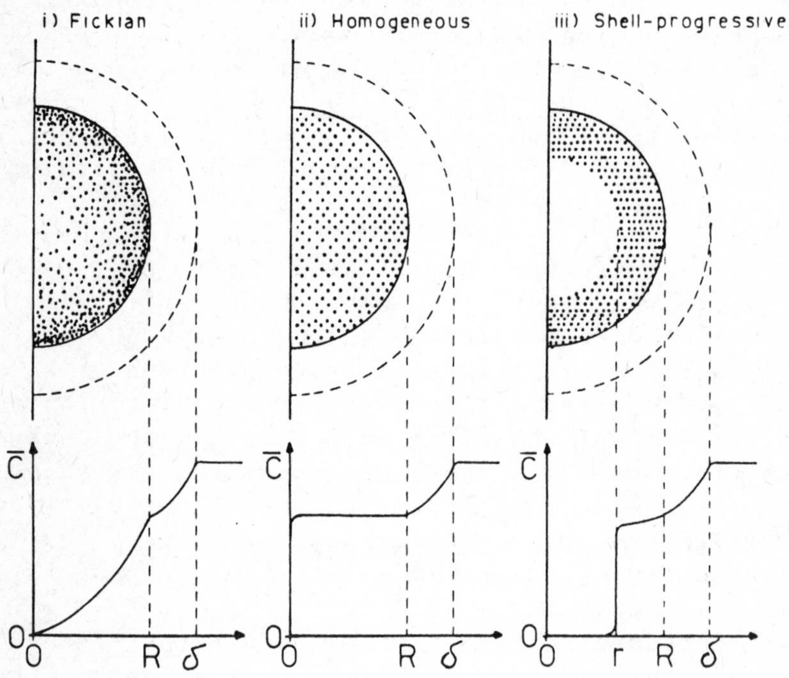

Fig.3. Fundamental questions to be solved in kinetic investigations.

more general 'finite' solution volume'(FSV) conditions must be done.

As summarized in Tab.II,indeed,only for 3 out of the 6 possible (relatively simple) situations where ion diffusion is controlling the exchange kinetics has a numerical solution been found for the rate laws of ion exchange(6).

Table II.Solutions to rate laws of ion exchange based on Nernst-Planck theory for diffusion controlled processes(*)

	ISV	FSV
Film diffusion control (fdc)	eqs.(6-46,48,49)	eq.(6-25)
Particle diffusion control (pdc)	eq.(6-41)	?
Film + particle diffusion control (f+pdc)	?	?

(*) equations code taken from ref.(6)

The picture is further complicated by the possibility (although reportedly remote) of chem-controlled,rather than diffusion controlled kinetics.In this latter case,the almost infinite analytical solutions for the n-th order kinetics of chemical reactions may be also considered.

Typically,this situation occurs for ion exchange processes accompanied by a chemical reaction (such as neutralization of a weakly basic anion resin by an aqueous acid).Helfferich(7) presented a theoretical analysis for 11 typical cases,grouped into 4 general classes,of ion exchange processes accompanied by chemical reaction and tentatively developed an analytical treatment for each class.Although efforts have been made by several researchers to confirm this analysis(8-18),disagreements about theoretical models for ion exchange kinetics accompanied by chemical reaction are still far from being resolved.

When,and if,the rate determining step has been found, the second question (ii),concerning the diffusion pattern of ions inside the resin,has to be solved for a full comprehension of the ion exchange process.Indeed,while ion diffusion in solution is acceptably explained by the well known Debye-Huckell theory(19),and the chemical exchange reaction at the resin active sites may be assumed to occur according to that of the corresponding inorganic species (i.e.,sulphuric or carboxylic acid for strong or weak cation resin,I^{ary} to IV^{ary} amines for weakly to strongly basic resins),at least 3 major different mechanisms may be hypothesized for ion diffusion in the resin. As shown in Fig.3,apart from the more common 'gradient' diffusion mechanism based on Fick' theory,two other possible mechanisms may be expected to occur,namely the 'homogeneous'(or'progressive conversion') mechanism,as discussed by Wen(20) and by Kunii and Levenspiel(21),or the 'shell progressive'(or 'unreacted core') mechanism,first develope by Yagi and Kunii(22).

As indicated by this short review,the situation of theoretical models for ion exchange kinetics is extremely complicated and theoretical solutions are available only in very few and particularly simple cases.

4.MONITORING TECHNIQUES AND OPERATING CONDITIONS

As shown in previous paragraph,selection of monitoring techniques and operating conditions is a crucial factor to meet one of the simple situations where kinetics are supported by a well defined theoretical background,which may then be used to interpret experimental results unambiguously.

4.1 Monitoring Techniques

A kinetic investigation usually requires the evaluatio of time variation of a suitable property of the system, followed either discontinuously or continuously (when possible).Depending on the chemistry of the exchange reaction,an electrochemical property such as pH or µS may be followed continuously and reproducibly when H^+ or OH^- ions are exchanged by other,less conductive species; or potentiometric measurements may be usefully performed when selective electrodes for one of the species exchanged are available. By means of reference curves, these properties are easily related to the actual concentration of the species exchanged so that the fractional attainement of equilibrium, U, can be obtained.

When the difference between the electric properties of the exchanging counterions is small, or when the extent of the exchange is limited (such as in 'differential' exchanges, where 10 to 20% of resin conversion is performed in each experiment, instead of 100% resin conversion from one ionic form to another obtained in 'integral' exchanges); or, more generally, whenever the ISV condition is to be applied, use of radioactive species appears extremely useful, with precise and affordable measurements being possible down to trace concentration of species in solution.

All the above mentioned techniques give an 'indirect' picture of what is effectively occurring in the resin phase, which is desumed through chemical balances from (dis)- appearance of ions in the external phase. In practice, experimental U vs t plots are compared with theoretical curves calculated for the few simple cases where analytical or numerical solutions exist, provided that a careful selection of operating conditions has been made previously.

If one wants to avoid the uncertainty always present when model equations are used and, moreover, to gain an insight into what is really occurring in the resin phase, other 'direct' monitoring techniques of investigation, such as autoradiography, X-ray microprobe analysis, etc., may be usefully used. Autoradiography, which requires the use of at least one radioactive species, consists of thin sections of resin particle, loaded with proper amounts of the isotope, being contacted in the dark with special films sensitive to isotopic radiation. The magnified picture of the resin section at different values of U permits one to have a visible verification of the mechanism of isotope distribution inside the resin particle (see Fig.4).

A detailed description of application of autoradiography to ion exchange investigation has been given recently by Petruzzelli and Boghetich(23).

Although α-emissions are reported to give strikingly linear tracks, and hence easily interpretable autoradiograms (17), satisfying applications of β-isotopes are largely described in the literature(24-28).

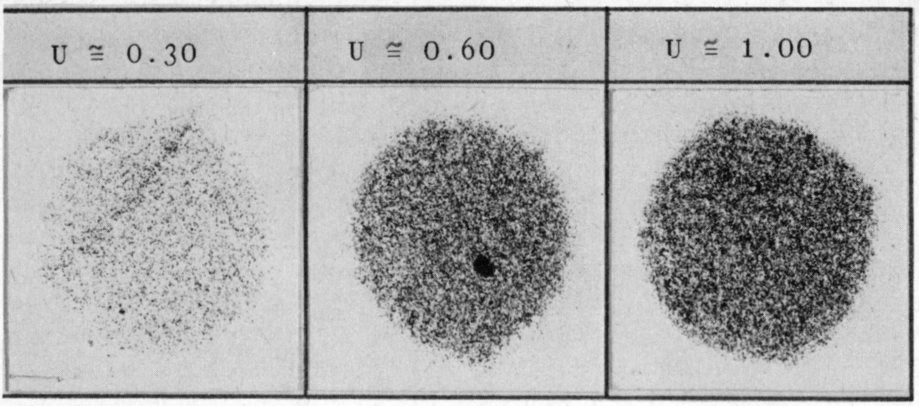

U ≅ 0.30 | U ≅ 0.60 | U ≅ 1.00

Fig.4. Typical autoradiogram for SO_4 invasion of resin Amberlite IRA 458 at different percentage of exchang (C=6x10^{-3}N)

X-ray microprobe analysis does not require the use of isotopes, as the energy of the electron beam of the instrument may be adjusted so as to reveal selectively almost any element present in the section of resin bead.

With X-ray microanalysis too it is possible to follow ion distribution within the resin phase (see an example in Fig.5)

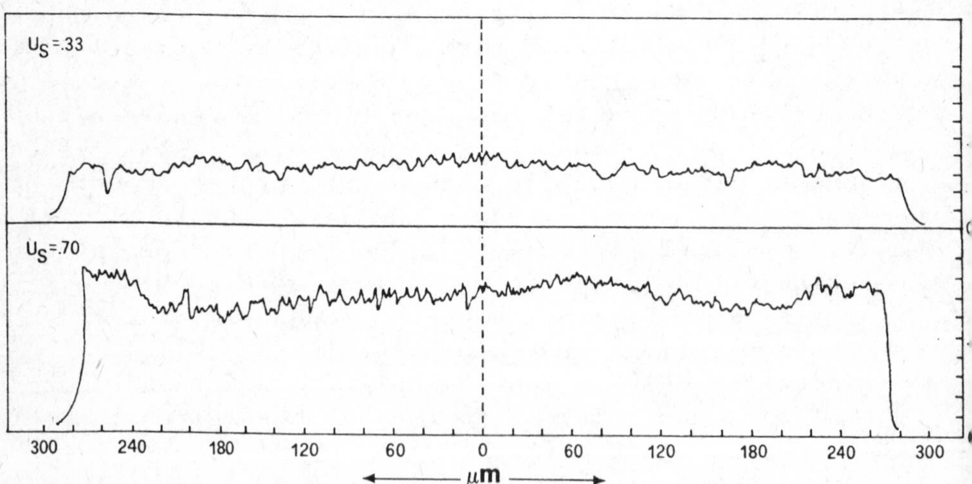

Fig.5. Sulphates distribution at approx. 33% and 70% resin conversion measured by X-ray microprobe analysis. (Resin Kastel A 102N; C=0.006 N)

Quite differently from autoradiography,however,which does not require expensive instrumentation (at least for qualitative measurements),an X-ray microanalyzer is very expensive (more than $ 100,000),which justifies its application to ion exchange investigation only under special circumstances.

Another very simple 'direct' investigation technique is that used by Holl(29),who exploits the variation of refractive index of,for instance,methacrylic resins when converted from H^+ to metal-alkali form.

4.2 Operating Conditions

It has been shown above that the two simplest limiting conditions are those for pure fdc-ISV and pure pdc-ISV kinetics,where a well established theoretical treatment of rate laws of ion exchange exists,based on Nernst-Planck theory.Relatively less simple to interpret may be the case of pure chem-controlled kinetics,while definitely more complicated (often impossible to analyse) are cases of mixed control kinetics.

Both theoretical and practical considerations allow one to expect a different dependence of exchange kinetics on experimental conditions,such as stirring rate,temperature,resin particle diameter,etc.,for the three simple limiting situations considered above,namely fdc,pdc and chem-control in ISV conditions.It has been already stated that a value $\gg$ 100 eqs in solution/eq in resin is usually considered sufficient to assume the ISV condition. Furthermore,it may be easily demostrated that low values of solution concentration,stirring speed,resin particle diameter,ion diffusion coefficient in solution and a high resin exchange capacity do facilitate the mechanism of fdc,while the contrary favors pdc mechanism.

By properly selecting the experimental parameters indicated,one is usually able to perform,or to anticipate, kinetic experiments occurring either in pure fdc or pdc conditions.Helfferich(6) has arranged semi-quantitatively the above considerations into a rate-determining criterion

$$\frac{\bar{C}\,\bar{D}\,\delta}{C\,D\,r}(5 + 2\alpha) \quad , \tag{1}$$

a value of which $\gg 1$ or $\ll 1$ does allow one to anticipate an fdc or a pdc mechanism respectively.

On the other hand, such an anticipation may be confirmed by comparing experimentally the influence of operating parameters on the rate of exchange, as resumed in the following Table.

Table III. Expected dependence of exchange rate on variation of operating conditions in limiting cases. (ISV)

	fdc	pdc	chem-control
resin particle	$\propto r$	$\propto r^2$	$\propto r$
temperature	low	low	high
stirring rate	high	none	low
interrupted test	none	discontinuity	none
selectivity	none	none	high

5. DATA INTERPRETATION

As anticipated in the first paragraph, interpretation and utilization of experimental kinetics may be used to accomplish different purposes for the pure and the applied researcher or for the design engineer. From a kinetic investigation the pure researcher usually requires fundamental information on the rate mechanism (controlling step, diffusion in resin) of unknown systems and/or values of interdiffusion coefficients in solution and in resin. To this aim a careful selection of operating conditions is extremely important to obtain a situation with a clearly defined theoretical background, such as in ISV-fdc or ISV-pdc kinetics.

In the case of ISV-fdc kinetics, for instance, with ions of similar mobility and different valence, exchange kinetics are expected to obey the following equation(6):

$$\ln (1-U) + (1-\frac{1}{\alpha})U = -\frac{3DCt}{\delta r \bar{C} \alpha} \quad , \qquad (2)$$

i.e.,a straight line when the left hand member is plotted vs t.If this is confirmed,then the ratio D/δ for the system investigated may be obtained from the experimental slope.

The applied researcher,on the other hand,may desire to compare,on a qualitative basis,equilibrium and kinetic features of a resin in a certain process,where the product of convenience has usually to show the best compromise between selectivity and rate.Accordingly,quantitative relations are no longer necessary,and different resins may be compared on the basis of their half-exchange times,$t_{0.5}$, determined in comparable kinetic experiments.

The design engineer,in turn,has normally little interest in interdiffusion coefficients and half-exchange times, preferring the less sophisticated column experiment results readily amenable to practical utilization.Nevertheless, at least for simple situations,design data may also be calculated to a good approximation through batch or shallow bed experiments.For pure ISV-fdc kinetics,for instance,it previously has been shown that application of eq.(2) permits one to obtain the interdiffusion coefficient D. This latter may be related to the liquid mass transfer coefficient,K_1a,by means of well known equations of the type(30)

$$K_1a = f(N_{Re}, N_{Sc}, D) \quad , \qquad (3)$$

and then,from K_1a,the height of a transfer unit,HTU,is readily obtained as

$$HTU = \frac{F/S}{K_1a} \quad . \qquad (4)$$

If,on the other hand,the number of transfer units,NTU, for the system investigated may be calculated by means of one of the well known graphical procedures,then the final depth of the bed

$$h_z = HTU \times NTU \qquad (5)$$

may be calculated with acceptable precision.

6. THE $Cl^-/SO_4^=$ EXCHANGE

As an example of rational application of previous considerations, the investigation of the $Cl^-/SO_4^=$ exchange kinetics on anion resins will be briefly illustrated.

This investigation, started in 1974 at IRSA-CNR and still in progress, had a triple purpose. First of all, it was aimed to select commercially available anion resins which exibit exchange rates comparable with high selectivity toward sulphate ions. Furthermore, the study was expected to give essential data for application of the DESULF process, i.e., the removal of sulphate ions from sea water feeding distillation plants to avoid formation of $CaSO_4$ scales. Finally, it was aimed to clarify the fundamental aspects of rate mechanism in this simple heterovalent system.

The first two objectives of the investigation were reached when certain resins, having weak basicity, were demonstrated to show a peculiar selectivity for sulphate over chloride ions and to be regenerable by the concentrated brine discharged by the evaporators with acceptable rates(31,32). Large evaporation plants with DESULF ion exchange pretreatment have already operated at Bari, Italy (3,000 m^3/d)(33), or are being put into operation at Doha East, Kuwait (11,000 m^3/d)(34), while a third one is under construction at Gela, Italy (28,000 m^3/d).

Full understanding of exchange mechanism is still far from having been achieved. Fig.1 shows a diagram of the apparatus utilized for fundamental study of this system. Electric potential variation with time (mV,t data), measured in solution by means of Cl-selective electrodes connected with an HP 2402A digital voltmeter, are continuously stored into an HP 91062A data acquisition system, controlled by an HP 2100S process computer, and automatically converted into U,t data by means of proper reference curve (see Fig.6). An HP 7210A digital plotter finally draws the kinetic plots, as shown in Fig.7.

The fundamental kinetic investigation, aimed to determine the rate controlling step, the $Cl^-/SO_4^=$ interdiffusion coefficients and the half-exchange times of several anion resins, was planned to occur under essentially 2 different operating conditions: high ($C \geqslant 0.6$ N) and low ($C \leqslant 0.006$ N) solution concentration.

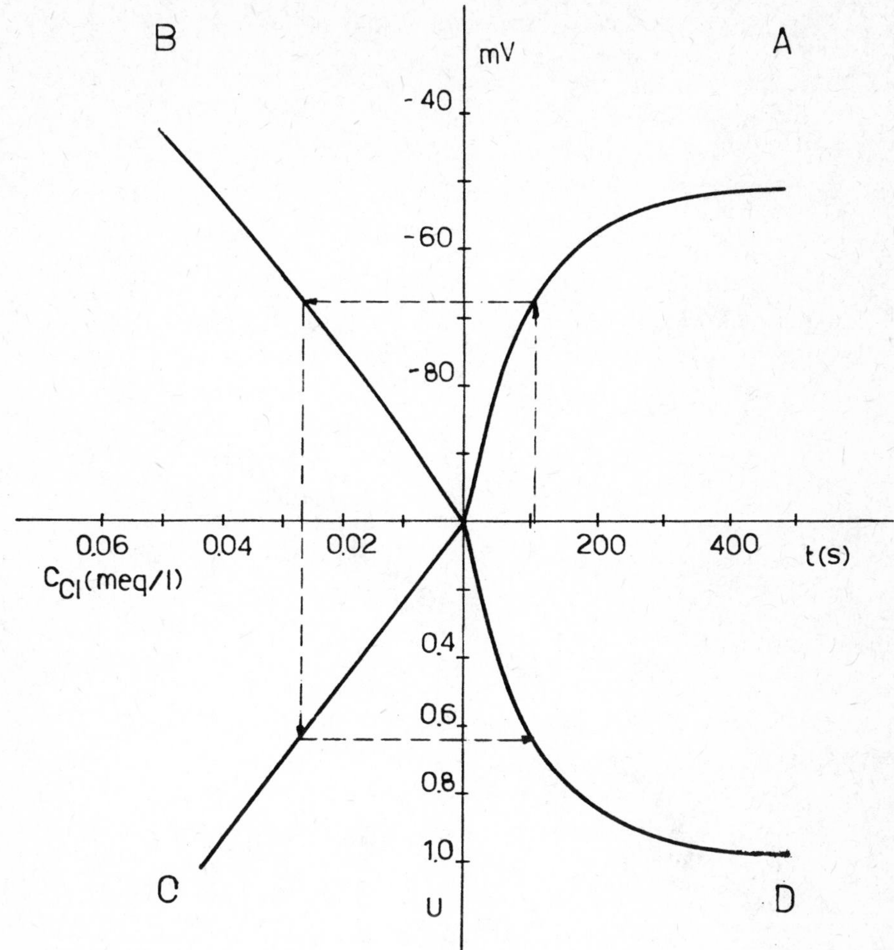

Fig.6. Nomogramme for automatic conversion of mv,t data into U,t data (follow the dotted line)

By using selected values for the remaining operating conditions,i.e.,3,000 rpm,20/30 U.S.mesh,25°C,high capacity synthetic resins,and with separation factors,α,ranging from ~10 to ~500 (sulphates always preferred) at the two solution concentrations,a pure ISV-pdc or a pure ISV-fdc rate mechanism was expected at high and low solution concentration respectively.

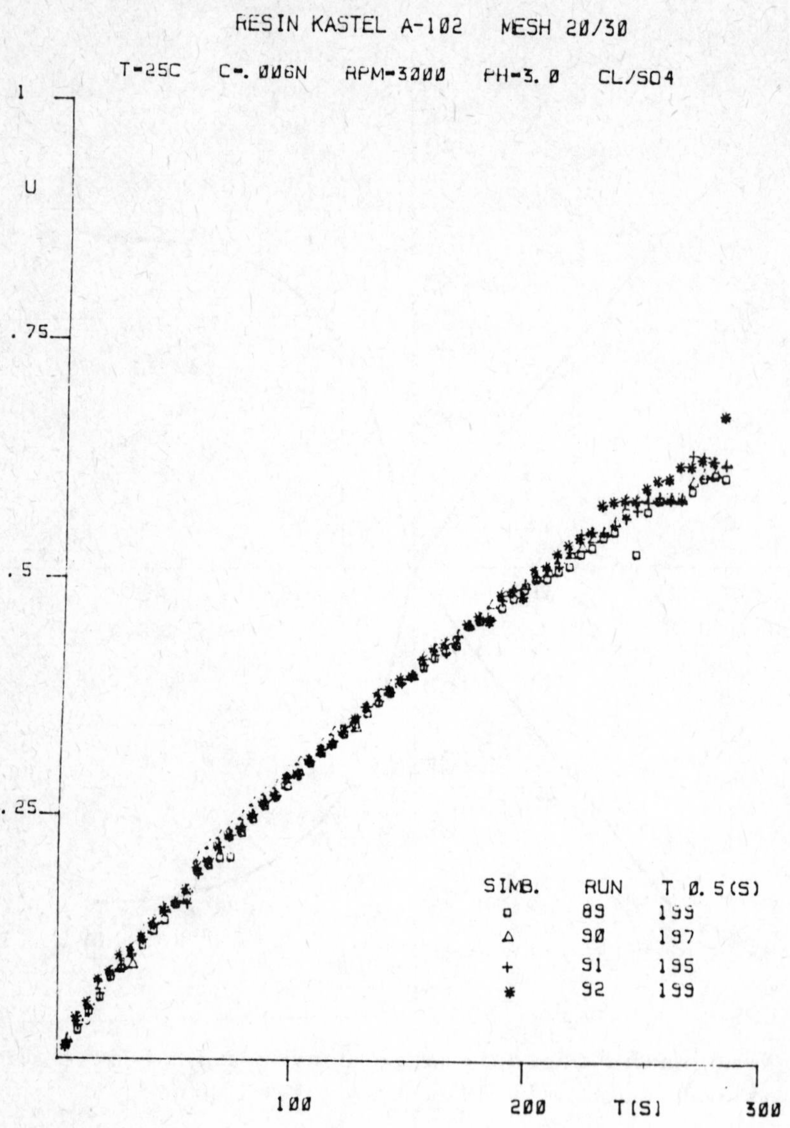

Fig.7. Typical experimental plot recorded automatically for the chloride/sulphate exchange kinetics.

In fact, the pure ISV-pdc mechanism, which according to the Nernst-Planck model is described by the following equation(6):

$$U = 1 - \frac{6}{\pi^2} \sum n^{-2} \exp\left(- \frac{\bar{D} n^2 t^2}{r^2}\right) \quad , \qquad (6)$$

was found in excellent agreement with experimental results at solution concentration equal to 1.8N (see Fig.8).

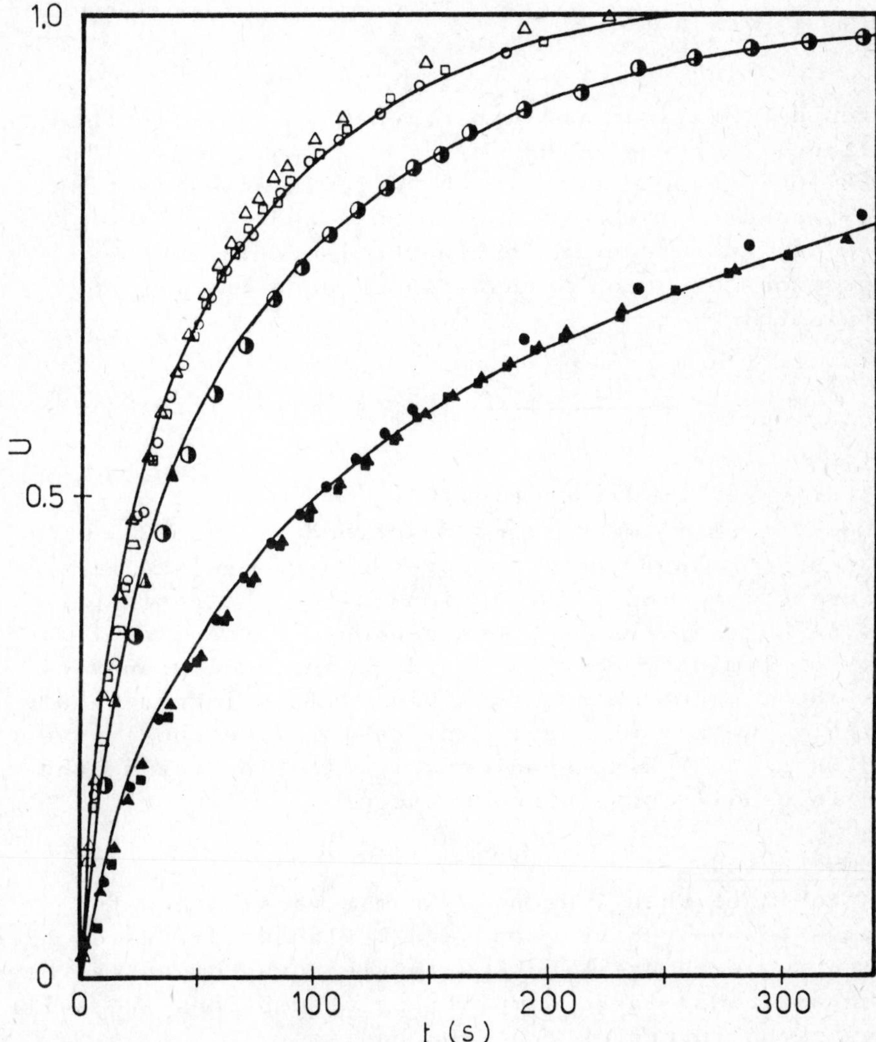

Fig.8. Exchange rate prediction, based on ISV-pdc mechanism and experimental data at 1.8N (Resin Wofatit AK 40)

On the other hand, theoretical expectations based on either a pure ISV-fdc or a less probable pure ISV-pdc mechanism were found not to agree with experimental results at low solution concentration ($C=1\div6x10^{-3}$ N), as shown in Fig. 9 A and B. Quite unexpectedly, a chem-control rate mechanism, developed on the basis of 2nd-order nucleophilic substitution reaction (SN2) in ISV conditions, described by

$$\ln\frac{1}{1-U} = \frac{3\ D^{o}_{Cl}\ C\ \beta\ t}{r\ \delta\ \alpha\ \bar{C}} \quad , \tag{7}$$

was found to explain the experimental results in these conditions (see Fig. 9 C).

Following these results, the kinetic investigation was extended to isotopic ($SO_4 \rightarrow SO_4$) and to reverse ($SO_4 \rightarrow Cl$) resin conversion. In the isotopic exchange, expectations based on Nernst-Planck equation for ISV-fdc mechanism(6)

$$\ln\frac{1}{1-U} = \frac{3\ D^{o}_{SO_4}\ C\ t}{r\ \delta\ \bar{C}} \tag{8}$$

were fully verified (see Fig.10).

None of the 3 mechanisms described by eqs.(2),(6) and (7) were found to apply for the reverse exchange (see Fig.11).

Furthermore, ion distribution inside the resin particle seems to obey always the 'homogeneous' pattern, as indicated by preliminary results based on autoradiography (Fig.4) and X-ray microprobe analysis (Fig.5). This latter result is highly unusual with gel-type resins, like those investigated, where a Fickian-gradient distribution inside the resin is usually considered to occur.

7. CONCLUSIONS

After more than 4 decades from Boyd's successful attempts to apply Fick' and Nernst-Planck' diffusion theories to ion exchange kinetics, little advancement has been gained and today large simplifying assumptions are still necessary to interpret rate mechanisms.

Theoretical models to predict ion exchange rates are essentially available, indeed, only in 2 ideal situations,

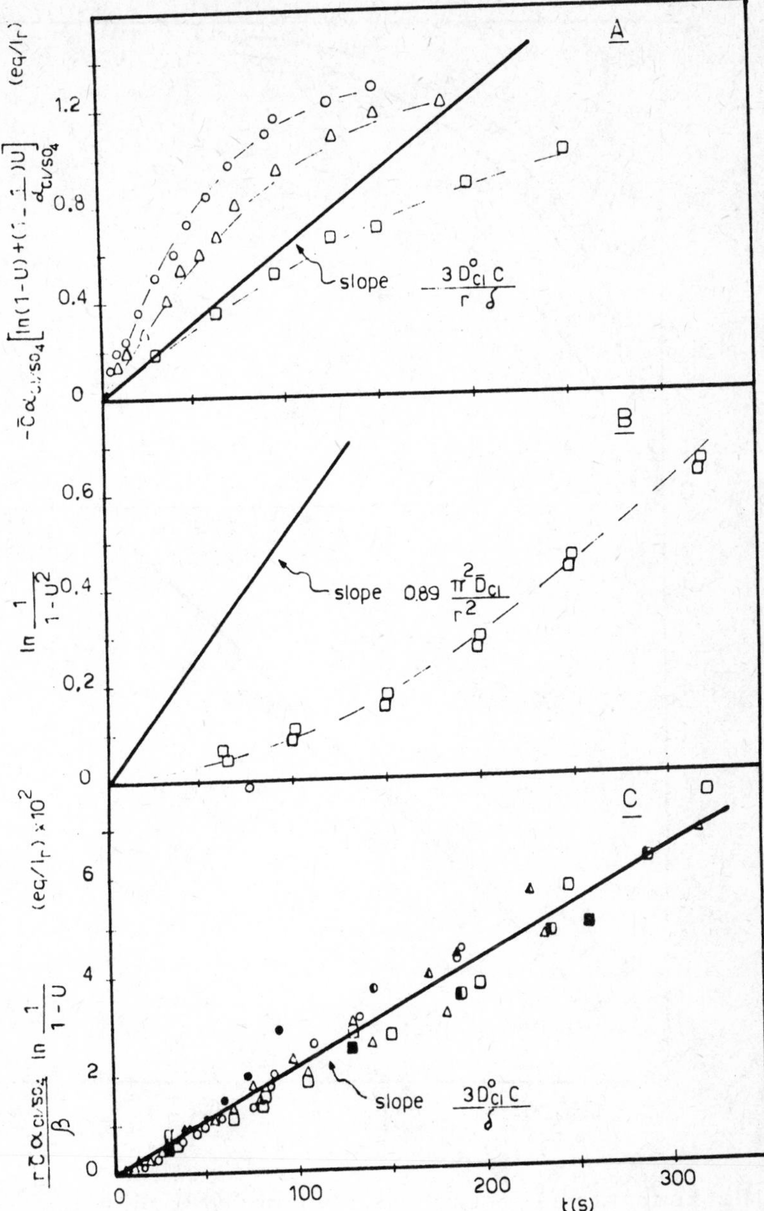

Fig.9.Experimental kinetics and model equations for direct ($Cl \rightarrow SO_4$) exchange on several anion resins(different symbols).
A:eq(2);B:eq(6);C:eq(7).

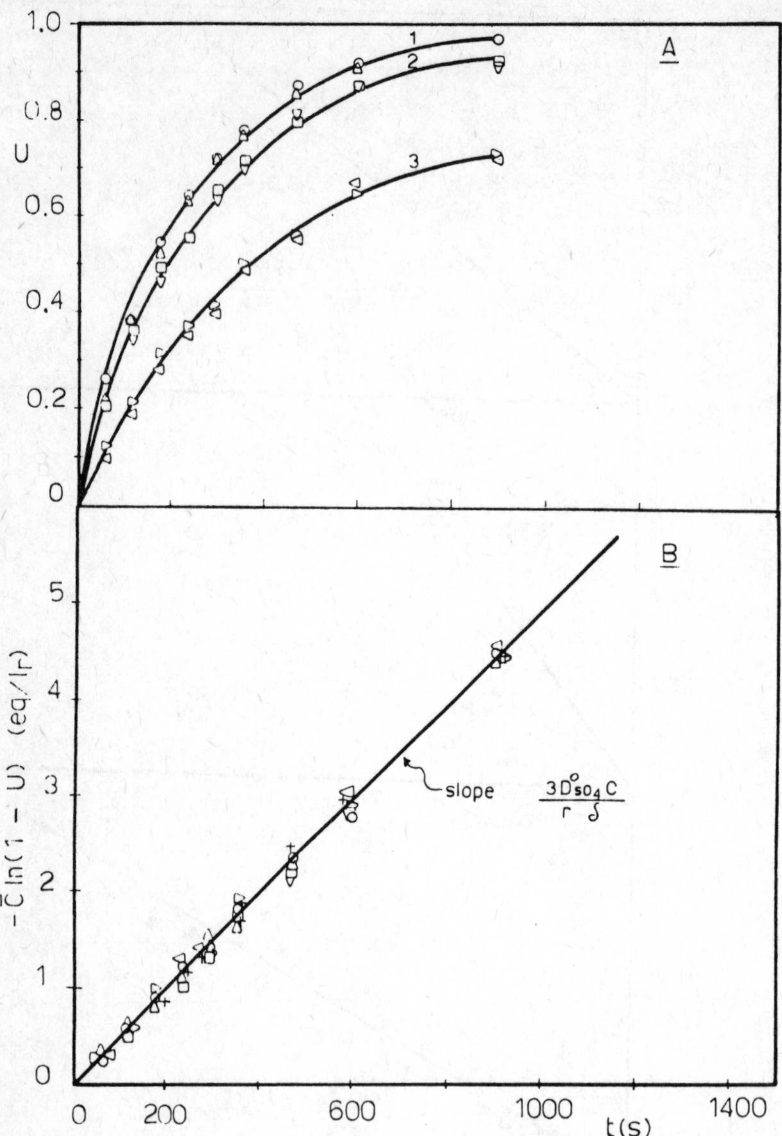

Fig.10. Experimental kinetics and model equation for isotopic ($SO_4 \rightarrow SO_4$) exchange on several anion resins (different symbols)
1:Amberlite IRA 458;2:Amberlite IRA 67;3:Kastel A102M
A:experimental rates
B:eq(8)

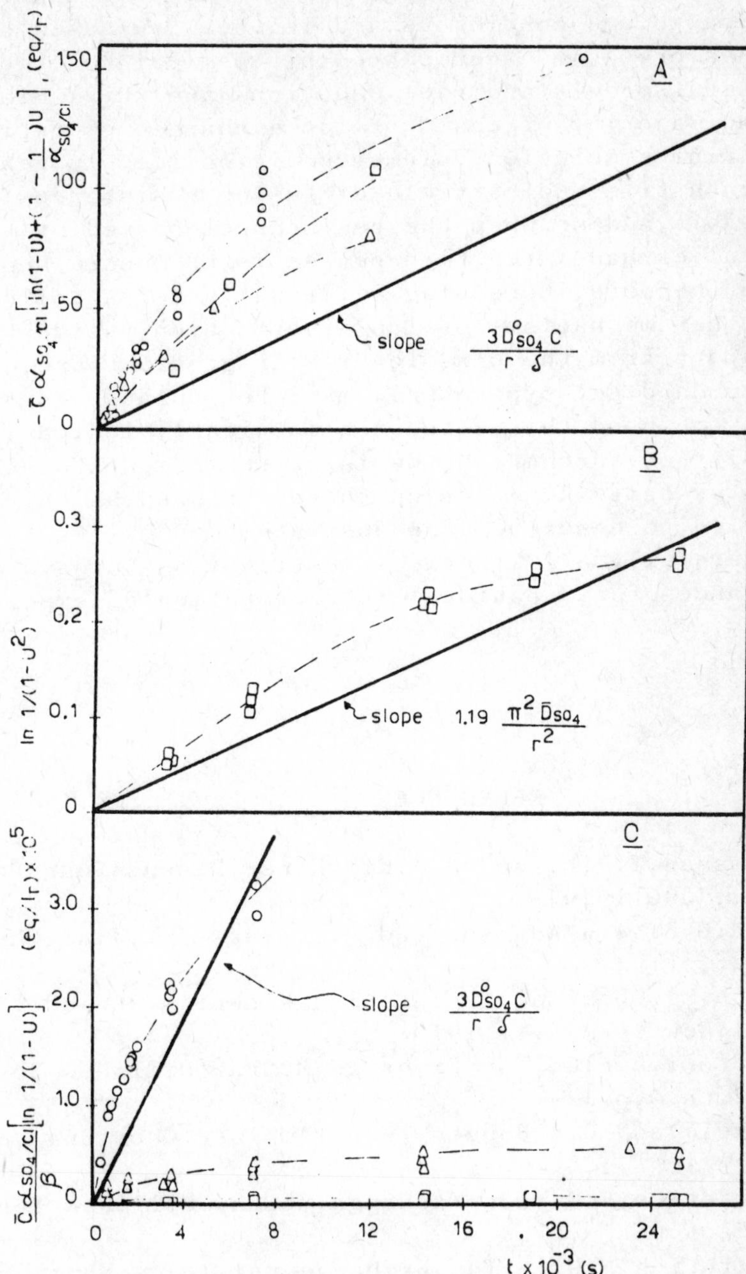

Fig.11.Experimental kinetics and model equations for reverse ($SO_4 \rightarrow Cl$) exchange.(see Fig.9)

i.e.,pure film or particle diffusion control under infinit solution volume conditions,and in very few cases of ion exchange processes accompanied by chemical reaction.

Useful models and rate equations are still missing for the majority of cases of ion exchange occurring under finite solution volume condition,i.e.,in real cases, when both film and particle diffusion resistances are noticeable,and/or when the resistance offered by the chemical exchange reaction per se is no longer negligible.

Furthermore,increasing evidence is being collected that other mechanisms of ion distribution inside the resin bead,apart from the classic Fickian-gradient one,exist for standard gel-type and new macroreticulate ion exchangers

On one hand this situation is clearly indicating the necessity of further and decisive refinements of theory of ion exchange kinetics.On the other hand,as this paper has tried to describe ,the absolute need exists of planning carefully all kinetic investigations to have acceptable chances for a rational interpretation of experimental results.

REFERENCES

1. Kressman,T.R.E. and J.A.Kitchener,Discussions Faraday Soc.,7,90(1949)
2. Boyd,G.E.,A.W.Adamson and L.S.Myers Jr.,J.Am.Chem.Soc., 69,2836(1947)
3. Smith,M.,Chemical Engineering Kinetics,2nd ed.,McGraw-Hill,New York,N.Y.,1970
4. Helfferich,F.G.,Ion Exchange,McGraw-Hill,New York,N.Y., 1962,ch.9,p.425
5. Wilke,J.and O.A.Hougen,Trans.Am.Inst.Chem.Eng.,41(1945) 445
6. Helfferich,F.G.,Ion Exchange,McGraw-Hill,New York,N.Y., 1962,ch.6
7. Helfferich,F.G.in Ion Exchange,vol.I,J.A.Marinsky,ed., M.Dekker,New York,N.Y.,1966,pp.86-93
8. Blickenstaff,R.A.,J.D.Wagner and J.S.Dranoff,J.Phys. Chem.,71(1967)1665
9. ibid.,1670

10.Adams,G.,P.M.Jones and J.R.Millar,J.Chem.Soc.,(1969)2543
11.Gupta,A.R.,Indian J.Chem.,8(11)(1970)1026
12.Schmuckler,G.,M.Nativ and S.Goldstein,Proc.Int.Conf. on Theory and Practice of Ion Exchange,S.C.I.,Cambridge, 17.1,1976
13.Helfferich,F.G.,J.Phys.Chem.,69(1965)1178
14.Dana,P.R. and T.D.Wheelock,Ind.Eng.Chem.Fundam.,13(1974) 20
15.Selim,M.S. and R.C.Seagrave,Ind.Eng.Chem.Fundam.,12 (1973)14
16.Efendiev,A.A.,A.T.Shahtahtinskaja and P.Meares,Proc. Int.Conf.on Theory and Practice of Ion Exchange,S.C.I., Cambridge,18.1,1976
17.Bahu,R.E.,M.J.Craske and M.Streat,Proc.Int.Conf.on Theory and Practice of Ion Exchange,S.C.I.,Cambridge, 19.1,1976
18.Nativ,M.,S.Goldstein and G.Schmuckler,J.Inorg.Nucl. Chem.,37(1975)1951
19.Debye,P. and E.Huckel,Phys.Z.,24(1923)305
20.Wen,C.Y.,Ind.Eng.Chem.,60(1968)34
21.Kunii,D. and O.Levenspiel,Fluidization Engineering, Wiley,New York,N.Y.,1969
22.Yagi,S. and D.Kunii,Chem.Eng.(Tokyo),19(1955)500
23.Boghetich,G. and D.Petruzzelli,ICP,XI,10(1981)
24.Madi,I.,T.Varro and E.Kovacs,Int.J.Appl.Radiat.Isotopes, 28(1977)473
25.Jarolska,H.,L.Rowinska and L.Walis,J.Radioanal.Chem., 38(1977)29
26.Varro,T.,G.Somogyi,Zs.Varga and I.Madi,Int.J.Appl. Radiat.Isotopes.,29(1978)381
27.Chikomska,K.,A.Lutze-Birk and L.Walis,Isotopenpraxis, 14(1978)224
28.Liberti,L.,I.Madi,R.Passino and L.Walis,J.Chromatogr., 201(1980)43
29.Hoell,W. and G.Geiselhart,Desalination,25(1978)217
30.Perry,J.H.,Chemical Engineers' Handbook,Int.Stud.Ed., 5th ed.,McGraw-Hill,New York,N.Y.,1973,sect.16
31.Boari,G.,L.Liberti,C.Merli and R.Passino,Desalination, 15(1974)145
32.Boari,G.,L.Liberti,C.Merli and R.Passino,Ion Exch.Membr., 2(1974)59
33.Boari,G.,L.Liberti,M.Santori and L.Spinosa,Desalination, 19(1976)283

34. De Maio, A., G. Odone and R. Zannoni, Proc. 7th Int. Symp. on Fresh Water from the Sea, Amsterdam, The Netherlands, 1980, vol. I, 375.

LIST OF SYMBOLS

a = resin surface per unit volume (cm^2/cm^3)
C = total solution concentration (eq/l)
$\bar{C}$ = total resin exchange capacity (eq/l_r)
D = interdiffusion coefficient in solution (cm^2/s)
D_i^{∞} = diffusion coefficient in solution of ith species at infinite dilution (cm^2/s)
$\bar{D}$ = interdiffusion coefficient in resin (cm^2/s)
F = volumetric flow rate (cm^3/s)
HTU = height of a transfer unit (cm)
h_z = height of the column (cm)
K_l = liquid phase mass transfer coefficient (cm/s)
N_{Re} = Reynolds number for porous systems
N_{Sc} = Schmidt number
NTU = number of transfer units
r = average radius of the resin bead (cm)
S = cross sectional area of the column (cm^2)
t = time (s)
$t_{0.5}$ = half exchange time of the reaction (s)
U = fractional attainment of equilibrium (i.e., conc. at time t/conc. at equilibrium)
α = separation factor
β = retardation factor
δ = thickness of the stagnant liquid film around the bead (cm)

NON-EQUILIBRIUM THERMODYNAMICS - A GENERAL FRAMEWORK TO DESCRIBE TRANSPORT AND KINETICS IN ION EXCHANGE. *)

Reinhard W. Schlögl

Max-Planck-Institut für Biophysik, Frankfurt/M, West-Germany

A common question pertinent to many of the talks presented at this meeting is the following: What is the range of validity of model theories, of kinetic descriptions, of the NERNST-PLANCK equations, of STEFAN-MAXWELL equations, of the variational principle proposed by DICKEL etc.? The most general formalism we have today is that provided by the so-called non-equilibrium thermodynamics. The fact it uses only general and unexceptionable "first principles" such as the conservation of mass, momentum and energy and the second law of thermodynamics is at once its strength and weakness; i.e., although it is of great generality, it yields only minimal information in a specific problem.

Non-equilibrium thermodynamics can be broadly decomposed into two completely different parts.

The first part, referred to as Part I here, consists of the "Balance equations". These were formulated by BERTRAND[1], JAUMANN[2], ECKART[3] and MEIXNER[4] and followed by PRIGOGINE[5], DE GROOT[6], MAZUR[7]

*) In response to the papers presented at the meeting, I made a few general remarks. The present version incorporates many of the points that I made not only during my talk but also during the discussions with the various other speakers.

and REIK[8] among many others. For our purposes, the dissipation function

$$\sigma_S = \vec{J}_q \cdot \nabla\left(\frac{1}{T}\right) + \sum_{\substack{i \\ \text{(species)}}} \vec{J}_i \cdot \left(\frac{\vec{F}_i}{T} - \nabla \frac{\tilde{\mu}_i}{T}\right) - \sum_{\substack{r \\ \text{(reactions)}}} \Gamma_r \frac{A_r}{T}$$

$$- \left(P - Ip\right) : \frac{\mathrm{Grad}\, \vec{v}}{T} \tag{1}$$

is of central importance. Here

σ_S = production of entropy per unit volume and unit time

$\vec{J}_q$ = heat flux

$\nabla\left(\frac{1}{T}\right)$ = conjugated force

$\vec{J}_i$ = $\rho_i(\vec{v}_i - \vec{v})$ = relative particle flux

ρ_i = specific density

$\vec{v}_i$ = average velocity

$\vec{v} = \sum \rho_i \vec{v}_i / \sum \rho_i$ = barycentric velocity

$\vec{F}_i$ = imposed driving force

$\tilde{\mu}_i$ = electrochemical potential

Γ_r = chemical production

A_r = affinity as defined by DE DONDER[9]

P = pressure tensor

I = unit tensor

p = "pressure" = $\frac{1}{3}$ × trace of P

The last term in the bilinear form (1) is the so-called RAYLEIGH dissipation function. It involves the overall viscosity effects. The balance equations need very few premises:

1) The system must contain a large number of particles.
2) It should not be so far from equilibrium that notions like temperature, pressure etc., lose their meaning. This part of the theory thus has a very wide range of validity.

Part II of the formalism postulates general relationships between the fluxes and the driving forces known as the "phenomenological laws". Here, for the first time, material and kinetic parameters come into play. Only in the vicinity of an equilibrium state can linear laws be postulated. To cite an example, for vanishing external forces $(\vec{F}_i = 0)$,

$$\vec{J}_i = \sum_k L_{ik} \nabla \frac{\tilde{\mu}_k}{T} + L_{iq} \nabla \frac{1}{T} \tag{2}$$

The flux of the species i and all the driving forces to which it is related have the same tensorial character (an immediate consequence of the CURIE principle as applied to an isotropic medium). It is this part of the theory where ONSAGER's name figures prominently. The L_{ik} are the elements of a non-negative definite symmetric matrix. VOIGT[10] first proved this symmetry property for the heat conductivity in crystals and ONSAGER[11] followed it up in the context of coupled chemical reactions. Later, ONSAGER recognized its general validity and gave a statistical justification. CASIMIR[12], CALLEN and GREENE[13] and others generalized and completed his proof.

As you see from the equation (2), the NERNST-PLANCK Ansatz follows directly if you set

$$T = \text{const.}, \quad L_{ii} \sim c_i D_i \;, \quad L_{ik} = 0 \;, \quad \vec{F}_i = 0$$

and neglect activity coefficients.

It is important to note that there is only *one* coefficient connecting the flux and the force. Thus you do not need EINSTEIN's relation connecting diffusion coefficient and mobility.

By again neglecting all the L_{ik} except L_{iL} (L = solvent) and correlating L_{iL} to an individual friction coefficient f_i, you can construct from equations (1) and (2) the LAGRANGE function which is the starting point for the variational principle of DICKEL.

If you relate the L_{ik} to individual friction coefficients (all of which, of course, must be positive) you are led to SPIEGLER's well-known transport equations.

If you interpret the L_{ik} in the sense of STEFAN, you get the description discussed by GRAHAM.

To describe transport across membranes, one can invoke the so-called discontinuous description. Here the membrane is regarded a "black box" and the driving forces are taken to be the differences of the intensive variables characterizing the outer phases; this description is more general than those of model theories which neglect some or all of the cross-coefficients L_{ik} . However, it is less general in that it needs expansions to higher powers in the forces for large deviations from equilibrium. The MICHAELI-KEDEM[14] equations, cited by SPIEGLER and the related KATCHALSKY-KEDEM[15] equations are derived from the dissipation function by somewhat different transformations (cp. in this connection SCHLÖGL[16]). These provide examples of the discontinuous description mentioned above restricted to linear relations between fluxes and forces. HOFF in his contribution discusses expansions to higher powers in the driving forces and focusses attention on the dependence of the transport coefficients on the reference state.

I am convinced that TURNER's title "Nernst-Planck or not" was said with tongue-in-cheek: of course, there are situations where NERNST-PLANCK equations provide an excellent description and others where they turn out to be almost totally invalid.

Thermodynamic considerations cannot serve to delimit the domain of validity of the linear laws. Whether the range of validity of the linear laws is "large" or "small" is a matter of definition as to what "large" means and is best left to the experimentalist. The only general statement that can be made, a negative one, can be phrased as follows: "When measurable changes of the intensive variables occur over dimensions which are smaller than the mean free path of the particles, the linear formulations are inadequate". I may be permitted to make three concluding remarks:

1) To the best of my knowledge, the first to formulate a principle of "least dissipation" was PRIGOGINE. Of course, the term "least" must be properly defined. It will be meaningless to point to a house and say "this *is* the smallest house" without stating by what measure it is the smallest (the height, the lay-out, the number of rooms etc.) and which part of the globe is meant. PRIGOGINE sought to compare dissipation in any possible non-steady state to that at the steady state. In this sense, the term "least" was well-defined. A "possible" state was any allowed by the conservation laws of mass, momentum and energy. MEIXNER and REIK showed that this principle holds only if a whole set of additional constraints were fulfilled and this drastically reduced its practical utility. Later PRIGOGINE and GLANSDORFF proposed a new, less general principle. For the time being, I feel that DICKEL's approach needs a better physical justification. Besides, the adjective "least" needs a sharp definition. Furthermore, the EULER-LAGRANGE-equations used by him allow for variations of fluxes but not of the concentrations profiles. The latter would be desirable.

2) Equations (2) contain neither the electric potential nor the swelling pressure explicity. Using non-polarizable, specific electrodes they allow the formulation of equations containing measurable quantities only as long as isothermal systems are discussed. If you do not like the splitting of electrochemical gradients into non-measurable quantities such as chemical, electrical and pressure terms, you are not forced to do so.

3) Very often the term "non-equilibrium thermodynamics" is used as if it were synonymous with the linear laws, the ONSAGER's reciprocity principle or both. This point of view is neither convenient nor warranted. Indeed, part II of the description envisages not only non-linear phenomenological laws but even allows for systems with "memory", where the fluxes may be related to the forces by an integral transformation. In addition, the balance equations contained in part I are not any less important.

Non-equilibrium thermodynamics cannot be considered a possible theory among many others. It is a mode of description in which almost all fields of classical physics are brought together. Any model theory, transport theory or kinetic theory has to fit into this framework. Indeed only "exotic" systems, as cited above, very far from equilibrium, may fall outside its purview. This discipline was not the outcome of any solitary individual, but that of a whole generation of physicists, physical-chemists, rheologists and others. Anybody who ignores this discipline and its history, places himself in a situation not unlike that of an electrical engineer who ignores MAXWELL's description of the electromagnetic field.

References

1. Bertrand, J.L.F., Thermodynamique, Paris 1887
2. Jaumann, G., Sitzungsber.Akad.Wiss. Wien, Math.-naturw.Kl. Abt.II A 120 (1911) 385.
3. Eckart, C., Phys.Rev. 58 (1940) 267, 924
4. Meixner, J., Ann.Physik {5} 35 (1939) 701; {5} 39 (1941) 333
5. Prigogine, I.: Etude Thermodynamique des Phénomènes irréversibles. Liège 1951
6. De Groot, S.R.: Thermodynamics of irreversibles Processes. Amsterdam 1951
7. Mazur, P. and Prigogine, I., J.Phys.Radium, 12 (1951) 616
8. Vgl. Meixner,J. und Reik, H.G., Thermodynamik der irreversiblen Prozesse, Handbuch der Physik (Flügge), III/2 (1959) 413
9. De Donder, Th., L'affinité (Gauthier-Villars) Paris 1927
10. Voigt, W., Nachr.Ges.Wiss. Göttingen, Math.Phys.Kl. 87 (1903)
11. Onsager, L., Phys.Rev. 37 (1931) 405; 38 (1931) 2265
12. Casimir, H.B.G., Rev.Mod.Phys., 17 (1945) 343
13. Callen, H.B., Greene, R.F., Phys.Tev. 86 (1952) 702
14. Michaeli, I., Kedem, O., Trans.Faraday Soc. 57 (1961) 1185
15. Kedem, O., Katchalsky, A., Biochem.biophys.Acta 27 (1958) 229
16. Schlögl, R.W., Stofftransport durch Membranen (Steinkopff) 44 (1964)

COLUMN DESIGN FOR SORPTION PROCESSES

Gerhard Klein

University of California, Berkeley
Water Thermal and Chemical Technology Center,
47th & Hoffman Blvd., Richmond, California 94804

ABSTRACT

Basic concepts, including the role of equilibrium, stoichiometric, and rate relations on the performance of fixed, monovariant sorption beds are reviewed, with emphasis on simple, idealized, linear-driving force rate models. A novel method is presented for combining film- and particle-diffusion mechanisms. The number of variables involved is reduced by combination into dimensionless groups, and the types of column performance are illustrated by generalized time-distance plots with concentration contours. A method for deriving rate constants and mechanism parameters from an ion-exchange isotherm and a constant-pattern effluent-concentration history is presented. Calculation methods are classified and outlined for the axis intercepts, the noncoherent, and the coherent portions of constant and proportional-pattern concentration contours. Column design is illustrated by example.

1 INTRODUCTION

The material presented here covers primarily design considerations for cylindrical fixed-bed ion exchangers. However, through the use of generalized variables, it is also applicable to adsorption processes, with minimal modifications. The term "sorption" is meant to apply to these and analogous processes. The concepts involved are useful in the analysis of a variety of phenomena of great practical or scientific interest, such as enhanced crude-oil recovery, underground chemical leaching, and some geochemical processes.

The transient phenomena of interest are governed by equilibrium, stoichiometry, and rate relations and involve a considerable number of variables, which increases with the number of components. For these reasons, a complete mathematical description is rarely feasible. The approach taken in chemical-engineering practice, and also here, is to adopt a number of simplifying premises, to combine several variables into dimensionless parameters, and to develop a clear understanding of typical column behavior under selected conditions, so that design can be guided by intuition, rather than merely by massive computations. This approach also conforms to the principle that the accuracy of a calculation need be no more than commensurate with that of the underlying data, which, in the present case, are numerous and laborious to accumulate.

The material presented comprises a selective review of relevant concepts, with emphasis on normalized time-distance diagrams, and illustrates their application in design. Also presented is a method for combining mass-transfer resistances and its implication for obtaining the rate parameters involved from constant-pattern effluent-concentration histories, alternative to the standard method. The treatment focusses primarily on two-component ion-exchange and analogous adsorption systems. For complementary material and data, the reader is referred to standard works on ion exchange, such as Helfferich (7), Kunin (9), Applebaum (2), the manuals published by some of the major manufacturers of ion-exchange resins, and, for a unified chemical-engineering treatment, the section by Vermeulen et al. (12) in the Chemical Engineers' Handbook. References to selected journal articles are given where applicable.

2 BASIC CONCEPTS

2.1 Variance

Sorption problems may be classified according to a variety of criteria. Perhaps the most fundamental of these is the number of variables affecting equilibrium that can be varied independently when the system is at equilibrium. We here designate this number as "variance". For simple ion exchange and isothermal adsorption, the variance is readily apparent, being n - 1 for an n-component ion-exchange system, and n for an n-component adsorption system, the concentrations being the variables in these cases. The situation becomes more complicated when such factors as chemical reactions, thermal or pressure effects, and the presence of additional phases enter the picture.

The present discussion is limited to monovariant problems. Systems of higher variance involve additional considerations, and their design requires a computational effort, which, as the variance increases beyond 2, rapidly tends to become unmanageable.

2.2 Concentration Variables

For adsorption systems, the concentration variables of interest then are the sorbate concentration (c) or pressure (p) in the fluid phase, and the local sorbate concentration (q) in the sorbent, here expressed in millimoles per gram of sorbent. For ion exchange, there are two exchangeable ion species of interest, but, with the premises of negligible solvent transfer between the fluid and exchanger phases, absence of chemical reaction, of negligible Donnan uptake of electrolyte by the exchanger, and of constant exchange capacity, the sums of the normal concentrations of both species in the fluid phase on the one hand, and in the exchanger phase, on the other, are constant, so that, provided that the total solution normality and the total exchange capacity are known, it suffices to specify the concentration of either species in both phases to define the composition. Here, in order to achieve analogy with adsorption systems, the concentration of that ion species which is higher in the feed than in the product stream will be selected for this purpose. This species, as well as the adsorbate, will be referred to generically as sorbate, and the medium into which it transfers, as sorbent.

It is now possible to represent the concentrations in both types of systems in normalized, dimensionless form, using the following transformations:

For the fluid phase,

$$X \equiv (c - c')/(c'' - c') = (p - p')/(p'' - p') \tag{1}$$

and for the sorbent phase,

$$Y \equiv (q - q')/(q'' - q') \tag{2}$$

In these relations, symbols without primes represent variable concentrations; q' is the uniform initial concentration in the sorbent, c' or p' concentration or pressure in the fluid phase in equilibrium with q', and c'' or p'', are the constant concentration or pressure of the sorbate in the feed, and q'' is the sorbate concentration in the sorbent in equilibrium with c'' or p''.

2.3 Types of Equilibrium Isotherms

As a convention, we shall here usually give compositions in terms of the relative concentrations of the component that is being taken up by the sorbent in the operating step under consideration; that is, in most simple cases, in terms of the concentrations of the feed component. (The wider definition given first above is required to avoid ambiguity in multivariant cases, and in monovariant systems with a two-component feed.) The respective concentrations X and Y in the fluid and sorbent phases then each range from zero to one. For monovariant systems, the isotherm is a line connecting points (X,Y) that represent equilibrium compositions.

For the purpose of the present discussion, monovariant isotherms can be classified qualitatively into (1) favorable, (2) unfavorable, (3) linear isotherms, and (4) isotherms exhibiting selectivity reversal. Favorable and unfavorable isotherms, respectively, do or do not favor uptake of the sorbate feed component over the entire range of concentrations considered. A favorable isotherm lies entirely above, and an unfavorable one, entirely below, the diagonal of the isotherm diagram, as shown in Figs. 1a and 1b. A linear isotherm is given by Y = X(Y), and an isotherm with selectivity reversal, illustrated by Fig. 1c, is composed of favorable and unfavorable regions, which implies that it crosses the diagonal of the diagram at least once.

Any isotherm determined experimentally can be represented graphically, but it is often convenient to model isotherms mathematically, and to represent them in terms of one or a few parameters, with varying degrees of realism and generality. While such models, for both ion exchange and adsorption, are quite numerous, we here give only three, of which the last two are subcases of the following mass-action expression for ion exchange:

$$\left(\frac{Y}{X}\right)^b \left(\frac{1-X}{1-Y}\right)^a = \underset{\sim}{K} = K(Q/C_0)^{a-b} \tag{3}$$

Here, a is the absolute value of the valence of the feed ion and b that of the ion initially on the sorbent. $\underset{\sim}{K}$ is a constant, dimensionless selectivity coefficient, which depends on the constant dimensional selectivity coefficient K, in $(ml/g)^{b-a}$, the exchange capacity Q, in meq./g, the total solution normality C_0, and the valences, as indicated in the second part of the equation. Eq. 3 applies only for c" = C_0, c' = 0, q" = Q, and q' = 0.

For exchanging ions of unit valence, the isotherm becomes a hyperbola, given by

$$\frac{Y}{X}\left(\frac{1-X}{1-Y}\right) = \alpha \tag{4}$$

where the constant separation factor α replaces $\underset{\sim}{K}$ or K for this special case. For this equation, the above restrictions on c', c", q' and q" do not apply, but the value of α depends on them, as well as on the exchanger and the ion species involved.

The isotherm for Langmuir adsorption equilibrium is given by

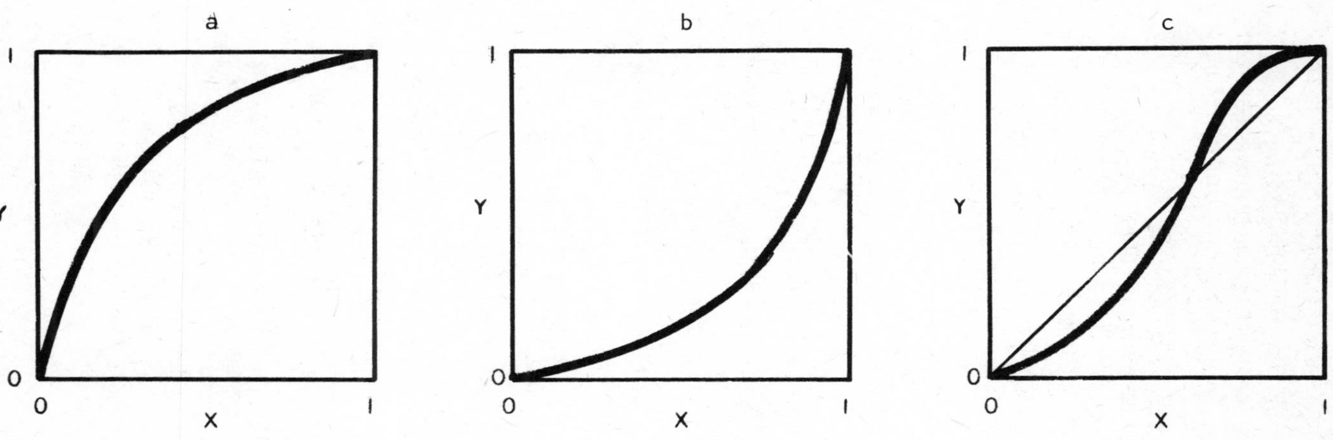

Fig. 1. a. Favorable hyperbolic isotherm ($\alpha = 5$); b. Unfavorable hyperbolic isotherm ($\alpha = 0.2$); c. Typical isotherm with selectivity reversal. X and Y are the relative sorbate concentrations in the fluid and sorbent phase, respectively.

$$q = \frac{QK_L c}{1 + K_L c} \tag{5}$$

where Q is the sorbtive capacity corresponding to full monomolecular occupancy, q, the sorbate concentration on the sorbent, both in the same units, and K_L, the Langmuir equilibrium constant, in units of reciprocal concentration.

With $X \equiv c/c_{ref}$ and $Y \equiv q/q_{ref}$, where c_{ref} is the feed concentration, and q_{ref} is the concentration in the sorbent in equilibrium with c_{ref}, Eq. 5 can be transformed into Eq. 4. It can also be shown that

$$a = K_L c_{ref} + 1 \tag{6}$$

2.4 Rate Considerations

The rate dY/dt of uptake of sorbate at a particular level of a fixed sorption bed is governed by the rate mechanism or combination of mechanisms applicable, by the relevant equilibrium isotherm, and by the values of the parameters entering the mathematical model that describes the mechanism. Usually, the better a model fits reality, the more complicated it is, the more parameters it involves, and the lengthier are the calculations required to utilize it. In practice, therefore, the engineer is willing, or forced, to make considerable compromises to keep the model manageable.

In ion exchange, the chief mass-transfer mechanism is diffusion in the fluid or solid phase, the diffusivity of the same ion in the latter being about an order of magnitude smaller than that in the former.

An ion about to enter a sorbent particle must first pass through a layer of fluid surrounding the particle, which is nearly stationary even if the fluid flows; the so-called Nernst film, or some equivalent, more recent and sophisticated, concept. The ion then passes through the overall particle surface either into the particle phase or into a pore. In the particle phase, or along the pore walls, the ion proceeds by "particle diffusion", which is governed by the diffusivity in the solid phase; in the pores themselves, by pore diffusion, which is governed by the diffusivity in the fluid phase, but independent of the film thickness. Film and particle diffusion proceed in series; pore diffusion and particle diffusion along the pore walls, in parallel.

The simplest two models are of considerable interest in that they permit manageable practical calculations involving the transient phenomena that occur in fixed-bed sorption even when the models need to be combined. These models adopt a number of grossly simplifying assumptions, the most salient of which are discussed in passing below. Nevertheless, they have in numerous cases been shown to be capable of satisfactory prediction of column behavior, perhaps because of compensation of errors and the ability of parameters obtained empirically to make up for the deficiencies inherent in the model.

In the first of these models, called "f-mechanism", the controlling resistance is in the film, and diffusion inside the particle is relatively very fast, so that the radial concentration gradient in the particle is virtually zero and the average concentration in the particle coincides with its concentration at the particle surface, where the fluid-phase concentration, X(Y), is in equilibrium with the concentration Y in the particle phase. These conditions are depicted in Fig. 2. Fig. 3 shows the corresponding X,Y-diagram. The ensuing rate law takes the form

$$\frac{dY}{dt} = \frac{k_f a}{\underset{\sim}{\Lambda}}\left[X - X(Y)\right] \qquad \text{(f-mechanism)} \qquad (7)$$

where the distribution coefficient $\underset{\sim}{\Lambda}$ is given by Eq. 30 further below, a is the surface area of the particles per unit volume of packed bed, in cm^{-1}, the bed being assumed to contain 40 percent voids, and where $k_f a$, in sec^{-1}, is given by

$$k_f a = 2.62(D_f u_o)^{0.5} d^{-1.5} \qquad (8)$$

Here, D_f is the diffusivity in the fluid phase, in $cm^2 sec^{-1}$; u_o, the approach velocity, in cm sec^{-1}; and d, the particle diameter, in cm.

The driving force [X - X(Y)] in Eq. 7 is called "linear" because it is proportional to the radial distance of the point in the film under consideration from the interface between fluid and particle.

Glueckauf and Coates' linear driving-force approximation (6), for relatively very fast diffusion in the fluid and the major resistance in the film-phase ("p-mechanism"), is entirely analogous to Eq. 7:

$$dY/dt = k_p a[Y(X) - Y] \qquad \text{(p-mechanism)} \qquad (9)$$

where

$$k_p a = 60 D_p d^{-2} \qquad (10)$$

D_p being the diffusivity in the particle, in cm^2 sec^{-1}. Here, the concentration in the fluid phase X is assumed to be uniform throughout and up to the particle surface, where the concentration

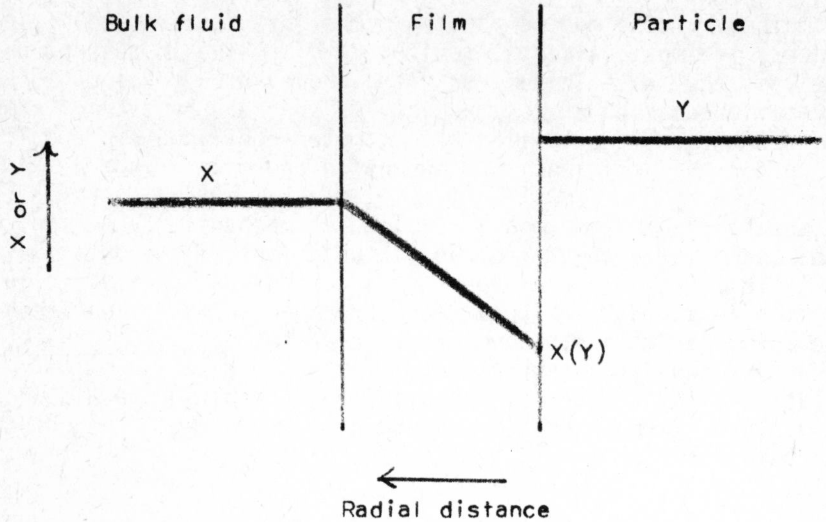

Fig. 2. Radial concentration profile for pure f-mechanism (large diffusivity in particle phase).

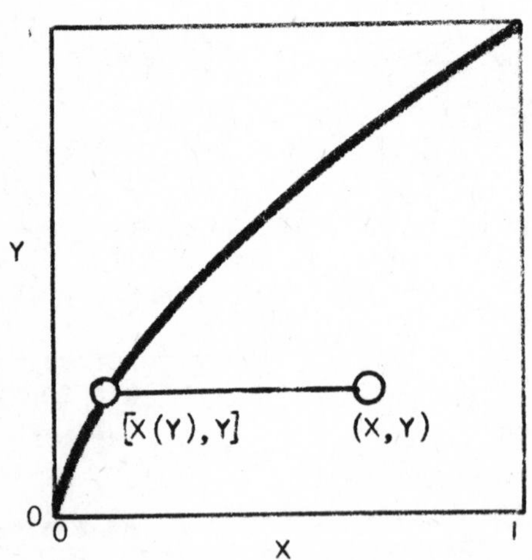

Fig. 3. X,Y-diagram corresponding qualitatively to concentration profile for f-mechanism in Fig. 2. [x - x(Y)] is the driving force in Eq. 7, X being the concentration in the bulk fluid phase. The curve represents an arbitrary isotherm.

$Y(X)$ in the sorbent is in equilibrium with X. In the particle, the concentration decreases linearly through an imaginary film, to the average concentration in the particle, Y. A typical radial concentration profile for this case is shown in Fig. 4, and the corresponding X,Y-diagram, in Fig. 5.

In this model, a purely conceptual solid film seems to be implied, and the spherical geometry of the particle is ignored. In actuality, the concentration profile in the particle is nonlinear and can, in principle, be calculated by numerical integration. This adds a major dimension to the calculations, so that, in spite of its deficiencies, the Glueckauf and Coates model is often invoked in practice, because it is readily integrable in many important cases, and can be used as part of a model for combined f- and p-mechanisms.

Both models consider the relevant diffusion coefficient invariant with respect to composition; a premise that is not justified, especially for heterovalent ions, as discussed by Helfferich (7). The effect of composition can be estimated with the aid of the Nernst-Planck equations, again with additional computational effort.

A third relatively simple model is the reaction-kinetic one, which considers the ion-exchange process as a second-order reaction, with the kinetics characteristic of such. This model is not physically realistic for ion exchange because the actual ion-exchange step is known to be rapid as compared to the one or more diffusion steps involved. Nevertheless, the model has value because, for systems approximately governed by hyperbolic isotherms, as are homovalent ion-exchange systems, it provides solutions over the entire range of possible separation factors, i.e., for systems that in many cases can be solved only numerically when other models are used. Equations for the reaction-kinetic and some other models are presented by Vermeulen et al. (12).

For systems in which both the f- and p-mechanisms are significant, radial profiles of the type shown in Fig. 6 arise, and the following considerations apply. X_I and Y_I are concentrations at the fluid-solid interface, in equilibrium with each other. It is seen that the driving force in Eq. 7 must be replaced by $X - X_I$, and the driving force in Eq. 9, by $Y_I - Y$. If the capacity of the film surrounding the particles is considered negligible, as is reasonable, the number of equivalents transported by the f- and p-mechanisms is virtually equal. Thus, with Eqs. 7 and 9,

$$\frac{dY}{dt} = \frac{k_f a}{\Lambda}(X - X_I) = k_p a(Y_I - Y) \quad \text{(f- and p-mechanism)} \qquad (11)$$

and

$$-\frac{k_f}{\Lambda k_p} = \frac{Y - Y_I}{X - X_I} \qquad (12)$$

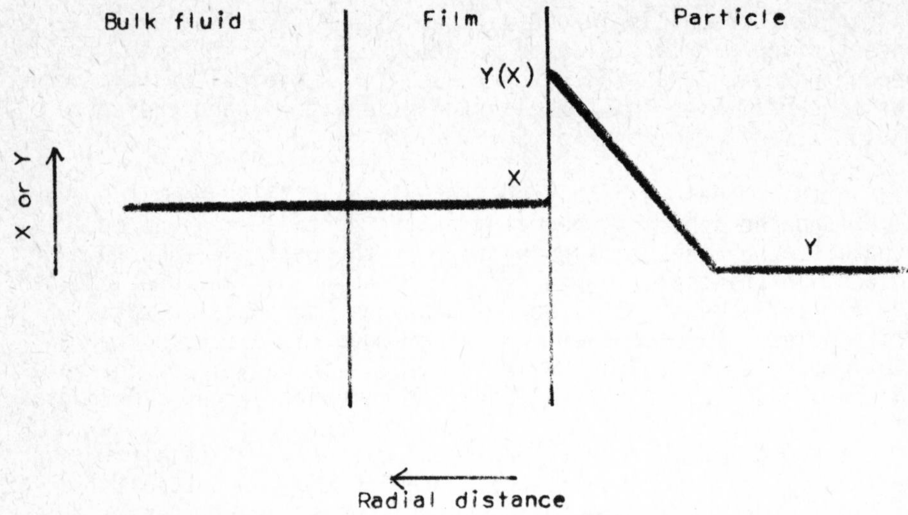

Fig. 4. Radial concentration profile for pure p-mechanism (large diffusivity in fluid film).

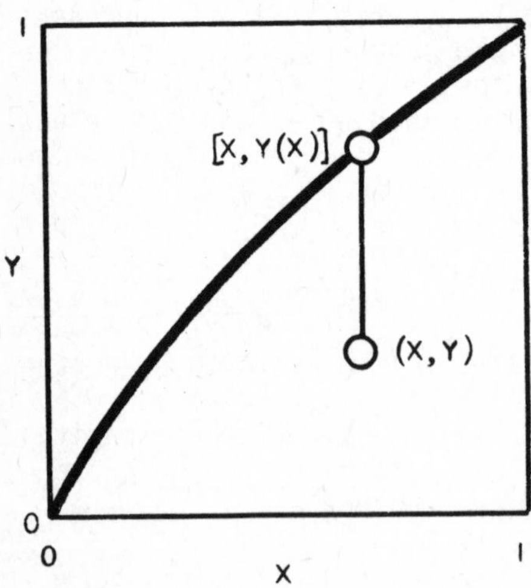

Fig. 5. X,Y-diagram corresponding qualitatively to concentration profile for p-mechanism in Fig. 4. $Y(X) - Y$ is the driving force in Eq. 9, Y being the average concentration in the sorbent phase. The curve represents an arbitrary isotherm.

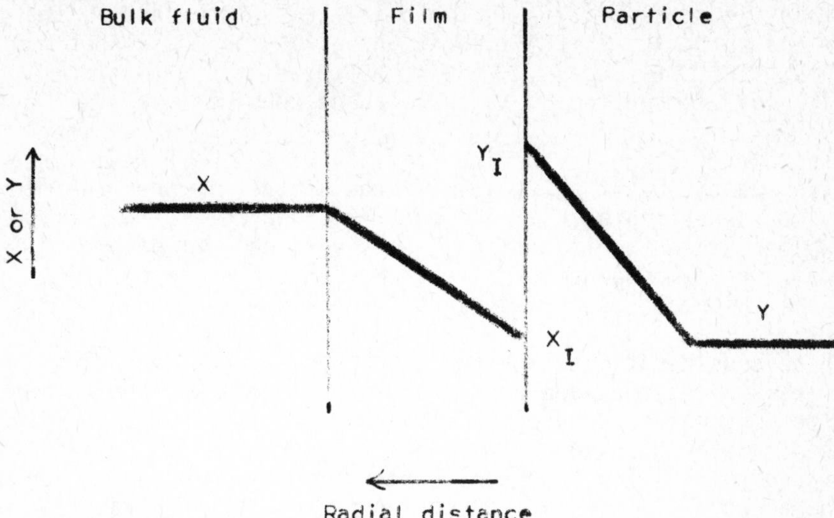

Fig. 6. Typical concentration profile for combination of f- and p-mechanisms. Significant concentration gradients in both phases. The concentrations X_I and Y_I at the interface are in equilibrium.

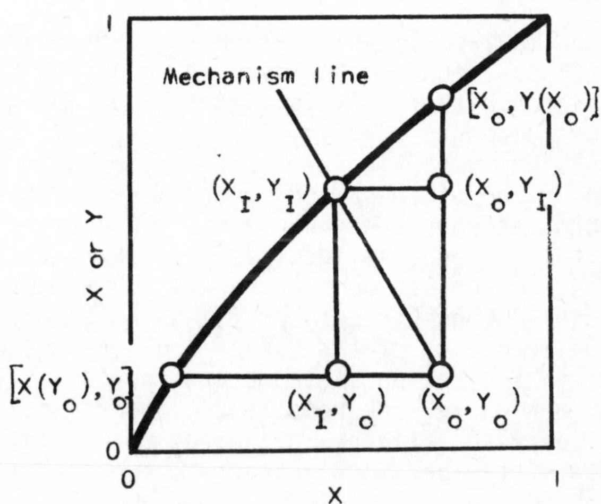

Fig. 7. Alternative method for combining f- and p-mechanisms. X_o and Y_o are the respective concentrations in the bulk fluid and sorbent phases. (X_I, Y_I) is the equilibrium composition at the interface (outside particle surface). This point is obtained as the intersection of the mechanism line, constructed with Eq. 12, and the isotherm (arbitrary curve shown).

where X is the concentration in the bulk fluid phase, and Y, the average concentration in the sorbent phase. Eq. 12 describes a "mechanism line" of slope - $k_f/(\underset{\sim}{\Delta} k_p)$, as shown in Fig. 7 for a specific fluid concentration X_0 and sorbent concentration Y_0.

The point (X_I, Y_I) is the intersection of the mechanism line, drawn through the point (X_0, Y_0), with the isotherm. When the mechanism line is horizontal, X_I becomes $X(Y_0)$, and Eq. 7, for f-mechanism, applies; when the mechanism line is vertical, Y_I becomes $Y(X_0)$ and Eq. 9, for p-mechanism, applies. For intermediate cases, it is immaterial which of Eqs. 11 is used.

These considerations constitute the basis for a method of combining resistances that appears to be more direct than the conventional method, which is based on the analogy between heat transfer and mass transfer governed by a linear isotherm, and which resolves the disparity resulting from nonlinearity in an ion-exchange isotherm with the aid of an empirical correction factor that can vary enormously over the possible range of concentrations. Application of the alternative method of combining the f- and p-mechanisms indicated here will be described further below.

3 CHARACTERIZATION OF FIXED-BED BEHAVIOR

3.1 General Remarks

Because of the multiplicity of variables that determine the response of a fixed sorption bed to given initial conditions and a given feed-composition history, it is important to use methods that will represent such response clearly, completely, and simply. Even for monovariant systems with single-step feed-composition change, the information given by the often-presented breakthrough curve, i.e., the effluent-concentration history up to the highest permissible concentration of the sorbate, is incomplete in that it gives no clue as to the remainder of the effluent-concentration history and the evolution of the axial concentration profiles in the bed with time. The reason for the frequency of the use of breakthrough curves is that they do give an indication of the operating capacity, and that they are readily obtainable with beds operating cyclically, while measurement of a complete effluent-concentration history requires uniform saturation of the bed with respect to the feed fluid.

The desired sequence of concentration profiles along the bed axis is difficult to obtain, even for the fluid phase only, because it must rely on samples taken along the column, and this not only requires elaborate apparatus, but also is likely to modify the results by affecting the hydrodynamics and continuity of the bed. Moreover, even such efforts will not yield the concommittant concentration profiles in the sorbent phase. To obtain them would require withdrawal of sorbent samples and their subsequent analysis; an undertaking generally practically out of the question.

While it is thus not easy to gather the desired information, it is nevertheless highly useful to find means of visualizing the phenomena inside the bed, and their dependence on time, and this can be based on tractable mathematical models of fixed-bed behavior. The "visible" performance of an actual column, in form of an effluent-concentration history, can then be compared to that predicted by the model. If the agreement is satisfactory, process design and optimization can then be undertaken.

The composition behavior in a fixed bed is transient in that both time and axial position must be specified to characterize it. In principle, then, a diagram at least two-dimensional in its independent variables is required to represent it. When a significant concentration gradient exists inside the sorbent particles, a third dimension is needed to characterize the distance of penetration into the particles, but in the present discussion, only time, t, and axial distance, s, are considered as independent variables. The corresponding performance diagrams can take a variety of forms, such as contour lines of constant axial distance s in a time-concentration (t,X or t,Y)-plane, or of X- and Y-contours in a t,s-plane. DeVault (5) has presented examples of the latter in perspective diagrams, with concentration as a third dimension.

3.2 Time-Distance Diagrams

Consider a time, distance- (t,s)-diagram for which we take as the origin on the t-axis the point corresponding to the time at which the feed fluid for the step under consideration begins to enter the column. At the inlet end (s=0), the time that the sorbent has been in contact with this feed fluid coincides with t. At any finite distance ($s > 0$) from the inlet end, the "local contact time", $\hat{t}$, will be t minus the time $\varepsilon s/u_0$ that it takes the feed fluid front to reach the distance s, ε being the void fraction of the bed, and u_0 the linear approach velocity of the feed fluid. Thus, if we denote local contact time by $\hat{t}$, we obtain

$$\hat{t} = t - \varepsilon s/u_0 \tag{13}$$

Fig. 8 shows a t,s-diagram with the locus $\hat{t} = 0$ as a straight line going through the origin and deviating from the s-axis by the angle $\theta = \arctan(\varepsilon/u_0)$. On this diagram, for any real time t and distance from the inlet end, s, the local contact time $\hat{t}$ is found as the horizontal distance of the point (t,s) from the line $\hat{t} = 0$.

In the following discussion, we assume plug flow of a fluid phase with negligible velocity variations with time, and along the

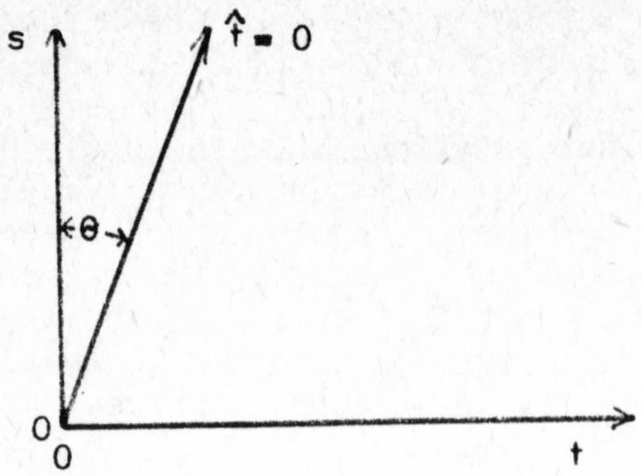

Fig. 8. t,s-diagram with locus at which local contact time $\hat{t}$ is zero. $\theta = \arctan(\epsilon/u_c)$.

column axis at any given time. We also assume absence of diffusion in the axial direction.

Under these premises, any interaction between the feed fluid and sorbent can begin only when the two are coming in contact, i.e., when $\hat{t} = 0$. The part of the diagram of Fig. 8 to the left of the line $\hat{t} = 0$ is therefore irrelevant and the ensuing mathematical treatment can be simplified by using a $\hat{t}$- instead of the t-axis, again in an orthogonal coordinate system.

With such a diagram, the initial axial concentration distribution can also be shown, as X- and Y-profiles plotted against the s-axis, and the influent-concentration history as a plot of X on an axis parallel to the t-axis and calibrated like the latter. In what follows, we will be concerned primarily with the important imposed condition of a single feed-concentration change, corresponding to

$$X = 1 \qquad (s = 0) \tag{14}$$

along the t-axis, and

$$Y = 0 \qquad (\hat{t} \leq 0) \tag{15}$$

along the s-axis, up to the line given by $\hat{t} = 0$.

3.3 Continuity

The differential continuity relation for an arbitrary sorbate species in a fixed bed implies that, in its course through the column, the species neither leaves nor is added through a side stream or a reaction accompanying the sorption process. This relation can be written in the form

$$u_0 \frac{\partial c}{\partial s} + \rho_b \frac{\partial q}{\partial t} + \varepsilon \frac{\partial c}{\partial t} = \varepsilon \frac{\partial}{\partial s}\left(D \frac{\partial c}{\partial s}\right) \tag{16}$$

where D is an effective coefficient for both diffusion and hydrodynamic dispersion in the axial direction.

The first term in this equation is convective; the second and third terms correspond to the time rate of accumulation in sorbent and fluid, with the second term usually preponderating considerably over the third. The dispersion term in the right-hand member reduces to $\varepsilon D\partial^2 c_i/\partial s^2$ if D is constant. Moreover, if this term can be neglected, Eq. 16 becomes

$$u_0 \frac{\partial c}{\partial s} + \rho_b \frac{\partial q}{\partial t} + \varepsilon \frac{\partial c}{\partial t} = 0 \tag{17}$$

If this equation is applied to a front in which sorbate is taken up, the first term will be negative and the second and third terms, positive. For desorption, these signs are reversed.

Because it is the local contact time $\hat{t}$ rather than t that, together with the appropriate rate law, initial axial concentration profile in the sorbent, and the sorbate-concentration history in the fluid, governs the rate of sorbate uptake, Eq. 17 can be reduced to a more manageable, two-term expression with the transformation given by Eq. 13. By considering c and q as functions of s and $\hat{t}$ instead of s and t, writing expressions for the partial derivatives in Eq. 17 in terms of s and $\hat{t}$, and substituting them back into Eq. 17, we obtain the desired result in the form

$$u_0(\partial c/\partial s)_{\hat{t}} + \rho_b(\partial q/\partial \hat{t})_s = 0 \tag{18}$$

3.4 Generalized Measurement of Effluent Quantity

The quantity of effluent from a fixed sorption bed is often measured in terms such as time, volume, or amount of feed component. To compare beds of different sizes and potential capacities, receiving feeds at different flowrates and potential concentration-reduction levels, it is desirable to apply certain transformations to the independent variables, time and axial distance, which will be discussed below. These transformations are based on the concept of a "differential equilibrium bed capacity" and the amount of fluid equivalent to this capacity. The resulting dimensionless variables can be used directly in conjunction with generalized models of wide applicability.

One distinguishes between specific and bed capacities, i.e., between capacities that apply to unit quantity of sorbent, and capacities that apply to a particular sorbent bed. Second, there are operating and equilibrium capacities.

Operating capacities depend on the contacting method, the breakthrough concentration specified, and kinetic factors. Moreover, because in practice complete regeneration is rarely used, the operating capacity predicted by a mathematical model, or measured in the laboratory with a fully regenerated bed, must be distinguished from the smaller cyclic operating capacity, which depends also on regeneration level and time, and can depend on the regenerant composition, including its total normality.

The concept of specific differential equilibrium capacity may be explained with the aid of Fig. 9, which applies to an ion-exchange bed uniformly presaturated to concentration q_0 in the sorbent phase, and receiving a feed having concentration c_0 of the component being sorbed. As shown in the figure, the fluid-phase concentration in equilibrium with $[q_0, c(q_0)]$, and the concentration in the sorbent in equilibrium with $[c_0, q(c_0)]$, lie on the equilibrium isotherm. $[q(c_0) - q_0]$ then is the specific differential equilibrium capacity of the sorbent, in milliequivalents per gram, and a bed of bulk density ρ_b and volume v may be said to contain $\rho_b v[q(c_0) - q_0]$ potential equivalents of sorbent, v being measured in liters. This differential equilibrium bed capacity is used as the unit for measuring effluent quantity in the method adopted here.

The concept can readily be extended to multicomponent systems, but to do so would exceed the scope of the present discussion. For an ion exchanger initially completely free of feed ion, and which is being equilibrated with a solution containing only the feed ion, the specific differential equilibrium capacity becomes the total

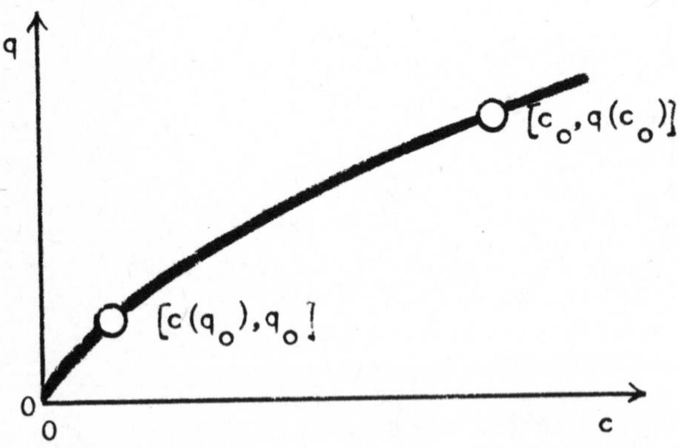

Fig. 9. Sorption isotherm with points $[c_0, q(c_0)]$ and $[c(q_0), q_0]$, for equilibrium with fixed-bed feed and presaturated sorbent, respectively. $q(c_0) - q_0$ is the specific differential equilibrium capacity. c = equivalent concentration of feed ion in solution; q = concentration in sorbent, meq. per gram.

specific capacity, Q, earlier referred to as "scientific operating capacity" by Helfferich (7).

To establish an equivalence between an exchanger bed and the fluid passed through it, consider a bed to be initially at the uniform concentration q_o, and finally at the uniform concentration $q(c_o)$, and the bed voids to be filled initially with incompressible fluid. Ideally, assume plug flow, and that sorbent and fluid are at equilibrium at each point along the bed axis; also, at least for the moment, that the isotherm is favorable. These conditions are known to lead to sharp and axially coincident fronts (mathematical discontinuities in the concentrations) in both phases that move more slowly (in most practial cases considerably more slowly) than the fluid front. Behind the front, the concentrations are c_o and $q(c_o)$ and ahead of it, they are $c(q_o)$ and q_o. (In the case of unfavorable equilibrium, a diffuse front would ensue, even with local equilibrium. This, however, does not invalidate the present derivation, which is based purely on stoichiometric considerations.)

If we let V_{stoich} be the influent volume at the time at which the front reaches the outlet end of the bed, then we can write the following material balance in terms of feed-component concentrations:

Input - output = final amount - initial amount

$$V_{stoich}c_o - V_{stoich}c(q_o) = v[\rho_b q(c_o) + \varepsilon c_o] - v[\rho_b q_o + \varepsilon c(q_o)] \quad (19)$$

or

$$(V_{stoich} - \varepsilon v)[c_o - c(q_o)] = \rho_b v[q(c_o) - q_o] \quad (20)$$

Here, the term $(V_{stoich} - \varepsilon v)$ is the volume of influent at breakthrough, reduced by the holdup volume, εv, that is equivalent to the differential equilibrium bed capacity.

In general, for any influent volume, V, the reduced volume $(V - \varepsilon v)$ is the volume of influent fluid with which the sorbent at the outlet end of the bed has come into contact. Using this variable, we can define a generalized throughput parameter T that measures the amount of potentially removable feed component, which varies with V,

$$T \equiv \frac{(V - \varepsilon v)[c_o - c(q_o)]}{\rho_b v[q(c_o) - q_o]} \quad (21)$$

In any actual case corresponding to a single-step feed-concentration change, breakthrough is never abrupt as in the ideal situation just considered, but instead, X rises gradually as V increases. For a complete effluent-concentration history, in which X rises from 0 to 1, Eqs. 20 and 21 then yield

$$T_{stoich} = 1 \tag{22}$$

where T_{stoich} is the value of T corresponding to $V = V_{stoich}$. It follows that, for such a history,

$$\int_0^1 TdX = 1 \tag{23}$$

Thus, when a complete effluent-concentration history is available in terms of X vs. (V- εv), the T-scale can be calibrated by integrating the area between the history and the X-axis graphically or numerically and then dividing the abscissa values by the value of the area to get abscissae in terms of T. This is illustrated in Fig. 10, where the shaded area was determined to be 25 liters, so that the unit point on of the T-scale corresponds to 25 liters on the (V - εv)-scale.

3.5 Generalized Time and Distance Variables

Further generalization of Eq. 17 is achieved through use of the concept of the number-of-reaction-units, N, defined in such a way that, with Eqs. 1 and 2, Eq. 17 takes the form

$$(\partial X/\partial N)_{NT} + (\partial Y/\partial NT)_N = 0 \tag{24}$$

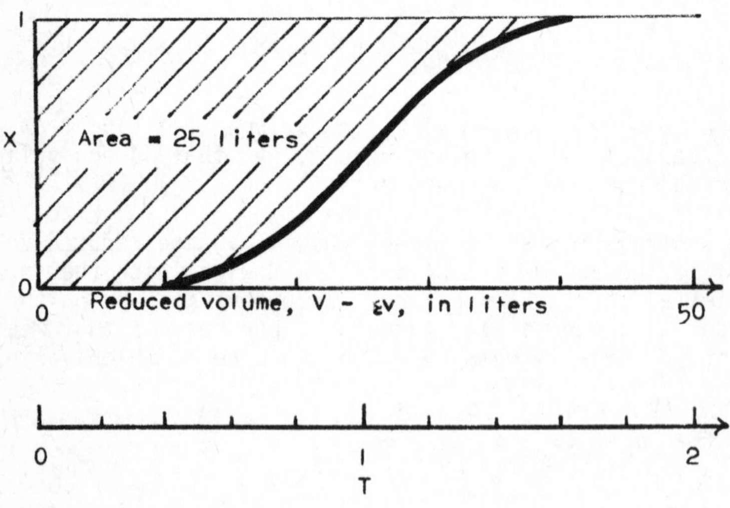

Fig. 10. Establishment of relation between V and T by graphical integration of complete effluent-concentration history for a single-step feed-composition change. The unit point on the T-scale corresponds numerically to the shaded area.

Although NT is the product of N and the throughput parameter T, the symbol NT here and in what follows is considered to represent a single variable, the meaning of which will be discussed further below.

Together with Eq. 24, the N-concept permits rate equations such as Eqs. 7 and 9 to be written in the generalized, dimensionless form applicable to a fixed bed,

$$(\partial X/\partial N)_{NT} = -\phi(X,Y) \qquad (25)$$

where $(\partial X/\partial N)_{NT}$ is the local slope of the generalized axial X-profile at a particular value of NT, and ϕ is a driving force that depends only on the local average concentrations X and Y.

For f-mechanism, the definition of N that satisfies these equations can be derived from Eqs. 7 and 21 to be that given in Table 1, so that N_f is seen to be the product of a rate factor, $k_f a$, and the apparent residence time, s/u_o. Moreover, with Eq. 8, N_f becomes inversely proportional to the square root of the approach velocity of the fluid, u_o. Other definitions of N, obtained similarly, are also given in the table. Also noted are the abbreviations for the mechanisms, which must be imagined to be subscripted to N for the corresponding definition. Where NT and N appear without subscript, the mechanism is indicated in the context, if necessary. $\underset{\sim}{\Delta}$ in the table is given by

$$\underset{\sim}{\Delta} \equiv \rho_b\left[q(c_o) - q_o\right] / \left[c_o - c(q_o)\right] \qquad (30)$$

L is an axial mixing length, and the ψ's are empirical correction factors.

In the normalized framework, NT becomes proportional to $\hat{t}$, as can be seen from Eqs. 26 through 29, and Eq. 21, and N to s, as seen from the expressions in Table 1. The nature of the proportionality between s and N is seen to presuppose constancy of the k's and of L, an assumption only approximately true.

Table 1

Definitions of N for Various Mass-Transfer Mechanisms

Mechanism	N	Equation No.
External diffusion (f)	$(k_f a/u_o)s$	(26)
Particle diffusion (p)	$(\psi_p \underset{\sim}{\Delta} k_p a/u_o)s$	(27)
Pore diffusion (pore)	$(\psi_{pore} k_{pore} a/u_o)s$	(28)
Axial dispersion (d)	$(1/L)s$	(29)

For f-mechanism, Eqs. 7, 13, 21, 24, 25 and 26 yield

$$\phi_f(X,Y) = X - X(Y) \qquad \text{(f-mechanism)} \qquad (31)$$

where ϕ_f is the dimensionless driving-force for this case. With Eq. 25, then,

$$(\partial X/\partial N)_{NT} = X(Y) - X \qquad \text{(f-mechanism)} \qquad (32)$$

Moreover, for the same mechanism, with Eq. 24,

$$(\partial Y/\partial NT)_N = X - X(Y) \qquad \text{(f-mechanism)} \qquad (33)$$

Similarly, Eqs. 9, 13, 21, 24, 25 and 27 yield, for p-mechanism,

$$\phi_p(X,Y) = Y(X) - Y \qquad \text{(p-mechanism)} \qquad (34)$$

$$(\partial X/\partial N)_{NT} = Y - Y(X) \qquad \text{(p-mechanism)} \qquad (35)$$

$$(\partial Y/\partial NT)_N = Y(X) - Y \qquad \text{(p-mechanism)} \qquad (36)$$

In Eqs. 32 through 36, N is understood to bear the subscript f or p, as appropriate.

3.6 NT, N-Diagrams

Representative X- and Y-contours on t,s- or NT,N-diagrams make it possible to describe the entire course of the transient phenomena occurring in fixed sorption beds. In what follows, some typical examples of such diagrams and their general features and classification are discussed for single-step feed-composition changes, and methods for calculating them and systems of other boundary conditions are outlined.

Figs. 11, 12 and 13 illustrate, for f-mechanism and hyperbolic isotherms with constant separation factor α, the typical cases of underlying isotherms favorable, linear, and unfavorable for the uptake of the feed component, in the form of (dashed) X- and (solid) Y-contours on NT,N-diagrams. The general features of these diagrams apply equally to other mechanisms.

3.6.1 General features. Eqs. 21 and 26, and Eqs. 14 and 15, respectively, establish

$$X = 1 \qquad (N = 0) \qquad (37)$$

as a boundary condition along the NT-axis, and

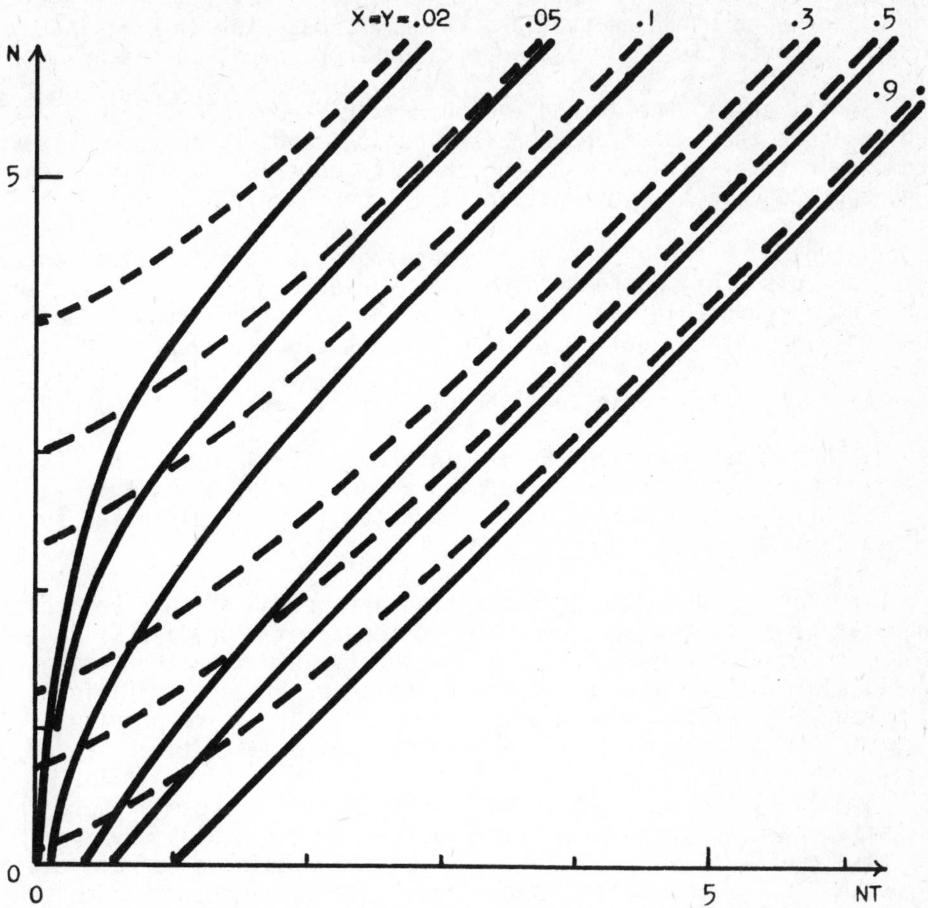

Fig. 11. Representative X-contours (dashed) and Y-contours (solid) in NT,N-plane, for single-step feed-composition change, hyperbolic isotherm strongly favoring uptake of feed component ($\alpha = 20$), and f-mechanism. Curved portions faired in between boundary-condition points and constant-pattern lines with unit slope.

$$Y = 0 \qquad (NT = 0) \tag{38}$$

as a boundary condition along the N-axis. The axes are thus seen to coincide with the concentration contours for the values just given. All X-contours intersect only the N-axis, and all Y-contours, only the NT-axis. Since NT is proportional to $\hat{t}$, and N to s, the slopes $(\partial N/\partial NT)_X$ and $(\partial N/\partial NT)_Y$ of X- and Y-contours represent normalized concentration velocities. For this type of diagram, a cross-plot of X and Y against N at a constant value of NT may be regarded as a normalized concentration profile, while a cross-plot of X against NT at constant N is a normalized effluent-concentration history.

In all cases, the X- and Y-contours are seen to cross in a region about the origin (noncoherent region) and, as they tend away from their intersections with the axes, to enter a region in which they approach a non-crossing condition (coherence).

3.6.2 Typical examples. Fig. 11 corresponds to a hyperbolic isotherm favoring uptake of the feed component. Here, and for any mechanism and form of favorable isotherm, the contours, as they tend to become coherent, approach parallel lines with unit slopes, along which

$$X = Y \quad \text{(favorable equilibrium; } N, NT \rightarrow \infty) \qquad (39)$$

As any such line is defined by its intercept N - NT on the N-axis, X (=Y) in the coherent region becomes a function of the single variable (N-NT) and this leads to ready tractability, as discussed further below.

If, for instance, we consider X as a function of this variable, then, as time increases, each point on the X- or Y-profile will move down the bed at a constant relative velocity, equal to the common (unit) slope of the asymptotic lines, so that, after a sufficient time required for its asymoptotic formation, the entire profile assumes a shape that virtually no longer changes with time.

Similarly, once the asymptotic shape has been closely approached, the effluent-concentration history in the form of a plot of X vs. NT exhibits constant shape, so that, for beds exceeding a certain minimum in depth, that shape becomes independent of bed length. For these reasons, the normalized profiles and effluent-concentration histories associated with parallel concentration contours and favorable equilibrium are said to be of the "constant-pattern" type.

The isotherm underlying Fig. 12 is linear. Here, the X- and Y-contours with the same concentration value also approach each other, and therefore coherence, but it is not apparent that they do so asymptotically and they never become rectilinear as they diverge from pairs of contours with other values. As coherence is being approached, the horizontal distance between two X- and Y-contours with particular concentrations widens approximately in proportion to the square root of N, so that this pattern is intermediate between the constant pattern and the proportional pattern explained in connection with the next figure.

A case of unfavorable equilibrium is illustrated by Fig. 13, where the coherent parts of the contours, when extrapolated toward the origin, are seen to radiate from the latter and where, in the coherent region, the local X- and Y-values are coupled through the isothermal equilibrium relation applicable. In this case, a normalized effluent-concentration history in the form of a plot of X vs. NT widens in proportion to N, so that this pattern is termed "proportional".

Asymptotic approach to the coherent pattern at increasing values of N and NT has been proved for monovariant systems for finite mass-transfer resistance without longitudinal dispersion, for equilibrium operation with longitudinal dispersion, and for finite mass-transfer

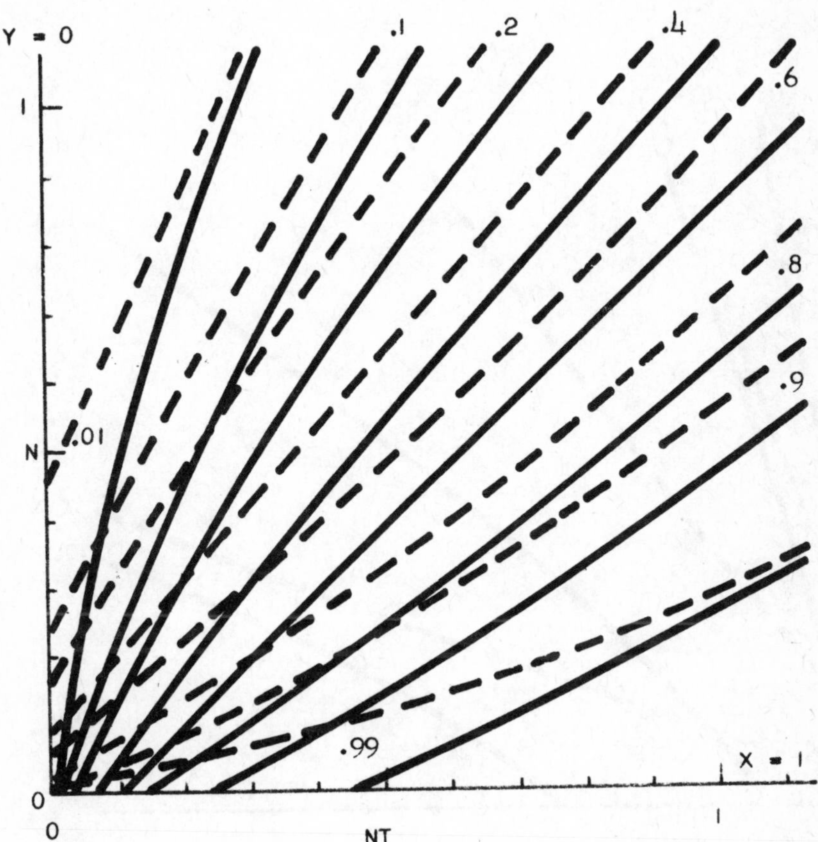

Fig. 12. Representative X-contours (dashed) and Y-contours (solid) in NT,N-plane, for single-step feed-composition change, linear isotherm ($\alpha = 1$), and f- or p-mechanism. Contour values are indicated between the X- and Y-contours to which they apply. Curves calculated from boundary-condition points and Klinkenberg's (8) approximation.

resistance and longitudinal dispersion by Cooney and Lightfoot (4). They have termed the corresponding coherent solutions "asymptotic", but this does not fit the case of infinitely favorable equilibrium with finite mass-transfer resistance but without longitudinal dispersion, where the analytical solutions available for this case indicate that the concentration contours actually merge into the corresponding coherent contours.

The question may be asked how NT,N-diagrams describing the dynamic response of a fixed bed and using coordinates defined in terms

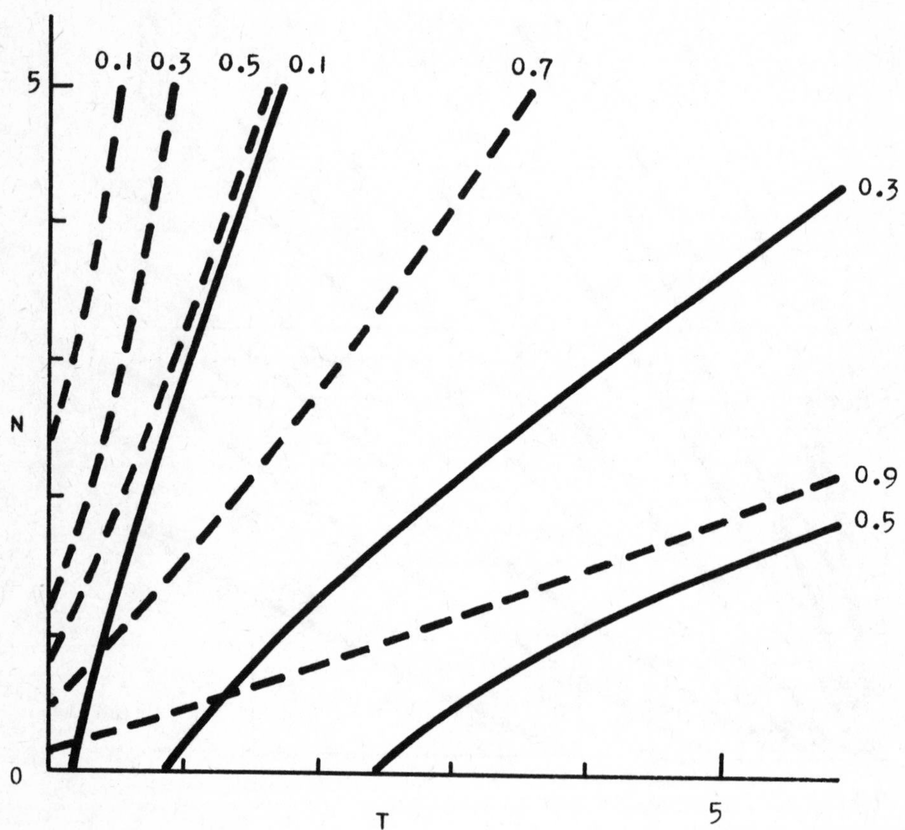

Fig. 13. Representative X-contours (dashed) and Y-contours (solid) in NT,N-plane, for single-step feed-composition change, hyperbolic isotherm unfavorable for uptake of feed component ($\alpha = 0.1$), and f-mechanism. Numbers inside diagram are contour values. Curved portions faired in between boundary-condition points and proportional-pattern lines radiating from origin.

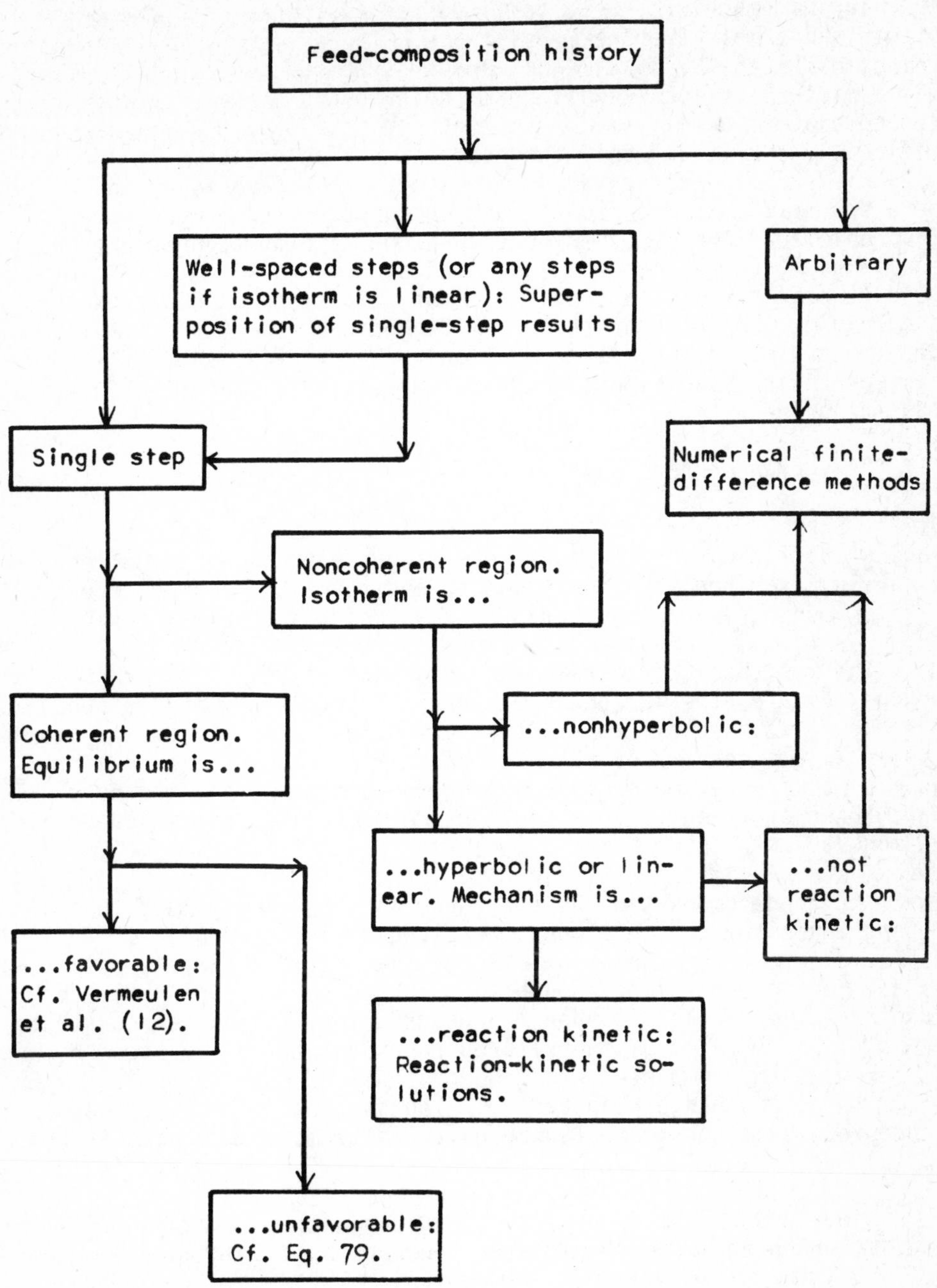

Fig. 14. Methods for predicting fixed-bed performance

of rate coefficients, may be related to a diagram describing local equilibrium behavior, i.e., to fixed-bed response under the premise of vanishing mass-transfer resistance. The answer is that at large values of NT and N, local equilibrium is approached, so that we can get a picture of local-equilibrium behavior by increasing the scale of the diagram until the noncoherent region around the origin becomes negligibly small, and only the coherent region counts. We then obtain either a set of parallel X- or Y-contours (for favorable equilibrium) or a set of X- and Y-contours fanning out from the origin (for unfavorable equilibrium). Moreover, in the case of favorable and linear equilibrium, because of the large scale of the diagram, the contours tend to coalesce into a single line of unit slope that passes through the origin. For both of these limiting cases, then, the actual values of NT and N no longer matter, but only their ratio, so that the rate part of their definition cancels out, as it should, for infinite mass-transfer rate.

3.7 Performance Prediction

3.7.1 Survey. An attempt is made here to classify the often confusing multiplicity of factors governing the transient phenomena that occur in fixed-bed sorption columns, and to relate them to the appropriate approach to a solution of the associated problems.

The most basic criterion for this purpose, that of variance, which indicates the irreducible degree of complexity of the problem. has been discussed already. Other criteria are the kinds of simplifying assumptions made, rate mechanism, feed-composition history, initial concentration profiles along the bed axis, and isotherm. The last three of these factors are considered in Fig. 14, which applies to monovariant systems.

The feed-composition history here is taken to include the time during which the bed developed the given "initial" profile. Thus, a single step change means that a feed stream of constant composition is applied at least long enough to saturate the bed uniformly and that this composition is then changed suddenly to another constant composition. Moreover, the reference component for any step change is taken as the component whose concentration in the feed is higher after the step change than before. The isotherm governing the sorption process is considered to be given in terms of the concentrations of this component.

Fig. 14 can best be understood in conjunction with Figs. 11, 12, and 13, which apply to single-step feed-composition changes.

Additional boundary conditions are discussed in Section 3.7.2. For the coherent region and the practically important constant-pattern case with favorable equilibrium, solutions are listed by Vermeulen et al. (12). For proportional-pattern cases, governed by unfavorable equilibrium, which sometimes cannot be avoided, Eq. 79 applies.

For the noncoherent region, and hyperbolic (including linear) isotherms and the reaction-kinetic mechanism, the reaction-kinetic treatment provides solutions. For other types of equilibria and rate-mechanisms, numerical methods can be resorted to. These methods are suitable also for arbitrary feed-composition changes and initial concentration profiles other than single-step feed-composition changes.

In cyclic operation, where the feed is suddenly changed between exhaustant and regenerant, if the resulting step changes are sufficiently separated, interference between the successive concentration waves resulting in the bed can be small enough so that superposition of the results for single-step changes can be applied. This method also becomes increasingly valid for more closely spaced steps as the isotherm approaches linearity.

3.7.2 Boundary conditions for single-step feed-composition change. The distribution of the variable concentration along the axes of an NT,N-diagram is of interest as a start for the numerical integration of equations such as Eqs. 25, 32, 33, 35, and 36, by the method of characteristics Acrivos (1). As an example, consider the distribution of Y along the NT-axis for a monovariant system governed by a curved hyperbolic isotherm and f-mechanism, with a single feed composition change.

Along the NT-axis, X = 1, so that Eq. 33, for f-mechanism, takes the form

$$(\partial Y/\partial NT)_N = 1 - X(Y) \qquad (N = 0;\ \text{f-mechanism}) \qquad (40)$$

where X(Y) is the value of X in equilibrium with Y. For hyperbolic equilibrium, with Eq. 4

$$X(Y) = Y/[(1 - \alpha)Y + \alpha] \qquad (41)$$

For the NT-axis, where N = 0, Eq. 40 can thus be written as the ordinary differential equation

$$dY/dNT = \alpha(1 - Y)/[(1 - \alpha)Y + \alpha] \qquad (N = 0) \qquad (42)$$

Integration now readily yields

$$1 + (\alpha - 1)Y - \ln(1 - Y) = \alpha NT + \text{const.} \qquad (N = 0) \qquad (43)$$

Since, for Y = 0, NT = 0, the constant assumes the value 1 and the distribution sought is given explicitly for NT but implicitly for Y by

$$NT = [(\alpha - 1)Y - \ln(1 - Y)]/\alpha \qquad (45)$$

which also satisfies $Y = 1$, $NT = \infty$.

Analogous procedures lead to the distribution along the axes for f- and p-mechanisms and for selected, simple isotherms, as given in Table 2. The conditions applying along each axis are noted below the axis-heading. Here, Y(X) is the value of Y in equilibrium with X, and X(Y), the value of X in equilibrium with Y. The symbols f and p in the mechanism column refer to the definitions of Eqs. 32, 33, 35 and 36; α in the isotherm column defines a hyperbolic isotherm as give by Eq. 4. However, $\alpha = 1$ and $\alpha = \infty$ refer to the respective limiting cases of a linear isotherm ($X = Y$), and of an isotherm infinitely favorable for the feed component ($X = 0$, $0 \le Y \le 1$; $0 < X \le 1$, $Y = 1$) The last column lists NT as a function of Y, or N as a function of X.

Table 2

Distribution of Compositions Along Axes of NT,N-Diagram

(For single feed-composition change, f- and p-mechanisms, and hyperbolic isotherms with constant separation factor α)

NT-AXIS

$N = 0$, $X = Y(X) = 1$

Mechanism	Isotherm	NT	Equation No.
p	any	$-\ln(1-Y)$	44
f	hyperbolic	$[(\alpha-1)Y-\ln(1-Y)]/\alpha$	45
f	infinitely favorable	Y	46

N-AXIS

$NT = 0$, $Y = X(Y) = 0$

Mechanism	Isotherm	N	Equation No.
f	any	$-\ln X$	47
p	hyperbolic	$[(\alpha-1)(1-X)-\ln X]/\alpha$	48
p	infinitely favorable	$1-X$	49

For p-mechanism, the Y-distribution along the NT-axis, and for f-mechanism, the X-distribution along the N-axis, are seen to be independent of the isotherm configuration, i.e., to be valid for any isotherm. The other distributions are isotherm-dependent. In genera the results for one axis are symmetric to the results for the same isotherm on the other axis in the sense that one result can be obtained from the other by replacing f by p, NT by N, and X by 1 - Y, and vice versa.

The results (not listed) for linear isotherms ($\alpha = 1$), and those (listed) for infinitely favorable isotherms ($\alpha = \infty$) are implicit in those for hyperbolic isotherms in general.

3.7.3 Constant-pattern contours. It can be shown that, at local equilibrium, higher concentrations of the feed component tend to travel faster than lower concentrations. This leads to the physical impossibility of several concentrations breaking through at the same time. What actually happens, is a compromise between the tendency of a concentration wave at local equilibrium to "break", and the effect of mass-transfer resistances. For a single-step feed-composition change, the result is a constant-pattern profile front with negative slope $(\partial X/\partial N)_{NT}$. If constant pattern has been reached, the X- and Y-profiles coincide, so that, for any bed level and time, X = Y (Eq. 33).

In order for the profiles to retain their shape, each concentration in them must move through the bed at the same velocity. But, since this velocity equals the slope of a concentration contour in a time-distance diagram, the concentration contours in the constant-pattern region must be parallel. In particular, in the constant-pattern region of an NT,N-diagram, the X- (and Y-) contours turn out to be straight lines with unit slope.

In such diagrams, the line N = NT corresponds to T = 1, i.e., to the average value $\bar{T}$ of T at constant N. This line therefore coincides with the contour of the concentration $X_{\bar{T}}$ which leaves the bed when $T = \bar{T} = 1$.

With Eq. 33, the dimensionless driving force, as well as the partial derivatives in Eqs. 24 and 25 can be expressed either as X or as Y throughout these equations, and they then become ordinary differential equations. For example, Eqs. 24, 25, and 33 yield

$$dX/dNT = \phi_1(X) \qquad (N = \text{const.}) \qquad (49)$$

where ϕ_1 is a dimensionless driving force in terms of X only, so that

$$\int [dX/\phi_1(X)] = NT + \text{const.} \qquad (N = \text{const}; 0 \le X \le 1, T \ge 0) \qquad (50)$$

or

$$f(X) = NT + \text{const.} \qquad (N = \text{const}; 0 \le X \le 1, T \ge 0) \qquad (51)$$

where

$$f(X) \equiv \int [dX/\phi_1(X)] \qquad (0 \leq X \leq 1) \qquad (52)$$

Where formal integration is not feasible, f(X) may be determined by graphical or numerical methods. Eq. 51, written for the point $(T = \overline{T} = 1,\ X = X_{\overline{T}})$, becomes

$$\text{const.} = f(X_{\overline{T}}) - N \qquad (N = \text{const.}) \qquad (53)$$

Substitution of this expression for the constant back into Eq. 51 yields

$$N(T - 1) = f(X) - \delta \qquad (N = \text{const.}) \qquad (54)$$

where

$$\delta = f(X_{\overline{T}}) \qquad (55)$$

An expression for the generalized effluent-concentration history may now be written in the form

$$T = 1 + [f(X) - \delta]/N \qquad (56)$$

δ may be evaluated by formal integration from the overall material balance of Eq. 22:

$$\int_0^1 T dX = \int_0^1 \left\{1 + [f(X) - \delta]/N\right\} dX = 1 \qquad (57)$$

In many cases, this integral is improper, and is more easily evaluated graphically or numerically. In the graphical method, the expression corresponding to f(X) is determined and values of it are plotted against N(T - 1). The integrated value of N(T - 1) is determined graphically as the area under the curve f(X). Finally, the origin of the N(T - 1)-scale is shifted to the point corresponding to the integrated average N(T - 1)-value on the original N(T - 1)-scale. The numerical procedure is analogous.

Since Eq. 54 gives X and Y (=X) implicitly as functions of N and NT, this equation predicts both the generalized effluent-concentration histories (X as function of T at constant N) and concentration profiles (Y as function of N at constant NT). It is thus not necessary to use the first of the partial derivatives in Eqs. 33. However, the same result can be obtained by starting with this derivative and using an analogous derivation.

If f(X) happens to be symmetrical about the point X = 0.5, δ vanishes. An example of this is the case in which the driving force is given by the reaction-kinetic expression, where

$$(\partial Y/\partial NT)_N = X(1 - Y) - Y(1 - X)/\alpha \tag{58}$$

With Eq. 39, ϕ_1 becomes

$$\phi_1(X) = (1-1/\alpha)X(1-X) \tag{59}$$

and f(X) assumes the simple form

$$f(X) = [\alpha/(\alpha - 1)] \ln [X/(1 - X)] \tag{60}$$

f(X) can be shown to be symmetrical about the point x = 0.5, and δ therefore to vanish. The solution is

$$NT - N = \left[\alpha/(\alpha - 1)\right] \ln \left[X/(1 - X)\right] \tag{61}$$

For f-mechanism, with Eq. 31, the dimensionless driving force, ϕ_f, is the horizontal distance between a point X,Y(=X) on the diagonal and the isotherm, as shown in Fig. 15. With Eqs. 52 and 54, one thus obtains

$$NT - N = \int [X - X(Y)]^{-1} dX - \delta \qquad \text{(f-mechanism)} \tag{62}$$

Similarly, from Eq. 34, one obtains for p-mechanism

$$NT - N = \int [Y(X) - Y]^{-1} dY - \delta \qquad \text{(p-mechanism)} \tag{63}$$

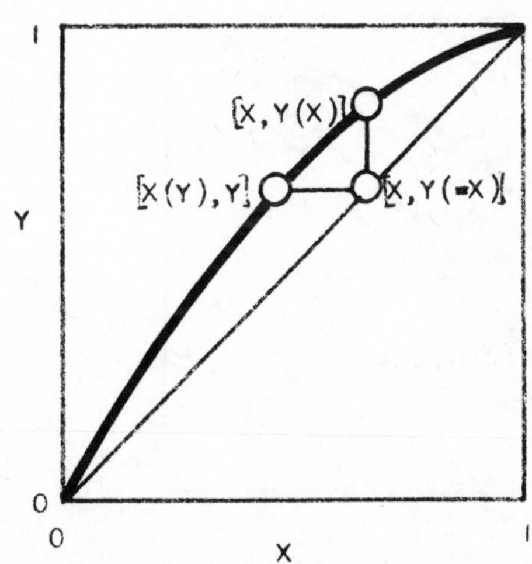

Fig. 15. Constant-pattern driving forces. Horizontal distance between isotherm and point [X,Y(=X)] for f-mechanism; vertical distance for p-mechanism.

where $Y(X) - Y$ is the vertical distance, also shown in Fig. 15. The difference between this figure and Figs. 3 and 5 is that, instead of a particular point (X_0,Y_0), which does not necessarily lie on the diagonal, a general point $[X,Y(=X)]$, on the diagonal, is shown.

For some isotherms, the integrals in Eqs. 62 and 63 can be evaluated formally. Where this is not possible or too cumbersome, graphical or numerical evaluation offers no significant difficulty.

For combinations of f- and p-mechanisms, the concept of the mechanism line as given by Eq. 12 is useful. In terms of dimensionless numbers only, this equation assumes the form

$$-(N_f/N_p) = (Y - Y_I)/(X - X_I) \quad \text{(f- and p-mechanisms)} \qquad (64)$$

i.e., that of a straight line of slope $-(N_f/N_p)$ and going through the point $[X, Y(=X)]$ on the diagonal. An example of such a line is shown in Fig. 16, which is similar to Fig. 7, except that the point (X_0,Y_0) has again been replaced by the point $[X,Y(=X)]$ and that the points $[X(Y_0),Y_0]$ and $[X_0,Y(X_0)]$ have been omitted, being no longer needed.

As can be seen, as the point (X,Y) moves along the diagonal, the points (X_I,Y) and (X,Y_I) trace the two curves shown as dashed lines and here termed f- and p-curves, respectively.

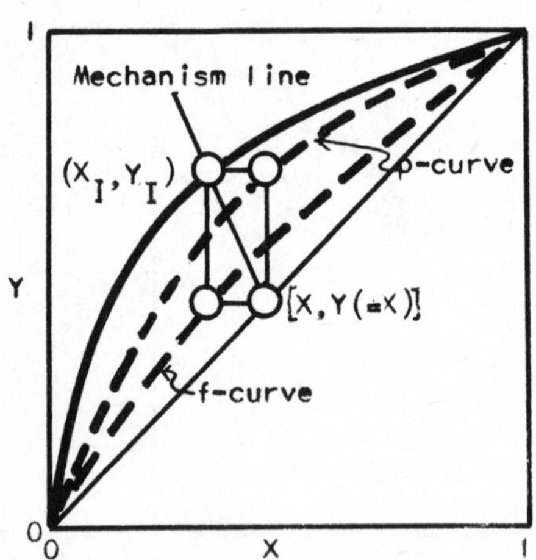

Fig. 16. Driving forces for constant pattern and combined f- and p-mechanisms as the horizontal distance between Point (X,Y) and the f-curve, or the vertical distance between that point and the p-curve.

With Eq. 11, the effective driving force is seen to be the horizontal distance between a point on the diagonal and the f-curve, or the vertical distance between that point and the p-curve, to be used in conjunction with N_f or N_p, respectively.

Eqs. 23 and 33, based on f-mechanism, in the present case yield

$$-(\partial X/\partial N)_{NT} = (\partial Y/\partial NT)_N = X - X_I \qquad (N = N_f) \qquad (65)$$

and Eqs. 35 and 36, based on p-mechanism,

$$-(\partial X/\partial N)_{NT} = (\partial Y/\partial NT)_N = Y_I - Y \qquad (N = N_p) \qquad (66)$$

The analogue of Eq. 62 becomes

$$NT - N = \int (X - X_I)^{-1} dX - \delta \qquad (N = N_f) \qquad (67)$$

and that of Eq. 63,

$$NT - N = \int (Y_I - Y) dY - \delta \qquad (N = N_p) \qquad (68)$$

The method for combining the mechanisms is now evident. It consists in plotting the isotherm and a sufficient number of parallel mechanism lines of the slope given by Eq. 64. Then, either the f- or the p-curve are constructed and the respective driving force is obtained as for f- or p-mechanism, only using the appropriate curve instead of the isotherm. In choosing either curve, the guiding consideration is that, if the mechanism line is nearly horizontal, the relative error in the driving force based on the p-mechanism will tend to become large, or the driving force will vanish because the rate coefficient is approaching infinity. It is therefore advantageous, if not mandatory, to use the f-curve in this case. Similarly, if the mechanism line approaches verticality, the p-curve should be used. In general, using the f- or p-curve according as N_f/N_p is smaller or larger than 1 should be the preferred procedure.

To construct specific concentration contours in an NT,N-diagram from equations of the type of Eq. 56, one may calculate their intercepts on the N-axis and draw lines of unit slope through them.

A number of constant-pattern solutions for specific isotherm types, mechanisms, and driving forces are given by Vermeulen et al. (12).

3.7.4 Acquisition of effective rate constants. The slope of the mechanism line can be evaluated with Eqs. 8, 10, 26, and 27 (with ψ_p=1) to be

$$-(N_f/N_p) = -0.044(D_f u_o d)^{0.5}/(D_p \triangle) \qquad (69)$$

and is thus seen to be proportional to the square root of the diffusivity in the fluid phase, the flow rate, and the particle diameter; and inversely proportional to the diffusivity in the sorbent and the distribution coefficient $\underset{\sim}{\Lambda}$.

The constants entering into the right-hand member of Eq. 69 are not always readily available. However, Eqs. 62, 63, 67, and 68 suggest a useful and simple method for evaluating N_f/N_p experimentally.

Consider for example, Eq. 62, which can be written in the form

$$T = 1 + \left\{ \int [X - X(Y)]^{-1} dX - \delta \right\} / N \qquad \text{(f-mechanism)} \qquad (70)$$

This, at a constant value of N, takes the form

$$T = (1/N) \int [X - X(Y)]^{-1} dX + \text{constant} \qquad \text{(f-mechanism)} \qquad (71)$$

Differentiation gives, after rearrangement,

$$dX/dT = N_f[X - X(Y)] \qquad \text{(f-mechanism)} \qquad (72)$$

A similar derivation for p-mechanism yields

$$dX/dT = N_p[Y - Y(X)] \qquad \text{(p-mechanism)} \qquad (73)$$

Since dX/dT is the slope of the constant-pattern effluent-concentration history, these two last equations permit evaluation of N_f or N_p, respectively, provided such a history, as well as the isotherm, are available. (With Eq. 33, Y in Eq. 73 equals the point-value of X in the effluent-concentration history.) However, this procedure presupposes that either the f- or the p-mechanism alone governs the column behavior. It is thus necessary to determine the slope of the mechanism line, which is again possible if the isotherm and an effluent-concentration history at constant pattern are available. For this purpose, differentiation of Eq. 72 yields,

$$d^2X/dT^2 = N_f[(dX/dT) - (dX/dY)(dY/dT)] \qquad \text{(f-mechanism)} \qquad (74)$$

At the inflection point $(T_\sim, X_\sim)$ of the generalized constant-pattern effluent-concentration history, $d^2X/dT^2 = 0$. Also, with X = Y. Eq. 74 reduces to

$$dY/dX = 1 \qquad (75)$$

The meaning of this expression is that, for pure f-mechanism, the slope of the isotherm is unity when the point value of X in the effluent is $X_\sim$. With Eq. 26, then, if

$$X_\sim = Y(X) \text{ at } dY/dX = 1 \qquad \text{(f-mechanism)} \qquad (76)$$

i.e., if X at the inflection point equals the ordinate of the point on the isotherm at which it has unit slope, f-mechanism governs exclusively.

Similarly, for pure p-mechanism,

$$X_{\sim} = X(Y) \text{ at } dY/dX = 1 \qquad \text{(p-mechanism)} \qquad (77)$$

X(Y) is the abscissa of the point at which the isotherm has unit slope.

These relationships are shown in Fig. 17.

The reasoning applied to pure f- and p-mechanism can be extended to cases with combined such mechanisms. The procedure is illustrated in Fig. 18, where P_3 is the point on the isotherm at which its slope is unity, and $X_{\sim}(= Y_{\sim})$ is the value of X at the inflection point of the generalized effluent-concentration history. The slope of the mechanism line therefore is

$$-(N_f/N_p) = (Y_3 - X_{\sim})/(X_3 - X_{\sim}) \qquad \text{(f- and p-mechanisms)} \qquad (78)$$

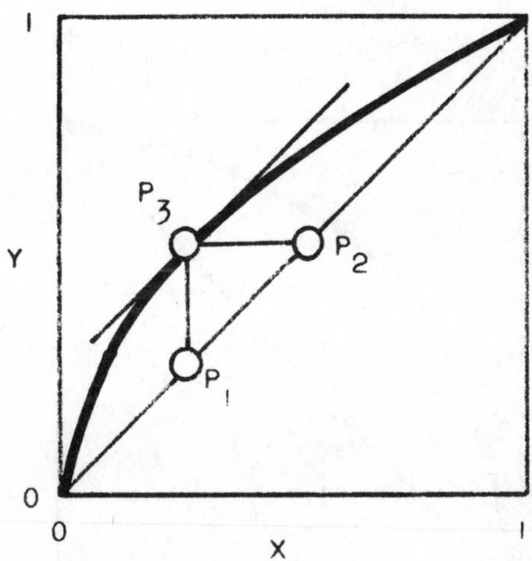

Fig. 17. Discrimination between pure f- and p-mechanisms, based on isotherm and inflection-point concentration of generalized constant-pattern effluent-concentration history. $P_3(X_{\sim}, Y_{\sim})$ is the point on the isotherm at which the tangent has unit slope. For f-mechanism, $Y_{\sim}$ is the abscissa of P_2; for p-mechanism, $X_{\sim}$ is the abscissa of P_1.

If there is reason to believe that the isotherm is of the mass-action type, its general equation, as suited for the particular combination of valences of interest, can be used in conjunction with Eq. 3 to provide diagrams from which the selectivity coefficient or the separation factor can be read, provided one has reasonable assurance of either complete f- or p-mechanism. An example is Fig. 19, which applies to hyperbolic isotherms. If the X-value of the inflection point of the generalized effluent-concentration history, $X_{\sim}$, and the mechanism are known, α can be found; if α and $X_{\sim}$ are known, either of the pure mechanisms can be established. Hyperbolic isotherms usually correspond to the exchange of ions of equal valence. Similar diagrams, not yet published, have been developed for other combinations of valences.

3.7.5 Proportional-pattern contours. In practice, one will try to avoid operation with unfavorable equilibrium because, for exhaustion, this would entail poor sorbent utilization at breakthrough, and, for regeneration, poor regenerant utilization. However, in the simplest case, involving the exhaustant and regenerant ions as the only species exchanging in the course of cyclic operation, if the equilibrium for either exhaustion or regeneration is favorable, and provided the other step is carried out at the same total solution concentration, the equilibrium for the other step will be unfavorable. A classic example of how this can be circumvented in the case of heterovalent

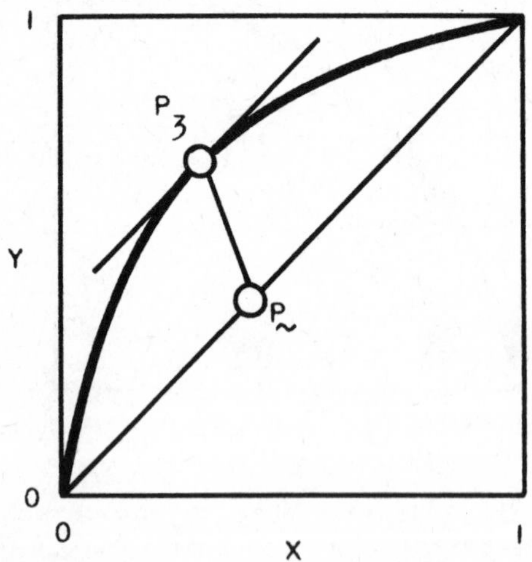

Fig. 18. Slope of mechanism line as that of the line connecting Points P_3, at which the isotherm has unit slope, and $P_{\sim}$, at which $X = Y = {}^3X_{\sim}$.

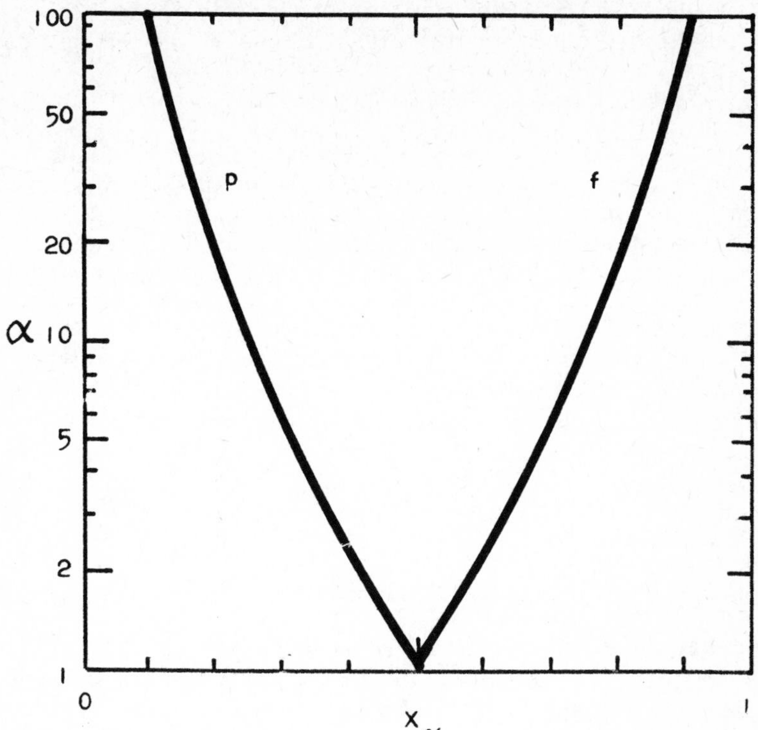

Fig. 19. Relation between concentration $X_{\sim}$ at inflection point of constant-pattern generalized effluent-concentration history (X vs. T) and constant separation factor, α. Curves marked f and p correspond to respective mechanisms.

exchange is that of water softening, where the uptake of calcium and magnesium ions proceeds in very dilute solutions, in which the equilibrium is highly favorable for this step. Regeneration, on the other hand, is done with a concentrated brine, in which the equilibrium is reversed, so that it is favorable for this step also. Other possibilities include the employment of weak-electrolyte exchangers and two-step regeneration involving acid or base. Reverse-flow regeneration can lead to improved regenerant utilization.

When unfavorable equilibrium for one of the operating steps cannot be avoided, performance prediction in the nearly coherent region of the NT,N-diagram can be made to a high degree of approximation with De Vault's equation (5) and a knowledge of the sorption isotherm. This equation states that the slope of the a coherent part of an X- or Y-contour in the NT,N-diagram equals the reciprocal of the slope of the isotherm at the corresponding composition point, and that the extrapolated coherent part of the contour passes through the origin of the NT,N-diagram. Formally, this relation can be stated as

$$\lim_{NT,N \to \infty} (\partial N/\partial NT)_{X \text{ or } Y} = dX/dY(X) \tag{79}$$

When the isotherm is available in form of an empirical curve, the coherent part of an X-contour corresponding to $X = X_0$ is obtained simply by measuring the slope of the isotherm at the point, calculating its reciprocal, and drawing a line through the origin of the NT,N-diagram with a slope of this value. If the isotherm is given in the form of Eq. 3, one first calculates $Y(X_0)$, algebraically, nomographically (12), or by a suitable zero-of-functions algorithm with the aid of a handheld electronic calculator. The slope of the X-contour is then given by

$$\lim_{NT,N \to \infty} (\partial N/\partial NT)_{X \text{ or } Y} = \frac{X}{Y} \frac{1 - X}{1 - Y} \frac{b + (a - b)Y}{b + (a - b)X} \tag{80}$$

with $X = X_0$ and $Y = Y(X_0)$.

Determination of the slope of a Y-contour for $Y = Y_0$ is analogous. Starting with an empirical isotherm, its slope is measured at $Y = Y_0$. If the isotherm is given by Eq. 3, one first calculates the X-value corresponding to Y_0, $X(Y_0)$, and then uses Eq. 80, with $X = X(Y_0)$, $Y = Y_0$.

3.7.6. Calculations for noncoherent regions. For a single-step feed-composition change, a crude approximation to the concentration contours in the noncoherent region may be obtained by calculating the intersections of the contours with the axes, as indicated in Section 3.7.2, constructing coherent contours for the same concentrations, and fairing in curves that go through the axis points and approach the coherent contours asymptotically.

For the reaction-kinetic rate model and hyperbolic and linear equilibrium isotherms, Thomas (11) has provided solutions that cover not only the noncoherent, but also the coherent, region. These solutions may be found in graphical form in Vermeulen et al. (12). For linear equilibrium, Klinkenberg (8) has developed a convenient nomograph.

For complex rate or equilibrium relations and irregular boundary conditions, numerical integration of Eqs. 24 and 25 may be carried out. The (explicit) method of characteristics, as described by Acrivos (1) is illustrated here for the simple case involving dimensionless rates and isothermal conditions. This method is not capable of accounting for axial dispersion and diffusion.

For this case, Eqs. 24 and 25 are used in finite-difference form, along characteristic lines, given by

$$NT = i\Delta \tag{81}$$

$$N = j\Delta \tag{82}$$

where i and j are integers and Δ is a suitably small constant. Eqs. 24 and 25 become

$$X_{ij} = X_{i,j-1} - \phi\Delta \tag{83}$$

$$Y_{ij} = Y_{i-1,j} + \phi\Delta \tag{84}$$

where the driving force, ϕ, is a function of X and Y. The flow of the calculations is shown in Fig. 20.

For a first approximation, ϕ is calculated using the X- and Y-values at the start of the computation interval; for successive approximations, ϕ is recalculated, using X and Y at the endpoint of the interval, and the average of the values of ϕ at the start and the end is used to get a new set of values of X and Y at the end point.

Application of the method to more complex problems (e.g. problems requiring enthalpy balances) is discussed by Acrivos (1).

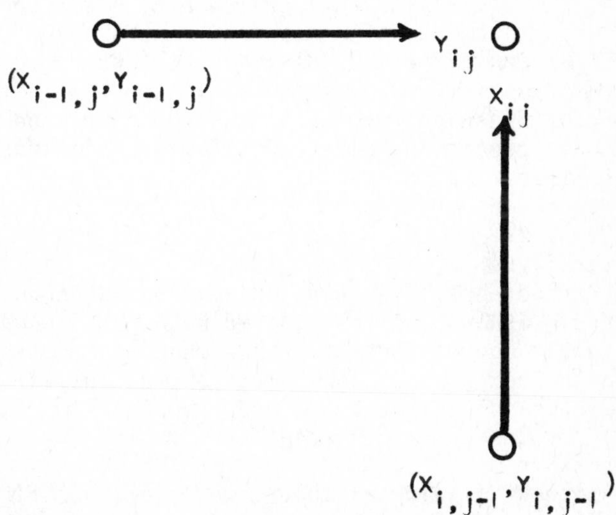

Fig. 20. Scheme for method of characteristics. X_{ij} and Y_{ij} are calculated from the X- and Y-values of the other two points.

4 EXAMPLES

Typical overall objectives of an industrial ion-exchange process are to reduce the concentration of an ionic constituent from a liquid stream at a given influent level to a specified (maximum or average) effluent concentration, or to produce a product, all at a specified rate. As this can often be achieved by a number of candidate methods, it is important not to overlook competing economically or environmentally preferable methods before going into detailed calculations. At the early stages of design, the ability to make rough estimates of process performance is especially valuable.

Even if ion exchange appears to be the method of choice, there may be several ion-exchange processes available or conceivable, from which the most promising must be selected. The minimum information that can serve as a basis for this step is the favorableness or unfavorableness of the equilibrium involved. If the equilibrium for the operating step under consideration is favorable, stoichiometric equality of the differential equilibrium bed capacity and the amount of the ion to be removed from the liquid stream can be assumed as a first approximation. This corresponds to the simple statement that the generalized throughput parameter T equals unity. If the equilibrium is unfavorable, DeVault's equation, used in conjunction with the applicable isotherm, can be expected to be valid to within a closer approximation.

Also to be considered are contacting methods and integral process design. The choice of contacting methods (batch, fixed-bed, counterflow regeneration, countercurrent, pulsed-bed, etc.) cannot be made without close consultation with equipment manufacturers. Here, only the still most common fixed-bed operation is considered. Integral process design takes into account the various operating steps, such as exhaustion, backwashing, regeneration and rinse, and an overall optimization of a process requires detailed calculations and judgment based on experience. Equipment size, resin inventory, controls, regenerant cost, and waste-disposal costs are some of the important cost items.

For the sake of brevity, only a single operating step can be considered here. Exhaustion is selected because it usually consumes most of the total operating time. To produce the desired average production rate, the rate during exhaustion must be adjusted to yield products at a rate sufficient to make up for the time during which the column is occupied by other steps.

While, in the following examples, we refer specifically to the removal of calcium from a solution of calcium and sodium chlorides, this is done merely to fix ideas. The data, conditions, and results are illustrative of the calculation methods only, and, in order to utilize existing graphs, little attempt has been made to simulate practice.

10^6 liters per day of a water are to be softened that is 0.002 normal each in calcium and sodium ions, and the maximum permissible point concentration of calcium in the effluent is 0.00002 normal. ($X = 0.01$). Laboratory experiments have yielded the isotherm shown in Fig. 1a for this case, and the effluent-concentration history of Fig. 10 has been obtained with a bed having a cross-sectional area of 1 cm^2 and a depth of 25 cm, operated at an approach velocity of 3 meters per hour. When the bed depth was increased to 40 cm, breakthrough at the same flowrate occurred later, but the shape of the effluent-concentration history (in terms of X vs. $V-\varepsilon v$) remained unchanged.

1. What is the specific differential equilibrium capacity of the exchanger?

2. If a cycle time of 24 hours is desired and 6 hours a day are to be used for subsidiary operations, what column diameter is needed if 3 columns are to be used at a flowrate of 3 m/hr.?

3. What is the bed depth required at this linear flowrate?

4. What would the results be at double the linear flowrate?

Solution to Problem 1. The bed volume of the laboratory column is 25 cm^3. From Fig. 10, the amount of calcium removed at complete exhaustion is equivalent to the calcium contained in 25 liters of reduced solution volume. the differential equilibrium bed capacity therefore is 25 x 0.002 = 0.05 equivalents, and the specific differential equilibrium capacity is 0.05/0.025 = 2 equivalents per liter of bed.

Solution to Problem 2. When the volumetric and linear flowrates, the fraction of on-stream time, and the number of columns are given, the bed diameter is independent of any factors related to the ion-exchange process. Thus, with 18 hrs/day on-stream time per day, the exhaustion rate required is $10^6/18$ = 55,556 l/hr. At the approach velocity of 3m/hr, we produce 3,000 l/(m^2hr). We therefore need 55,556/3,000 = 18.52 m^2 total cross-sectional area, or 18.52/3 = 6.17 m^2 per column, corresponding to a diameter of 2.8 m for each column.

Solution to Problem 3. From Fig. 10, breakthrough from the laboratory column occurs at a reduced volume of 10 l, which, at the approach velocity of 3 m/hr, corresponds to 10,000/300 = 33 hrs. As this is nearly twice the desired exhaustion time of 18 hours, the bed height should be shortened. This is undesirable not only because bed heights below 25 cm are not used in practice, but also because there is no assurance that constant-pattern behavior, on which the present calculations are based, will prevail down to the correspondingly smaller bed height. However, the calculation will be carried through for illustration.

18 hrs corresponds to a reduced volume of 18 x 300/1000 = 5.4 l. This means that the X-vs.-(V-εv)-curve, whose shape is invariant under the constant-pattern assumption, is moved 10 - 5.4 = 4.6 l to the left. This also shifts the stoichiometric point, at X = 0.5, to 25 - 4.6 = 20.4 l. Since the equivalent amount of exchanger is proportional to the bed height, the latter must be adjusted to 25 x 20.4/ 25 = 20.4 cm.

It will be noted that the product volume per cycle has been reduced out of proportion to the reduction in bed depth. This is due to the marked reduction in specific operating capacity that an exchanger undergoes during further reduction of an already relatively small bed depth.

This calculation is analogous to the exchange-zone concept of Michaels (10) and the length-of-unused-bed (LUB) method of Collins (3), which are based on the invariance of the depth of a constant-pattern zone with bed diameter and bed depth.

Solution to Problem 4. The cross-sectional area needed per column is 6.17/2 m^2 = 3.08 m^2, corresponding to a diameter of 1.98 m.

The symmetry of hyperbolic isotherms like that of Fig. 1a, in conjunction with the symmetry of Fig. 10, for which $X_{\sim}$ = 0.50, and Fig. 16, make it readily apparent that the slope of the mechanism line, - (N_f/N_p), equals - 1 for the 25-cm bed depth at u_0 = 3m hr^{-1}. From Fig. 10, also, one can then read, or from Eq. 4, calculate, that, for $X = X_{\sim} = 0.500$, $X_I = 0.309$. The effective dimensionless driving force based on f-mechanism thus is 0.500 - 0.309 = 0.191 at this point For the same flowrate and bed depth, from Fig. 10, dX/dT (= dY/dT) at $X = X_{\sim} = 0.500$ is 1.25, so that, with Eq. 65, at constant pattern, $N_f = (dY/dT)/(X_{\sim} - X_I) = 1.25/0.191 = 6.54$.

Eqs. 8 and 26 show N_f to be inversely proportional to u_0, so that N_f, at u_0 = 6 m hr^{-1}, will be 6.54/ 2 = 4.62. With Eq. 69, the slope of the mechanism line for the new flowrate is proportional to the square root of the ratio of the new and old flowrates; hence, for u_0 = 6 m hr^{-1}, - (N_f/N_p) will equal - 6/3 = - 1.41. With this slope and the construction indicated in Fig. 16, an f-curve can now be drawn, and with it and the analogue of Eq. 67 for mixed mechanisms, X established as a function of $N_f(T - 1)$ for 25-cm bed depth and the doubled flowrate. The result indicates breakthrough at $N_f(T - 1)$ = - 7.3.

With u_0 = 6 m hr^{-1} and a 1-cm^2 bed cross-section, the (reduced) throughput volume in the desired 18-hr exhaustion time will be 1x600x 18/1000 = 10.8 l. Hence, with Eq. 21, T = 10.8 x 0.002/(2 x 0.001s) = 10.8/s, where s is expressed in cm.

From Eq. 27, N_f is directly proportional to the column length, s. Hence, because of the invariance of $N_f(T - 1)$ as a function of X, $N_f(T-1)$ = - 7.3 = 4.62x(s/25)x(10.8/s - 1), and s = 49.5 cm. Doubling the flowrate thus is accompanied very nearly by a doubling of the bed depth. Since the total cross-section of the exchangers was halved, the bed volume has remained virtually constant. The small decrease it has undergone will probably be more than compensated for by an increase in pumping cost.

SYMBOLS

a	outside surface area of sorbent particles per unit bed volume; valence of ion being sorbed
b	valence of ion being desorbed
c	normality of feed component in fluid phase
C_0	total solution normality
d	diameter of sorbent particles
D	diffusivity
k	mass-transfer coefficient
K	dimensional selectivity coefficient
$\underline{K}$	dimensionless selectivity coefficient
$\tilde{K}_L$	Langmuir equilibrium constant
L	axial mixing length
n	variance
N	number of transfer units (dimensionless distance)
NT	dimensionless time
p	partial pressure
q	concentration of feed component in sorbent, equivalents per unit mass
Q	total specific equilibrium exchange capacity, equivalents per unit mass
s	axial distance
t	time
$\hat{t}$	local contact time
T	dimensionless throughput parameter
$\bar{T}$	integrated average value of T
u_0	linear approach velocity of fluid
v	bed volume
V	feed volume
X	normalized concentration in fluid phase
Y	normalized concentration in sorbent

Subscripts

b	bulk sorbent
f	film mechanism
I	interface
o	definite point value,, specifically initial or feed
p	particle mechanism
ref	reference
stoich	stoichiometric
$\sim$	value at inflection point

Greek Letters

α	separation factor
δ	integration constant
ε	void fraction
Δ	computation interval
θ	angle between t- and $\hat{t}$- axes
$\underset{\sim}{\Delta}$	distribution coefficient (cf. Eq. 30)
ρ	density
ϕ	dimensionless driving force
ϕ_1	dimensionless driving force in terms of X only
ψ	correction factor

ACKNOWLEDGMENT

Thanks are expressed herewith to the University of California Water Thermal and Chemical Technology Center for providing time and facilities, to Judy Sindicic and Gloria Arneson for assistance in typing and editing, and to the Center's director, Professor Theodore Vermeulen, for council and encouragement.

REFERENCES

1. Acrivos, A. Method of Characteristics Technique. Application to Heat and Mass-Transfer Problems. Ind. Eng. Chem. 481 (1956) 703-710.
2. Applebaum, Samuel B. Demineralization by Ion Exchange (New York, Academic Press, 1962).
3. Collins, J.J. The LUB/Equilibrium Section Concept for Fixed-Bed Adsorption. Chem. Eng. Progress Symp. Ser. 74, 63 (1967) 31-35.
4. Cooney, D.O. and E.N. Lightfoot. Existence of Asymptotic Solutions to Fixed-Bed Separations and Exchange Equations. Ind. Eng. Chem. Fundls. 4 (1965) 233-236.
5. DeVault, D. The Theory of Chromatography. J. Am. Chem. Soc. 65, No. 4 (1943) 532-540.
6. Glueckauf, E. and J.I. Coates. Theory of Chromatography. Part IV. J. Chem. Soc. (1947) 1315-1321.

7. Helfferich, Friedrich. Ion Exchange (New York, McGraw-Hill, 1962).
8. Klinkenberg, A. Numerical Evaluation of Equations Describing Heat and Mass Transfer in Packed Solids. Ind. Eng. Chem. 40 (1948) 1992-1994.
9. Kunin, Robert. Ion Exchange Resins, 2nd ed. (New York, Wiley, 1958).
10. Michaels, A.S. Simplified Method of Interpreting Kinetic Data in Fixed-Bed Ion Exchange. Ind. Eng. Chem. 44 (1952) 1922-1930.
11. Thomas, H.C. Chromatography: A Problem in Kinetics. Ann. New York Acad. Sciences. 49, Art. 2 (1948) 161-182.
12. Vermeulen, T., G. Klein and N.K. Hiester. Ion Exchange and Adsorption. Section 16 in Chemical Engineers' Handbook, 5th ed., R.H. Perry and C.H. Chilton, eds. (New York, McGraw-Hill, 1973).

DYNAMICS OF ION EXCHANGE PROCESSES

Alírio E.Rodrigues

Department of Chemical Engineering
University of Porto,Porto,Portugal

1 INTRODUCTION

Ion exchange is an old process as we can believe from Aristotle's statement that water loses part of its salt when flowing through sand. However,only in 1850,Thomson and Way stated the principles governing the interaction of fertilizers with soil (exchange Ca^{++}/NH_4^+)and the first industrial unit for water softening using aluminosilicates was developed by Gans fifty years later.A spectacular development began in 1935 with the synthesis of organic ion exchanger resins by Adams and Holmes.The 1950 were the "Golden Age of Ion Exchange" (1) because this operation was considered as an unit operation as a consequence of the work of Vermeulen et al.,at Berkeley.

At this moment you can ask:what is an ion exchanger?what is ion exchange? The first part of the question was already answered in this Institute:in short an ion exchange resin consists of polymer chains, to which ionizable groups are chemically bound,interconnected into an insoluble three dimensional structure.Basically the ion exchange resin behaves as an electrolyte in which an ion is bound to an insoluble matrix.This matrix can carry ionic groups,such as — COO^- in cationic resins or —NH_3^+ in anionic resins.The mobile ions,called counter-ions,have opposite charge to fixed ions and ensure electroneutrality.The concentration of fixed charges is just the ion exchange capacity.The ions of the solution having the same sign as the fixed groups are the co-ions.

The ideal ion exchanger (2) should have good thermal,chemical and physical stability properties,good ion exchange capacity and a structure allowing fast kinetics as sketched in Figure 1.

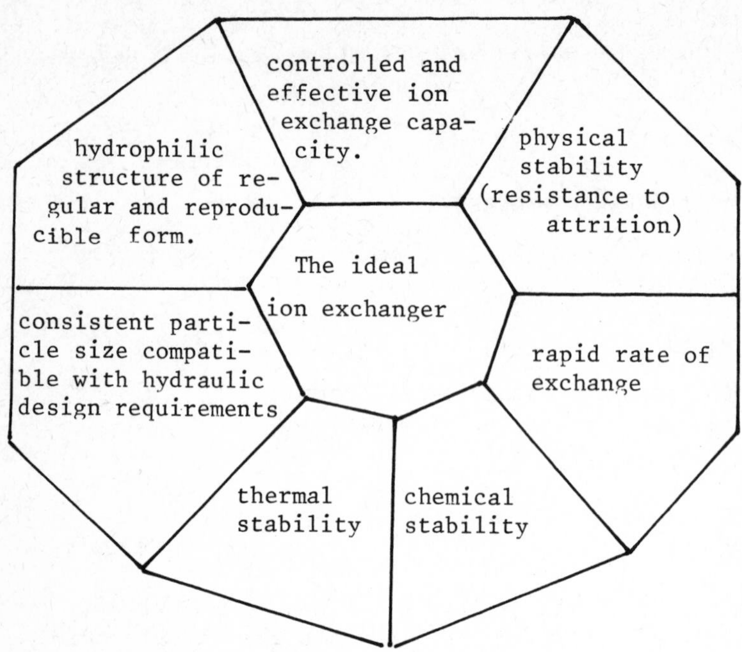

Figure 1 - The ideal ion exchanger

Let us now look at the ion exchange phenomenon.Three aspects should be stressed:

i) stoichiometric nature of ion exchange,i.e.,every ion removed from the solution is replaced by an equivalent amount of another ionic species of the same sign;this property explains why ion exchange is more attractive economically for the removal of heavy metals than the lighter ones (3).

ii) concentrating action of ion exchange,i.e.,if a diluted solution is contacted with a resin the solute is transferred to the resin and then eluted in a regeneration step at much higher concentration.

iii) selectivity,i.e.,a resin presents different affinities for different species and this can be used to separate them.

We talked above about "transfer of solute from the solution to the resin".This brings us to chemical engineering: ion exchange is a mass transfer operation between a liquid and a solid phase;an ionic species is transferred from the solution to the resin provided a driving force exists.Consequently two aspects should be discussed: what is the maximum transfer which can be achieved? how fast is mass transfer? Answers to these questions require information on equilibrium and kinetics of ion exchange.

Ion exchange is then an unit operation in which a fluid phase flows through a packed bed of beads exchanging mass/heat with the solid phase;these operations have been called percolation processes or chromatographic processes.However,coming from the reaction engineering side,ion exchange can also be viewed as an heterogeneous reactor.

In any case the following objectives can be distinguished when using ion exchangers (4):

i) separation of products,e.g.,separation of aminoacids
ii) recovery of solutes,e.g.,recovery of gold from liquors
iii) purification of diluents,e.g.,removal of pollutants (heavy metals from wastewater)
iv) production of chemicals,e.g.,dehydration of methyl lactate to get acrylic acid.

Whatever the application is the design of ion exchange equipment for separation or reaction purposes involves : mass and energy balances,equilibrium isotherms at the solid/fluid interface,kinetic laws of mass/heat transfer and/or reaction,boundary conditions and optimization criterion.

Fixed bed ion exchange becomes a complex problem due,in part,to the dynamic nature of this process.However,even in multicomponent cases one can get valuable information on how the different species leave the bed by using the so-called equilibrium theory.The framework of this theory was built by Tondeur(5),Klein and Helfferich (6) and Rhee (7). Obviously real concentration histories will deviate more or less from this ideal picture due to dispersive effects : hydrodynamics and kinetics of mass transfer in the film and inside particles.

Consideration of hydrodynamics and diffusional effects will lead to complex model equations which require a lot of computational work to be solved,if possible.If we add the influence of electric field (8,9,10) the degree of complexity increases.But if we want to analyse the contribution of each factor it is necessary to start with a model to correctly describe particle structure;in fact diffusion equations in homogeneous or gel type,porous or bidisperse (macroreticular) resin beads are different and then it is meaningless to study the importance of one phenomenon starting with a wrong particle diffusion equation.

2 HOW TO DESIGN ION EXCHANGE EQUIPMENT?

Let us now discuss design methods for ion exchange equipment.We assume that an "a priori" decision about the mode of operation (batch, fixed bed,fluidised bed,moving bed) was made.Then we should be able to predict the effluent concentration as a function of time (breakthrough curve) which depends on various factors:flowrate,particle diameter,feed concentration,etc.

This brings us to the problem of modelling an ion exchange system. A realistic model provides a tool for the scale-up from laboratory to large scale plants, prediction of the dynamic behavior of the unit and optimization of operating conditions.

We will discuss different approaches to modelling and design of ion exchange equipment.

2.1 Empirical Approach to Design

This can be illustrated by the following example: we want to design an ion exchange column in order to remove an ionic species at concentration c_o in a feed with flowrate U. Our goal is to get an effluent free of that ion (we accept a breakthrough concentration, $c_{Bp} < 0.05c_o$) during a time t_{Bp} of operation using an ion exchanger of capacity Q. The question is: what volume of resin should be used in a fixed bed?

Some people would simply make an overall mass balance and say:

$$(v_r)_1 = c_o \xi U\, t_{st}/(1+\xi)Q \qquad |1|$$

where

$t_{st} = \tau(1+\xi)$, stoichiometric time

$\xi=(1-\varepsilon)Q/\varepsilon c_o$, mass capacity factor

$\tau=\varepsilon v/U$, space time

and v is the bed volume with porosity ε.

Equation (1) gives a volume which is only true <u>if</u> the ion exchange isotherm is favorable (since then $t_{Bp}=t_{st}$) and no dispersive factors exist.

However, as it will be shown later, we know that for unfavorable isotherms of "constant separation factor type" (K< 1) the history of concentrations is:

$$\frac{c}{c_o} = \frac{\sqrt{\dfrac{\xi K}{(1+\xi)\theta-1}}-1}{K-1} \qquad |2|$$

where $\theta=t/t_{st}$. For very diluted solutions, $\xi\gg 1$ and then

$$\frac{c}{c_o} = \frac{\sqrt{K/\theta}-1}{K-1} \qquad |2a|$$

At the breakthrough point we get from Eq. (2):

$$t_{Bp}=(1+K\xi)t_{st}/(1+\xi) \qquad |3|$$

For large values of the mass capacity factor, $t_{Bp}=Kt_{st}$ and then:

$$(v_r)_2 = (v_r)_1/K \qquad |4|$$

We need a larger volume of resin, for a given time of operation, in the case of an unfavorable isotherm: or with the same volume of resin we can operate during

$$(t_{Bp})_2 = K(t_{Bp})_1 \qquad |5|$$

This calculation shows that we have to consider not only catalog information on ion exchanger capacity but the whole ion exchange equilibrium isotherm for the design of fixed bed equipment.

Since the breakthrough time is reduced, relatively to the equilibrium case, by dispersive effects some people just use a "safety factor", S_f and then:

$$v_r = S_f\,(v_r)_1 \qquad |6|$$

Instead of this empirical approach let us discuss now the use of the concept of width of mass transfer zone for the design of ion exchange equipment.

2.2 Semi-Empirical Approach : The Concept of Mass Transfer Zone

When we look at an ion exchange column at a given time we can distinguish three zones: the first one almost saturated, the second where exchange is taking place and the third almost free of feed ionic species. The second region is called Mass Transfer Zone and its width measures the efficiency of the ion exchange process; let us call Z_e the width of the MTZ. Figure 2 shows the concentration profile in the bed when the effluent concentration begins to increase (breakthrough time, t_{Bp}).

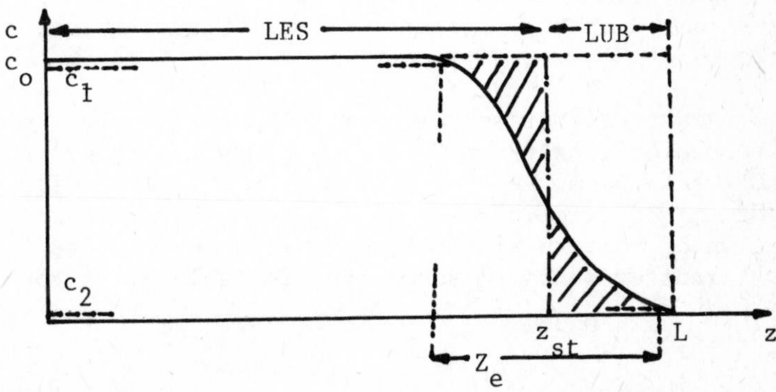

Figure 2- Concentration profile at the breakthrough time

The Unused Bed Length (LUB) is simply:

$$LUB = L - z_{st} \qquad |7|$$

and

$$L = LUB + LES \qquad |8|$$

where LES is the Length of the Equilibrium Section. Taking into consideration that:

$$L = \frac{u_i}{1+\xi} \; t_{st} \qquad |9a|$$

$$z_{st} = LES = \frac{u_i}{1+\xi} \; t_{Bp} \qquad |9b|$$

we finally get:

$$LUB = L\left(1 - \frac{t_{Bp}}{t_{st}}\right) \qquad |10|$$

Equation(10) provides a tool for calculating LUB from an experimental breakthrough curve, simply by measuring t_{Bp} and t_{st}.

For symmetric fronts, $Z_e = 2$ (LUB) and then

$$Z_e = L\,(\theta_2 - \theta_1) \qquad |11|$$

where θ_1 and θ_2 are the reduced times at which outlet concentrations are $c_1 = .05\ c_o$ and $c_2 = .95\ c_o$, respectively. Again from a breakthrough curve we can easily get the width of mass transfer zone.

If we run several experiments at various flowrates we obtain a relationship between Z_e and the superficial velocity, u_o (or the Reynolds number) which enables us to design an ion exchange column. We can say that all the mechanisms contributing to the ion exchange process (diffusion in the film, internal diffusion, axial dispersion, etc) are lumped into the concept of Mass Transfer Zone and then in the measure of Z_e.

As an example we present some results obtained at the laboratory scale for anionic exchange Cl^-/OH^- accompanied by neutralization. A column of L=15 cm, section of 5.3 cm2 , filled with particles (d_p=0.06 cm) of anionic resin in the form ROH^- is percolated with a solution of HCl. Experimental conditions and results for the width of mass transfer zone are summarized in Table I. We got:

$$Z_e = 14.55\ u_o^{\,0.362} \qquad |12|$$

with the superficial velocity, u_o in cm/sec and Z_e in cms.

TABLE I - Determination of the width of MTZ, Z_e

Run	Flowrate(ml/min)	c_o(meq/ml)	$\theta_2 - \theta_1$	Z_e(cm)
1	4.76	1.53×10^{-2}	0.212	3.18
2	10.60	1.47×10^{-2}	0.283	4.24
3	45.50	1.53×10^{-2}	0.480	7.20

Michaels (11) used the width of mass transfer zone, Z_e to develop an analogy between fixed bed and counter-current operations. While in distillation the Height Equivalent to a Theoretical Plate, HETP is the ratio between the column length, L and the number of plates, N, i.e., HETP=L/N he defined, for fixed bed processes, the Height of Transfer Unit, HTU as:

$$HTU = Z_e / NTU \quad |13|$$

where NTU is the Number of Transfer Units.

However, different expressions for NTU can be derived according to the mechanisms of mass transfer considered in the analysis. Michaels assumed that film mass transfer was the controlling mechanism; then for the moving bed sketched in Figure 3 we have:

- Mass balance around the bed

$$U_s(c_o - c_\infty) = U_r(Q - q_\infty) \quad |14a|$$

where U_s, U_r are the flowrates of solution and resin, respectively, and Q is the resin concentration in equilibrium with c_o. In a point of the bed we get:

$$U_s c = U_r q \quad |14b|$$

- Mass balance in a volume element of thickness, dz

$$-U_s dc = (k_f a) A \, (c - c^*) \, dz \quad |14c|$$

where A is the column section, and $(k_f a)$ the film mass transfer coefficient in sec^{-1}.

- Equilibrium law

$$K = q(c_o - c^*)/c^*(Q - q) \quad |14d|$$

Solving the system of Eqs. (14a to d) we get, after integration on z from 0 to Z_e and on c from c_1 to c_2:

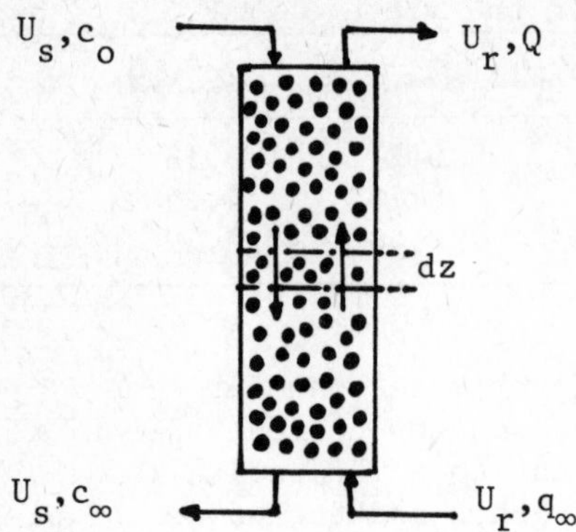

Figure 3- Sketch of a moving bed

$$NTU = \frac{(k_f a) A\, Z_e}{U_s} = \frac{K}{K-1} \ln \left\{ \frac{c_o - c_1}{c_o - c_2} \, \frac{c_2}{c_1} \right\} - \ln \left\{ \frac{c_o - c_1}{c_o - c_2} \right\} \qquad |15|$$

Usually $c_1 = 0.05\, c_o$ and $c_2 {*} 0.95\, c_o$ so:

$$NTU = \frac{K}{K-1} \ln(19^2) - \ln(19) \qquad |15a|$$

If another kinetic law is used a different expression for NTU is obtained. Moison et al. (12) considered a kinetic law of Thomas type and arrived to:

$$NTU^* = \frac{K}{K-1} \ln' \left\{ \frac{c_o - c_1}{c_o - c_2} \, \frac{c_2}{c_1} \right\} \qquad |16|$$

or

$$NTU^* = \frac{K}{K-1} \ln(19^2) \qquad |16a|$$

with

$$NTU^* = \frac{(K_L a) A Z_e}{U_s}$$

and $1/K_L a = (1/k_f a) + (c_o / KQ\, k_r a)$. This additivity of resistances in the fluid film and inside the particle is controversial for nonlinear systems. If K is close to 1, then $\ln K \simeq (K-1)/K$ and Eq. (16) becomes the Fenske equation, well-known in distillation at total reflux.

Now we can calculate for the runs of Table I the values of HTU according to Michaels and Moison. The results are shown in Table II.

TABLE II - Height of Transfer Unit

Run	u_o(cm/sec)	$Re=u_o d_p/\nu$	HTU(Michaels) (cm)	HTU*(Moison) (cm)
1	0.015	0.089	0.81	0.41
2	0.033	0.197	1.08	0.54
3	0.143	0.850	1.88	0.94

We get

$$HTU = 3.69\ u_o^{0.362} \qquad |17|$$

and finally k_f=1.3 x10^{-3} cm/sec.This value is lower than that predicted from Carberry correlation (k_f=5.9x 10^{-3} cm/sec) since all resistances were lumped into the kinetic law for film mass transfer.

Several investigators have tried to find an analogy between chromatographic and counter-current processes.One equation largely used but never proved,relating the number of plates in chromatography,N_c and in distillation,N_d is:

$$N_c = N_d^2 \qquad |18|$$

Conder and Fruitwala (13) revisited this question and showed that for $K-1 \ll 1$,

$$N_c \leqslant 0.20\ N_d^2 \qquad |19|$$

Equation (18) arises because N_d was probably underestimated by assuming total reflux.It is probably because reflux was not considered and comparisons were made on wrong basis that discrepancies between various derivations occur. Valentin (14) states that comparison should be made on the basis of equivalent separation quality and same average residence time in chromatographic and countercurrent processes.

He defined the residence time for a species A in chromatography as:

$$\bar{t}_A = \frac{H(1+\xi)}{u_i}\ N \qquad |20a|$$

where H is the HETP and N is the number of theoretical plates.For counter-current processes he wrote:

$$\bar{t}_A = \frac{H(1+\xi)}{u_i}\ \frac{N^+\rho^+ - N^-\rho^-}{1-R} \qquad |20b|$$

where N^+ and N^- are the plates in the extraction and rectification sections, ρ^+ and ρ^- are the fractions of entering flow of A collected at the top and at the bottom,respectively.Noting that the residence time per plate is:

$$1/E = H(1+\xi)/u_i$$

then,in general,we have:

$$\bar{t}_A = (1/E)\ f(N,R) \qquad |21|$$

In chromatography without reflux or recycling ,f =N :but if reflux is used then a lower number of plates is required for a given separation,so:

$$f = \frac{N'}{1-R'} \qquad |22a|$$

In distillation, $\rho^- << \rho^+$, $\rho^+ \simeq 1$, $N^+ \simeq N^-$ and then:

$$f = \frac{N^+}{1-R} \qquad |22b|$$

The form of Eqs(22a) and (22b) is similar;in chromatography without recycling we need a much higher number of theoretical plates than in countercurrent processes where reflux is always relatively high.

2.3 Scientific approach:modelling and simulation

A scientific approach to design of ion exchange equipment starts with a model of the process,which enables us to predict breakthrough curves,and thus,process efficiency.Many philosophies of modelling are encountered;we use both differential or staged approach,appropriate kinetic laws for description of transport phenomena and a model for the particle structure .Model parameters are then obtained from well-designed experiments. However,some authors simply lump all mechanisms in a kinetic term,$\mathcal{R}$;this leads to what I call "Models of chemical kinetic type law",in which $\mathcal{R}$ is described by a kinetic equation similar to a rate of chemical reaction.

Thomas model developed in 1948 (15) assumes a kinetic law equivalent to a reversible second order reaction rate equation,i.e.,

$$\mathcal{R} = k_1 \ \{c(Q-q) - \frac{1}{K}\ q(c_o-c)\} \qquad |23|$$

The mass balance equation,for plug flow of the fluid phase,is:

$$u_i \frac{\partial c}{\partial z} + \frac{\partial c}{\partial t} + \frac{1-\varepsilon}{\varepsilon} \ \frac{\partial q}{\partial t} = 0 \qquad |24|$$

where $\partial q/\partial t \equiv \mathcal{R}$ .For a step input and assuming that no solute is in the bed at the beginning of the operation,we get:

$$\frac{c}{c_o} = \frac{J(\tilde{N}r,\tilde{N}T)}{J(\tilde{N}r,\tilde{N}T)+\{1-J(\tilde{N},r\tilde{N}T)\}\exp\{(r-1)\tilde{N}(T-1)\}} \qquad |25|$$

where $r=1/K$, $T=c_o(V-\varepsilon v)/(1-\varepsilon)Qv$ is the throughput parameter and the number of reaction units is $\tilde{N}=k_1 Qz/\varepsilon u_i$. The function J is defined as:

$$J(x,y)=1-\int_0^{\infty} \exp(-y-\lambda)\ I_o(2\sqrt{y\lambda})\ d\lambda$$

with I_o -modified Bessel function of zero order and first kind.

From this model we get solutions for a number of models from the literature,which use other kinetic laws (4):

Bohart model (r=0)

$$c/c_o = \exp(\tilde{N}T)/\{\exp(\tilde{N}T) + \exp(\tilde{N}) - 1\} \qquad |25a|$$

Walter model (r=1)

$$c/c_o = J(\tilde{N},\tilde{N}T) \qquad |25b|$$

Walter model (r high)

$$c/c_o = \{\sqrt{K/T} - 1\} / (K-1) \qquad |25c|$$

Klinkenberg model (r=1,$\tilde{N}$ and $\tilde{N}T$ high)

$$c/c_o = \frac{1}{2}\{1+ \mathrm{erf}(\sqrt{\tilde{N}T} - \sqrt{\tilde{N}})\} \qquad |25d|$$

Other authors use models of "Physical kinetic type law" in which kinetic laws describing mass tranfer in the film and inside the particles are introduced.For linear ion exchange equilibrium isotherm these models have reached a high degree of complexity but allowing analytical solutions by the use of Laplace transform.

Rosen model (16) assumes linear equilibrium,plug flow of the fluid phase but considers film diffusion and particle diffusion;the particle structure considered is the homogeneous one.

The mass balance equation is:

$$u_i \frac{\partial c}{\partial z} + \frac{\partial c}{\partial t} + \frac{1-\varepsilon}{\varepsilon}\frac{\partial \bar{q}}{\partial t} = 0 \qquad |26a|$$

The equilibrium law:

$$q = \frac{Q}{c_o}\ c_i^* \qquad |26b|$$

The kinetic law for film mass transfer:

$$(\partial q/\partial t)_{\rho=R} = (k_f a_p)(c - c_i^*) \quad |26c|$$

and the diffusion equation for an homogeneous particle is:

$$\frac{\partial q(\rho,z,t)}{\partial t} = D_e\left(\frac{\partial^2 q(\rho,z,t)}{\partial \rho^2} + \frac{2}{\rho}\frac{\partial q(\rho,z,t)}{\partial \rho}\right) \quad |26d|$$

Since $\bar{q}(z,t)$ is the average over the sphere of $q(\rho,z,t)$ we get, for a step input and no solute in the bed at time zero:

$$c/c_o = 1/2 + 2/\pi \int_0^\infty \exp(A)\sin(B)\, d\lambda/\lambda \quad |26e|$$

where A and B are functions of the model parameters.

If the internal diffusion is negligible the Rosen model becomes the Anzelius or Schuman model ,which has the solution:

$$c/c_o = J(\tilde{N}_f, \tilde{N}_f T) \quad |27|$$

with $\tilde{N}_f = k_f a\, z/\varepsilon u_i$ (number of film mass transfer units for the length z of the bed).

The transfer function of the column,using transforms with regard to $t - z/u_i$ is,for the Rosen model:

$$G(s) = \exp\{-\frac{1-\varepsilon}{\varepsilon}\tau\, Y_T(s)\} \quad |28|$$

where

$$Y_T(s) = \frac{Y_D(s)}{1 + R_f Y_D(s)}$$

and

$$R_f = 1/k_f a_p$$

$$Y_D(s) = \frac{6D_e\,\xi}{R^2}\sum_{n=1}^{\infty}\frac{s}{s + \frac{D_e\,\pi^2}{R^2}n^2}$$

For the Anzelius model the tranfer function is simply:

$$G(s) = \exp\left(-\frac{N_f \tau s}{N_f/\xi + \tau s}\right) \quad |29|$$

where $N_f = k_f a_p \frac{1-\varepsilon}{\varepsilon}\tau$ is the number of film mass tranfer units,calculated for the whole bed.

Comparing the moments of the impulse responses from the transfer functions of Rosen and Anzelius models,we get:

$$N_f(\text{Anzelius}) = \left\{\frac{N_f}{1+ Bi_m/5\xi}\right\}_{\text{Rosen}} \qquad |30|$$

where $Bi_m = k_f R/D_e$ is the mass Biot number.

From the Rosen model we can derive simplified models or extend it ,by including axial dispersion effects:

Wicke model

This model includes axial dispersion but neglects mass transfer resistances and accumulation in the fluid phase.The solution is:

$$c/c_o = \frac{1}{2}\left\{2-\text{erf}\sqrt{Pe}\,\frac{z^*+\theta}{2\sqrt{\theta}} - \text{erf}\sqrt{Pe}\,\frac{z^*-\theta}{2\sqrt{\theta}}\right\} \qquad |31|$$

For deep beds the solution becomes:

$$c/c_o = \frac{1}{2}\left\{1- \text{erf}\sqrt{Pe}\,\frac{z^*-\theta}{2\sqrt{\theta}}\right\} \qquad |31a|$$

where $z^*=z/L$ and $Pe=u_i L/D_{ax}$ is the Peclet number.

Kawazoe and Babcock model

This is an extension of Rosen model ,which simply includes axial dispersion;for large z the asymptotic solution is:

$$c/c_o = \frac{1}{2}\left\{1+\text{erf}\,\frac{u_F\sqrt{t} - z/\sqrt{t}}{2\sqrt{D_{ov}}}\right\} \qquad |31b|$$

where $u_F=u_i/(1+\xi)$ is the velocity of the stationnary front and D_{ov} is an overall diffusion coefficient based on the additivity of resistances due to film mass tranfer,particle mass transfer and axial dispersion.

For nonlinear isotherms it is more difficult to get analytical solutions;however there are some which use simplified equations for particle diffusion and the hypothesis of a stationnary front.Let us recall some solutions:

Glueckauf and Coates

For Freundlich isotherm, $q=ac^b$ and linearized particle diffusion equation, $\partial q/\partial t=(k_p a_p)(q_i-q)$ we get:

$$(c/c_o)^{1-b} = 1- \exp\{(b-1)N_d(\theta-\theta_{Bp})\} \qquad |32|$$

with $N_d=k_p a_p \tau(1+\xi)$.For irreversible isotherms,b=0 and:

$$c/c_o=1-\exp\{-N_d(\theta-\theta_{Bp})\} \quad |32a|$$

For isotherms of "mass action law" type we get for external diffusion control :

$$\frac{1}{r-1}\ln(c/c_o) + \frac{r}{r-1}\ln(\frac{c}{c-c_o})=1-N_f(\theta-1) \quad |33|$$

For internal diffusion control we have:

$$\frac{r}{r-1}\ln(c/c_o)+\frac{1}{r-1}\ln(\frac{c}{c-c_o}) =-\{1+N_d(\theta-1)\} \quad |34|$$

Vermeulen solutions

For irreversible isotherms Vermeulen expressed the original solution of Drew et al. in terms of parameters N_f and T (film diffusion) and N_d and T (particle diffusion),respectively:

$$c/c_o=\exp\{N_f(T-1)-1\} \quad |35a|$$

$$c/c_o=1-\exp\{-N_d(T-1)-1\} \quad |35b|$$

Hall model

The model considers pore diffusion ,film diffusion and a shell-progressive boundary;the result is:

$$N_{pore}(T-1)=\phi(c/c_o) + \frac{N_{pore}}{N_f}(\ln\frac{c}{c_o} +1) \quad |36|$$

where $\phi(c/c_o)=2.39 -3.59\sqrt{1- c/c_o}$ and $N_{pore}=15D_{pore}(1-\varepsilon)\tau/\varepsilon R^2$. The model also assumes irreversible equilibrium.

Some authors approach a nonlinear isotherm by a bilinear isotherm, then solve,analitically,some models and get the solution for the irreversible case as a limiting situation.

However,the important case of irreversible isotherm has been tackled in an elegant way by Acrivos and later completed by Quilici and Vermeulen.Guided by the same ideas,Rodrigues developed a model which considers that the first part of the front is controlled by axial dispersion and film diffusion and the second part by axial dispersion and internal diffusion.The model equations are (17):

<u>mass balance</u> :

$$D_{ax}\frac{\partial^2 c}{\partial z^2}=u_i\frac{\partial c}{\partial z}+\frac{\partial c}{\partial t}+\frac{1-\varepsilon}{\varepsilon}\frac{\partial q}{\partial t} \qquad |37a|$$

<u>kinetic laws</u>:

$$\partial q/\partial t=k_f(c-c^*) \quad , \quad 0\leqslant c\leqslant c_i \qquad |37b|$$

$$\partial q/\partial t=k_p(Q-q) \quad , \quad c_i\leqslant c\leqslant c_o \qquad |37c|$$

The basic hypothesis is that of stationnary front,with velocity given by $u_F=u_i/(1+\xi)$;then a moving coordinate is introduced which follows the front:

$$\bar{z}=(z-u_F t)\frac{u_i}{D_{ax}}$$

for large capacity factors; using $x=c/c_o$,$y=q/Q$ we get:

$$\frac{d^2x}{d\bar{z}^2}=\frac{dx}{d\bar{z}}-\frac{dy}{d\bar{z}} \qquad |38a|$$

$$-\frac{dy}{d\bar{z}}=\beta(x-x^*) \qquad |38b|$$

$$\frac{dy}{d\bar{z}}=\delta(y-1) \qquad |38c|$$

where $\beta=N_f/Pe$,$\delta=N_d/Pe$ and $x^*=ry/\{1-(1-r)y\}$;r=0 for irreversible equilibrium.

For this case,r=0 we can integrate Eq.(38a) to get:

$$\frac{dx}{d\bar{z}}=x-y \qquad |39|$$

From Eqs(39) and (38b) we have,at x=0 after application of L'Hopital's rule :

$$\left.\frac{dy}{dx}\right|_{x=0}=(1\pm\sqrt{1+4\beta})/2$$

and finally :

$$y=(\lambda+1)\,x \quad ,0< x < (\lambda+1)^{-1}$$

$$y=1 \quad , \quad x > (\lambda+1)^{-1}$$

with

$$\lambda= \frac{1}{2}(\sqrt{1+4\beta} -1)$$

Integrating now Eq(39) and taking into account that at $\bar{z}=0$, $x=x_i$ we get:

$$x=x_i \exp(-\lambda\bar{z}) \qquad |40|$$

For the second part of the front we get similarly:

$$x=1+ \frac{(\delta-1)(x_i-1)+(y_i-1)}{\delta-1} \exp(\bar{z}) - \frac{y_i-1}{\delta -1} \exp(\bar{z}\delta) \qquad |41|$$

where

$$x_i= \frac{\delta}{\beta(1+ \delta/\lambda)}$$

In the time domain,we get finally:

$$x= x_i \exp\{-\lambda Pe(\theta_i -\theta)\}$$

$$x=1+ \frac{\lambda\delta(\delta-1)/\beta +\delta(1-\lambda)-\delta^2}{(\lambda+ \delta)(\delta-1)} e^{-Pe(\theta-\theta_i)} + \frac{\lambda}{(\lambda+\delta)(\delta-1)} e^{-\delta Pe(\theta-\theta_i)}$$

where

$$\theta_i=1- \frac{1}{Pe} \left\{\frac{\delta\beta(\lambda+\delta)-\delta^2(\lambda+1) +\lambda\beta}{\beta\delta(\lambda+\delta)}\right\}$$

Most of the models developed above consider simplified kinetic equations for particle diffusion and the structure of the resin is assumed to be homogeneous.Nonlinear equilibrium cases coupled with complete diffusion equation in realistic situations (homogeneous, macroporous) will lead to complex problems which can only be solved by numerical techniques.

Models presented above are differential models but we can develop staged models,in which the bed is viewed as a series of perfectly mixed cells;for a large number of cells again the numerical handling is time consuming.However the approach can be of interest to simulate ion exchange in fluidised beds and to learn the effect of independent parameters on the behavior of the process.

The dynamic behavior of ion exchange processes is determined by equilibrium factors ,hydrodynamics and kinetics of mass transfer (film diffusion,pore diffusion,particle diffusion).In the next section we will discuss the importance of the ion exchange equilibrium isotherm and concepts issued from the equilibrium theory.

3 FUNDAMENTAL NOTIONS

The behavior of fixed bed processes can be predicted if we understand how concentration and/or heat waves propagate through the column.Let us then start with a simple model for sorption (adsorption, ion exchange,etc) processes,which assumes instantaneous equilibrium in every point of the bed,plug flow for the fluid phase and isothermal operation.This is the so-called equilibrium model;for a single solute adsorption or binary ion exchange the mass balance in a volume element of the bed is:

$$\frac{\partial x}{\partial z^*} + \frac{\partial x}{\partial \theta^*}\{1+ \xi y'(x)\} =0 \qquad |42|$$

where θ^* is the time reduced by the space time and $y'(x)$ is the slope of the equilibrium isotherm.From Eq(42),using the cyclic relationship between partial derivatives,we obtain:

$$u_x = (\partial z^*/\partial \theta^*)_x = 1/\{1+\xi\ y'(x)\} \qquad |43|$$

Equation(43) gives the velocity of a concentration x;in the absence of adsorbent $u_x=1$ (note that this velocity is reduced by the interstitial velocity,u_i).

3.1 Dispersive and compressive waves

According to the nature of the equilibrium isotherm we get different shapes of the concentration wave:

a) dispersive waves

It occurs for unfavorable isotherms ($y''(x)>0$)since then higher concentrations travel slower than lower concentrations.

b) compressive waves

It happens for favorable isotherms ($y''(x) < 0$) since then higher concentrations travel faster than lower concentrations.

The shape of the concentration wave travelling through the bed is important since it determines the time at which the front arrives at the bed outlet: breakthrough time, t_{Bp} or θ_{Bp} in reduced coordinates. Usually it is defined in a arbitrary way as the time at which the outlet concentration is 5% of the inlet concentration. The breakthrough time is an important design parameter since it is the useful time of operation for a given set of operating conditions.

Whatever the shape of the equilibrium isotherm is, the overall mass balance should be satisfied; if a feed contains a solute to be adsorbed or an ion to be exchanged at concentration c_o and the corresponding solid capacity is Q then we get:

$$t_{st} = \tau(1+\xi) \qquad |44|$$

or in reduced variables: the stoichiometric time is equal to one.

Equation(43) enables us to find the histories of concentration for any shape of the equilibrium isotherm; we will use Eq(43), also called De Vault equation, to obtain the histories for the cases of unfavorable and favorable equilibrium. However, because of its interest when leading with quasi-linear hyperbolic equations (which often occur in equilibrium models of ion exchange columns) let us briefly introduce the method of characteristics.

3.2 The Method of Characteristics

A system of $\underline{n}$ first order partial differential equations, with unknowns $x_1, x_2, \ldots, x_n$ can be written as:

$$\underline{A}\frac{\partial \underline{x}}{\partial z} + \underline{B}\frac{\partial \underline{x}}{\partial \theta} = \underline{H} \qquad |45|$$

where the matrixes are :

$\underline{A}(nxn)$: $a_{ij}(x,z,\theta)$

$\underline{B}(nxn)$: $b_{ij}(x,z,\theta)$

$\underline{H}(nx1)$: $h_{ij}(x,z,\theta)$

$$\underline{x} = \begin{bmatrix} x_1 \\ x_2 \\ \vdots \\ x_n \end{bmatrix}$$

The system is linear if $\underline{A}$ and $\underline{B}$ do not depend on x_i and $\underline{H}$ is linearly dependent on x_i; if $\underline{A}, \underline{B}$ and $\underline{H}$ depend only on x_i the system is quasi-linear.

Introducing $\underline{M}=\underline{A}^{-1}.\underline{B}$ and $\underline{R}=\underline{A}^{-1}.\underline{H}$ we get from Eq(45):

$$\frac{\partial \underline{x}}{\partial z} + \underline{M}\,\frac{\partial \underline{x}}{\partial \theta} = \underline{R} \qquad |46|$$

with

$\underline{R}(nxn)$: r_{ij}

This system is hyperbolic if $\underline{M}$ is diagonalizable.Then it exists a matrix $\underline{P}(nxn)$ such as:

$$\underline{M}=\underline{P}^{-1}.\underline{D}.\underline{P}$$

with

$\underline{P}$:p_{ij}

$$\underline{D}=\begin{bmatrix} \sigma_1 & & \\ & \ddots & \\ & & \sigma_n \end{bmatrix}$$

where $\sigma_i(i=1,2,\ldots,n)$ is the family of eigenvalues of $\underline{M}$ and the familly of eigenvectors of $\underline{M}$ is

$$\begin{pmatrix} p_{1j} \\ \vdots \\ p_{nj} \end{pmatrix} \quad j=1,\ldots,n$$

The method of characteristics enables us to replace the original system by two systems of ordinary differential equations;the characteristic directions of this system are σ_i and we call characteristic curves in the plan (z,θ) the equations $d\theta/dz=\sigma_i$.

If D_i is the directional derivative along σ_i,i.e.,

$$D_i= \frac{\partial}{\partial z} + \sigma_i\frac{\partial}{\partial \theta}$$

we get from Eq.(46):

$$\sum_j p_{ij}D_i(x_j) = \sum_j p_{ij}r_j \qquad |47|$$

Eq.(47) is the equation associated with the characteristic direction, $d\theta/dz=\sigma_i$.Through a point P in the plan (z,θ) pass $\underline{n}$ characteristic curves and the $\underline{n}$ relationships,Eqs(47) should be verified to find the solution of the system.

The determination of the characteristic directions and associated equations needs the following steps: inversion of $\underline{A}$, calculation of $\underline{M}=\underline{A}^{-1}.\underline{B}$ and finally, determination of the eigenvalues and eigenvectors of $\underline{M}$.

In practice we add to the system (45) the system:

$$d\underline{x}= \frac{\partial \underline{x}}{\partial z} dz + \frac{\partial \underline{x}}{\partial \theta} d\theta$$

and we get:

$$\underline{A} \frac{\partial \underline{x}}{\partial z} +\underline{B} \frac{\partial \underline{x}}{\partial \theta} =\underline{H} \qquad |48a|$$

$$\underline{I}_n dz \frac{\partial \underline{x}}{\partial z} +\underline{I}_n d\theta \frac{\partial \underline{x}}{\partial \theta} =d\underline{x} \qquad |48b|$$

If we put the determinant of this system equal to zero, i.e.,

$$|\underline{A}.\underline{I}_n \frac{d\theta}{dz} -\underline{B}| =0$$

which is equivalent to:

$$|\underline{I}_n \frac{d\theta}{dz} - \underline{A}^{-1}.\underline{B}| =0$$

we note that $d\theta/dz$ are the eigenvalues of $\underline{A}^{-1}.\underline{B}$ and then the characteristic directions of the system (45). The system (48a,b) has a solution if its determinant is zero; it is sufficient then to write that the $\underline{n}$ characteristic determinants of

$$\begin{bmatrix} \underline{H} \\ d\underline{x} \end{bmatrix}$$

are zero.

From Eq. (42) and considering a parameter $\underline{s}$ running along σ_i the method of characteristics leads to:

$$\frac{dz^*}{1} = \frac{d\theta^*}{1+ \xi y'(x)} = \frac{dx}{0} =ds \qquad |49|$$

or

$$\frac{dz^*}{d\theta^*} = \frac{1}{1+\xi y'(x)} \qquad |50a|$$

$$\frac{dx}{d\theta^*} =0 \qquad |50b|$$

where $y'(x)$ is the slope of the ion exchange equilibrium isotherm

$$y = \frac{Kx}{1+(K-1)x}$$

Eq.(50a) is the characteristic direction along which Eq.(50b)is integrated;in this case x=const. along the characteristics.Equation (50a) tells us that the velocity of a concentration $\underline{x}$ is inversely roportional to the slope of the equilibrium isotherm at that point. The general equation of characteristics is then:

$$\theta^* = \left\{ 1 + \frac{\xi K}{|1+(K-1)x|^2} \right\} z^* + \text{const.} \qquad |51|$$

For a step input in concentration since at $\theta^*=0, x=0$ the characteristics leaving the z^*-axis are:

$$\theta^* = (1+\xi K) z^* + \text{const} \qquad |52|$$

and since at $z^*=0, x=1$ the characteristics leaving the θ^*-axis are:

$$\theta^* = \left(1 + \frac{\xi}{K}\right) z^* + \text{const} \qquad |53|$$

The diagram of characteristics and the history of concentrations $x(1,\theta^*)$ are shown in Figure 4 ($\xi=1, K=0.1$) while in Figure 5 is represented the diagram of characteristics and the history for favorable isotherms (K=10).

The interesting situation which occurs with favorable isotherms is due to the formation of a shock;the shock velocity is:

$$\left(\frac{dz^*}{d\theta^*}\right)_s = \frac{1}{1 + \frac{y^- - y^+}{x^- - x^+}} \qquad |54|$$

where superscripts - and + denote concentrations behind and ahead the shock.The shock velocity is inversely proportional to the slope of the chord between points representing the feed state (x^-, y^-) and the presaturation state of the bed (x^+, y^+).

The practical implications of these results are:for favorable isotherms the equilibrium model predicts a breakthrough time equal to the stoechiometric time,$t_{Bp}=t_{st}$ while for unfavorable isotherms $t_{Bp}=(1+K\xi)t_{st}/(1+\xi)$.In Figure 6 we plot θ_{Bp},predicted by the equilibrium theory,as a function of K,with the capacity factor as a parameter.

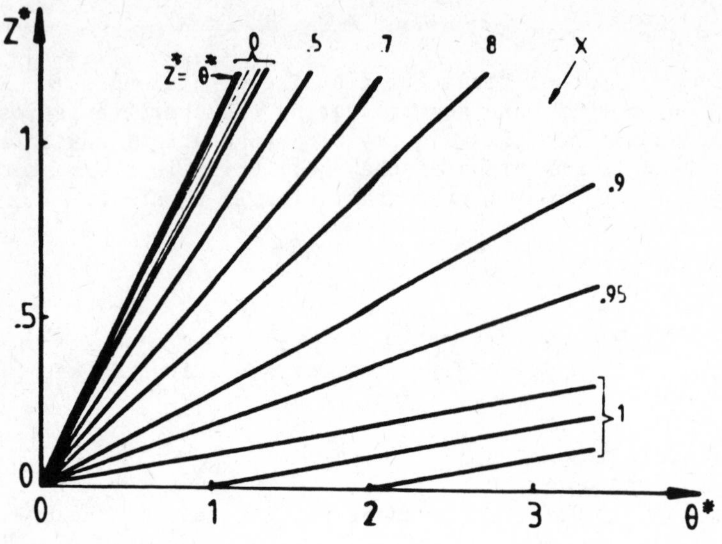

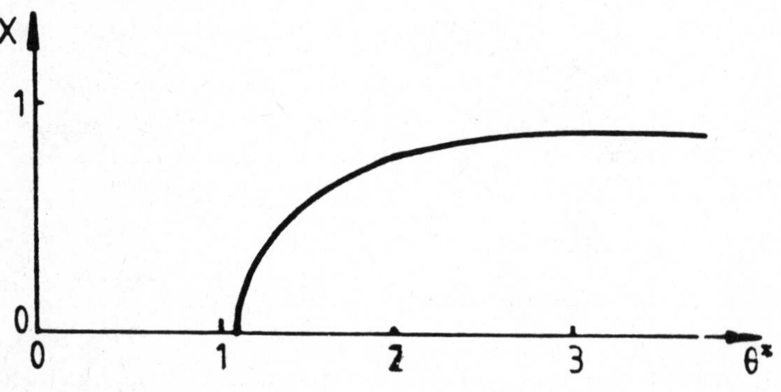

Figure 4- Diagram of characteristics and history of concentrations for unfavorable isotherm($K=0.1, \xi=1$).

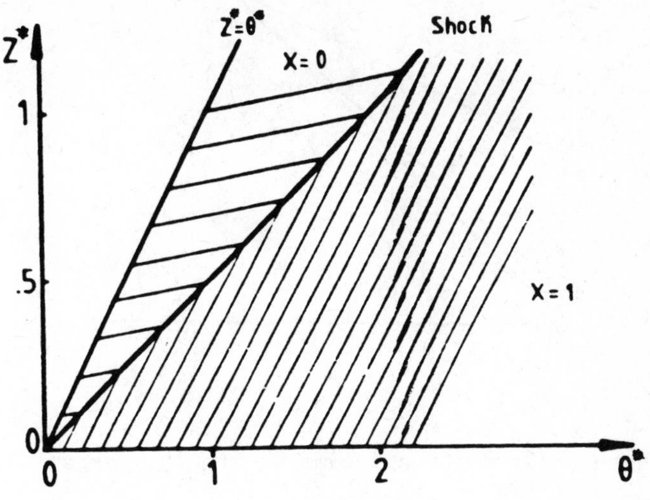

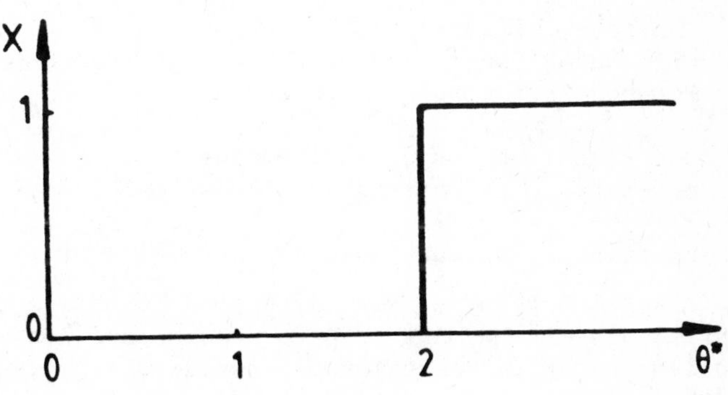

Figure 5- Diagram of characteristics and history of concentrations for favorable isotherms ($K=10, \xi=1$).

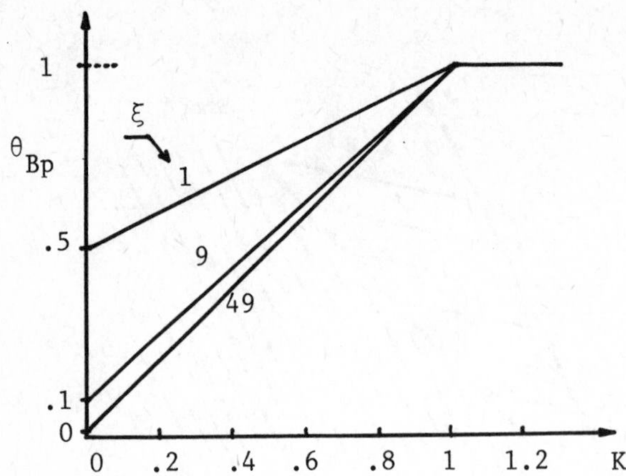

Figure 6- Breakthrough time, $\theta_{Bp}=t_{Bp}/t_{st}$ versus K, with ξ as a parameter; equilibrium model.

Responses of the bed to other inputs: pulse, Dirac can be derived by using the same methodology (18). As an example for a pulse injection of height $x_o=2$ and duration of the injection $\theta^*=a/x_o=1/2$ we get the diagram of characteristics shown in Figure 7. The history of concentrations for pulse and Dirac inputs is the same provided we plot the reduced outlet concentration as a function of $\theta^*- a/x_o$ (Figure 8).

We said above that the equilibrium model of a fixed bed in which favorable ion exchange is carried out leads to the notion of compressive wave. The physically limiting situation is the formation of a shock or discontinuity which propagates at a velocity given by Eq(54).

In real practice we never find discontinuities due to dispersive effects (axial dispersion, mass transfer resistances); however there is an opposite effect of the favorable nature of the equilibrium isotherm and the result is, that after a certain time of formation, the front will propagate through the bed keeping its shape. This is the so-called stationary front. Mathematically we just have:

$$c(z+dz,t+dt)=c(z,t) \qquad |55|$$

and then a <u>shock layer</u> propagates with the velocity of the discontinuity. For plug flow systems it is easily shown that the stationary front hypothesis leads to x=y, which is a result found in other areas as deep bed filtration.

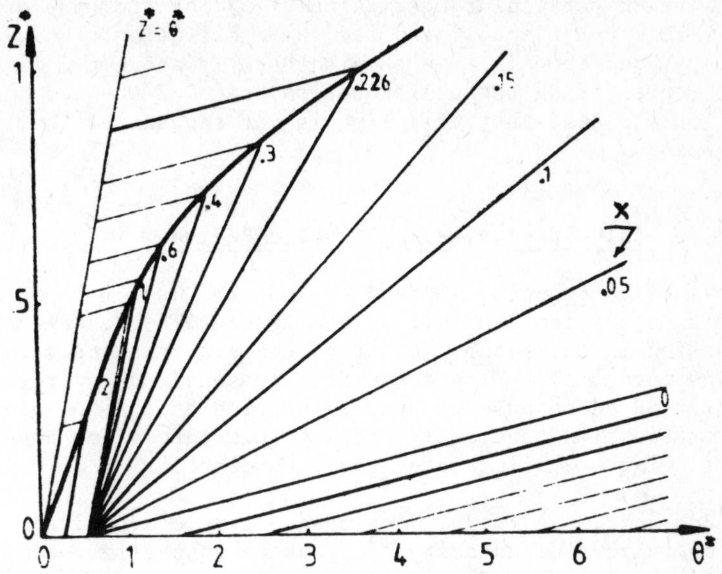

Figure 7- Diagram of characteristics for a pulse input of concentration of height x_o=2

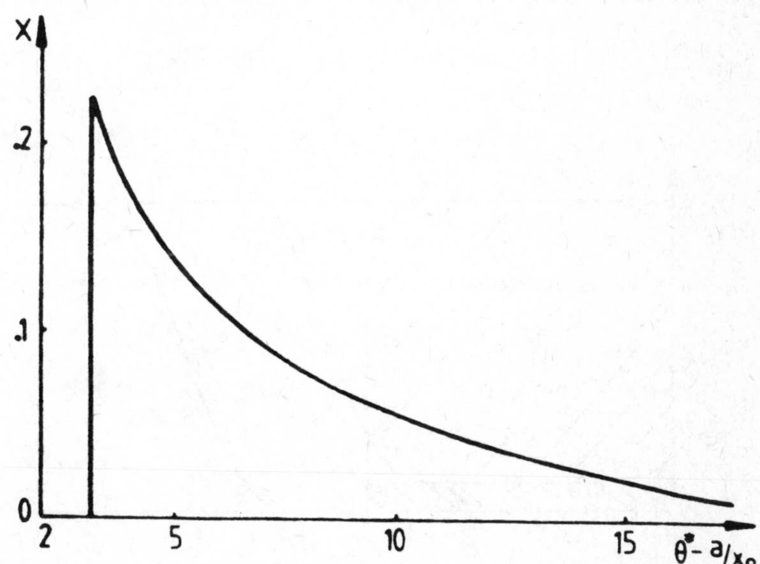

Figure 8- History of concentrations for pulse or Dirac inputs

The criterion for the existence of stationary fronts has to be found in each particular case;Safonov(19) for the case of plug flow of the fluid phase and linearized solid diffusion concluded that the necessary condition is that the isotherm is above the line between points representing the presaturation state of the bed and the feed condition and also that this line is not tangent to the isotherm in those points.

3.3 Useful Capacity.Leakage.Cyclic Regime.

For design purposes we need to know the useful capacity under operating conditions as well as the leakage.The leakage of the ionic species fed in the saturation step occurs because previous regeneration was not complete (by economical reasons);moreover the capacity is also reduced because we stop saturation at a fixed breakthrough concentration (and obviously breakthrough curves are never discontinuities).Figure 9 illustrates these concepts.

Dodds and Tondeur (20) developed a simple method for the prediction of the useful capacity and leakage in cyclic regime based on overall mass balances.Considering the ion exchange between ion A in solution and ion B initially in the resin we have:

saturation step

accumulation of A in the resin= $A_{in}-A_{out}$= B which left the resin

$$Y_{sj}-Y_{si} \qquad =\tau_{si}-X_{si}\tau_{si} \qquad =Y_{ri}-Y_{rj} \quad \triangleq F_{si}Y_{ri}$$

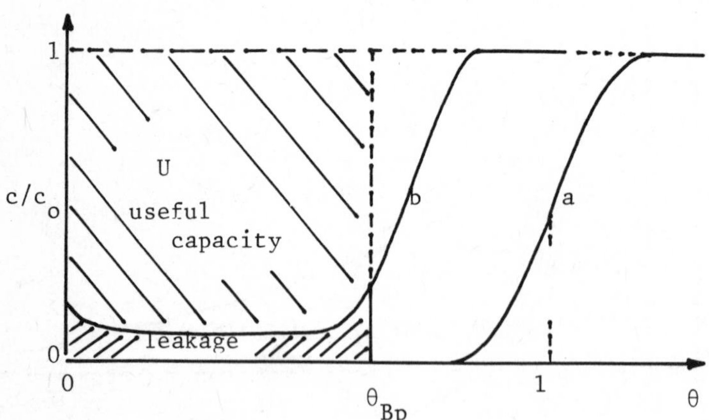

Figure 9- Definitions of useful capacity and leakage
a) saturation after a complete regeneration
b) saturation after partial regeneration in cyclic regime.

<u>regeneration step</u>

accumulation of B in the resin=$B_{in}-B_{out}$=A which left the resin

$$Y_{rk}-Y_{rj}=\tau_{rj}-X_{rj}\,\tau_{rj}=Y_{sj}-Y_{sk} \quad \triangleq F_{rj}Y_{sj}$$

where s and r denote saturant (ion A) and regenerant (ion B),respectively,and

$$X_s=\frac{\text{equivalents of A in the effluent}}{\text{total equivalents entered in the bed during saturation}}$$

$$Y_{ri}=\frac{\text{equivalents of B in the bed at the begining of step i}}{\text{total capacity of the bed}}$$

$\tau_s=c_sV_s/Q$,$\tau_r=c_rV_r/Q$ (V is the volume passed through the bed)

$$F_{si}=\frac{\text{equivalents of B which comes out during saturation i}}{\text{equivalents of B in the bed at the beginning of saturation i}}$$

$$F_{rj}=\frac{\text{equivalents of A leaving during regeneration j}}{\text{equivalents of A in the bed at the begining of regenerationj}}$$

The above mass balances contain the hypothesis that the accumulation of saturant (or regenerant) depends only on Y_s (or Y_r),the proportionality constants being the saturation factor (or the regeneration factor).

We also have in each step:

$$Y_{ri}+Y_{si}=1$$

In cyclic regime F_si and F_{rj} are constants and the useful capacity ($Y_{sj}-Y_{si}$) is simply:

$$U=F_sY_r=F_rY_s \qquad |56|$$

with

$$Y_s=\frac{F_s}{F_s+F_r-F_rF_s}$$

$$Y_r=\frac{F_r}{F_s+F_r-F_rF_s}$$

The method only requires the experimental determination of curves F_s(or F_r) versus τ_s(or τ_r).

The leakage is X_s given by:

$$X_s = 1 - \frac{U}{\tau_s} \qquad |57|$$

and the regenerant efficiency,E_j is:

$$E_j = F_r Y_{sj} / \tau_{rj} = 1 - X_{rj} \qquad |58|$$

3.4 Multicomponent Equilibrium Model

Equilibrium models are really interesting for predicting breakthrough curves in the case of multicomponent feed.Before stating some rules to draw these histories we should keep in mind that the first step is the determination of ion exchange equilibrium isotherms.This can be done either by batch equilibration or by dynamic methods. Some authors used regeneration curves to get equilibrium data,stating that equilibrium prevails during regeneration step (unfavorable equilibrium).Other dynamic methods involve step-by-step saturations ,in which total concentration is kept constant but the equivalent fraction of an ionic species is increased.Alternatively if we run several experiments on a column RB with feed containing different proportions of A and B we get information on equilibrium isotherm.

In Figure 10 we present the equilibrium isotherm at 20C for the system Cl^-/OH^- obtained by Diaprosim and in the laboratory using a dynamic method.The column was filled with particles of d_p=450μ and had L=10 cm and internal diameter of 0.3 cm;the flowrate used was 6 ml/min and total concentration 0.01 N.Solutions of NaCl and NaOH in different proportions (90%,75%,50%,25%,10% Cl^-) were passed through the bed.

Experimental results for those runs are shown in Figure 11. It is interesting to notice that the throughput parameter T has the following values for runs 1 to 4:

Run	Feed (NaCl+NaOH) Cl%	T
1	25	2.083
2	50	1.295
3	75	1.060
4	90	1.037

and $T=\Delta y/\Delta x$ for stable fronts.T is proportional to the slope of the chord linking points (0,0) and (x,y) and then the stoichiometric points of the fronts are directly proportional to those slopes as predicted by Eq.(54).

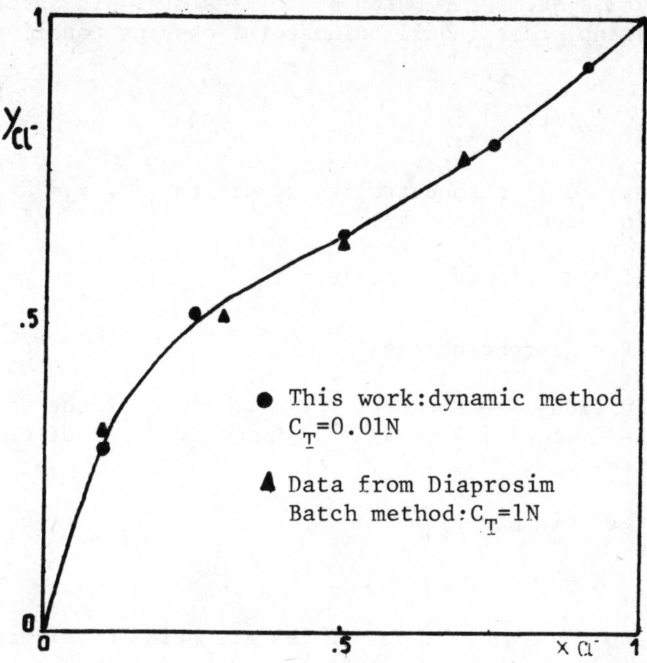

Figure 10- Ion exchange equilibrium isotherm Cl^-/OH^- (20C, Na^+ as co-ion)

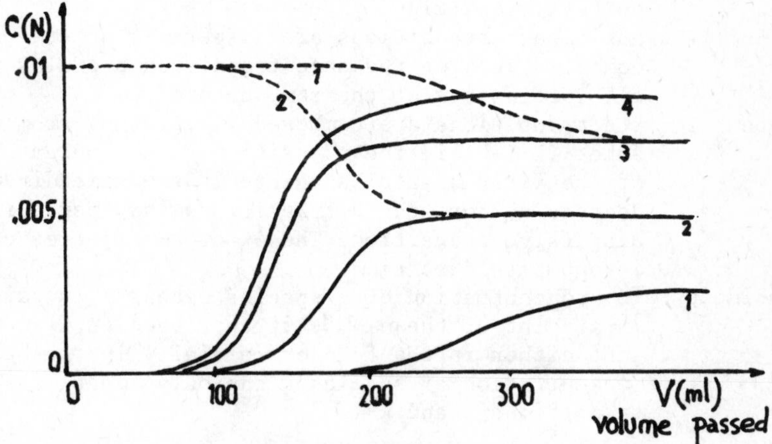

Figure 11- Breakthrough curves for a column ROH^- fed with mixtures of NaCl+NaOH (C_T=0.01N);——Cl^- ---OH^- 1-25%Cl^-, 2-50% Cl^- ,3-75%Cl^- ,4-90%Cl^-

For multicomponent feeds Tondeur (21) stated a number of rules for "well-behaved systems",in which affinities of ionic species are in a given constant order .Equilibrium relationships considered were of the type:

$$(y_i/x_i)^{\nu_i}/(y_j/x_j)^{\nu_j} = K_j^i$$

with K_j^i constant and electroneutrality condition is $\sum_i x_i = \sum_i y_i = 1$. Mass balances for each species are:

$T = dy_i/dx_i$ (for continuous functions)

and

$T = \Delta y_i/\Delta x_i$ (for discontinuities)

The initial conditions are the presaturation state of the bed at T=0 and the feed composition at 1/T=0.A profile can be defined in terms of y_i versus 1/T.

The characteristic equation is:

$$\sum_i \nu_i \frac{x_i y_i}{Tx_i - y_i} = 0$$

which has n-1 positive roots if n species are being considered.The roots $T_1 \ldots . T_{n-1}$ and each one defines an exchange path.The basic rules for drawing a profile are:

Zone rule : a profile for a system of n species has n plateau and n-1 transitions

Nature of transitions: transitions are dispersive or stable(or compressive);for the same transition profiles for all species are of the same nature

Slope rule: the transitions are numbered 1,2,...,n-1 from the inlet (1/T=0);in the transition k the concentrations of the first k species change in the same direction, decreasing for stable transitions and increasing for dispersive transitions.The other n-k species change in opposite direction.

Annulment rule: concentration of a species necessarily cancels in a point of the profile if that species is not present either in the feed or initially in the bed. Concentration of species k can only cancel in the transitions k and k-1.

Let us consider an hypothetical example: species are numbered in order (decreasing) of affinities - 1,2 and 3.The bed is initially saturated with species 3 and the feed contains all species. The profile is shown in Figure 12.

For more details on multicomponent equilibrium theory the reader should look at Helfferich and Klein book (6).

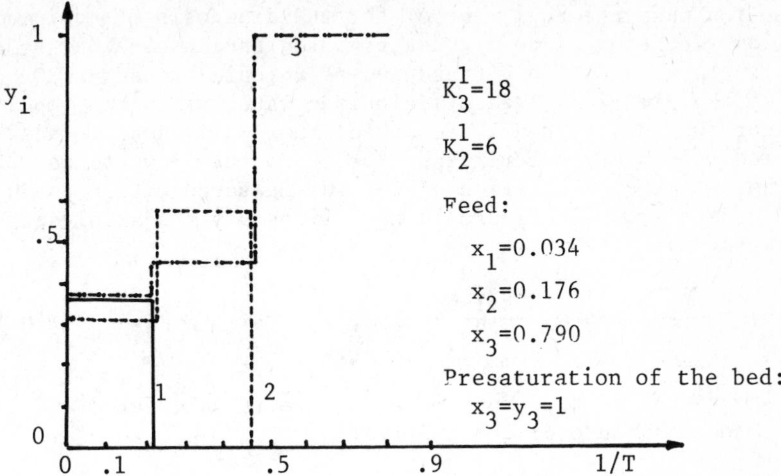

Figure 12- Profile y_i versus 1/T for multicomponent ion exchange: equilibrium theory

4 HYDRODYNAMICS.KINETICS AND MASS TRANFER.

One of the factors governing the dynamic behavior of ion exchange column is the nature of the fluid flow through the bed,i.e.,hydrodynamics.One way of describing fluid flow is by using the concept of residence time distribution,$E(t_r)$ in which the residence time is the time a fluid element spent in the system from the moment it entered until it left the bed.The distribution or density function of the character,t_r is called RTD,$E(t_r)$ such as $E(t_r)dt_r$ gives the fraction of fluid elements at the outlet which has residence time between t_r and t_r+dt_r.

This theoretical distribution can be related to experimental information arising from the use of tracer methodology.A tracer is any substance,injected in the system,which doesn't modify the hydrodynamics and is easily detected at the outlet.

The normalised response to a Dirac input is the C-curve of Danckwerts while the normalised response of the system to a step input is the F-curve.We can show that:

$$E(t)=C(t)/\tau \quad |59|$$

and

$$E(t)=dF(t)/dt \quad |60|$$

These experiments can be used to test models for fluid flow through ion exchange equipment or for the diagnosis of equipment operation (detection of dead zones,by-pass).Let us consider some experiments carried out in a fixed bed of anionic exchanger ROH^- of L=15.5cm,A=5.3cm^2;in a feed of deionized water we inject impulses of tracer (which is linearly adsorbed):NaOH 0.1N :Runs 1 and 2 used 0.3ml and 0.2 ml of tracer,respectively and the flowrate was 23 ml/ /min.The outlet concentration of OH^- was measured with a conductivity meter;data shown in Figure 13 were fitted by the axial dispersion model,which has a RTD:

$$E(\theta)= \frac{\sqrt{Pe}}{2\sqrt{\pi\theta^3}} \exp\left\{-\frac{Pe(1-\theta)^2}{4\theta}\right\} \qquad |61|$$

The variance of the experimental curve is calculated and compared with the variance of the RTD,i.e.,

$$\sigma^2=2/Pe \qquad |62|$$

and then the model parameter,Pe is obtained.

The literature provides a number of correlations for the estimation of axial dispersion in liquid-solid systems,in the form of Pe versus $Re'=u_o d_p/\nu(1-\varepsilon)$.For Re' < 20,Pe≃0.2.Recently ,Wesselingh and his group (22) published interesting work on axial dispersion in staged fluidised beds,namely on the influence of column diameter.

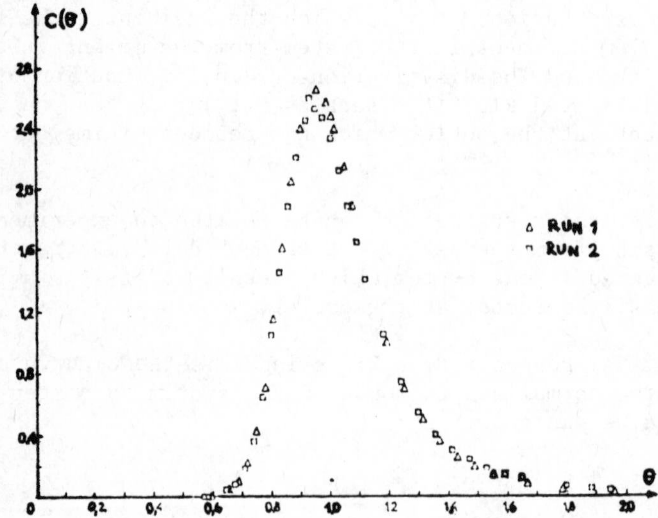

Figure 13- Impulse response of an ion exchange column C(θ)

Let us now move to other dispersive phenomena produced by the resistances to mass transfer through the film around beads and inside the particles.

4.1 Film Mass Transfer

We will start with liquid-solid mass transfer in stirred tanks. Several correlations have been suggested for the estimation of k_f (film mass transfer coefficient) such as the equation due to Ranz and Marshall:

$$Sh_p = \frac{k_f d_p}{D} = 2 + 0.6\, Re_p^{1/2}\, Sc^{1/3} \qquad |63|$$

where $Re_p = Ud_p/\nu$ and $Sc = \nu/D$. The question is the calculation of the fluid-solid relative velocity U involved in the Reynolds number. Some authors state that, as the solid is completely suspended in the liquid, Stokes law can be used to get U, i.e.,

$$U = g d_p^2 (\rho_s - \rho_\ell)/18\mu_\ell \qquad |64|$$

Levins and Glastonbury (23) presented a correlation derived from the analysis of 400 experiments :

$$Sh_p = 2 + 0.47 \left\{ \frac{d_p^{4/3}\, \bar{\varepsilon}^{1/3}}{\nu} \left(\frac{D_s}{T}\right)^{0.28} \right\}^{0.62} Sc^{0.36} \qquad |65|$$

where D_s/T is the ratio between stirrer and tank diameters and $\bar{\varepsilon}$ is the energy dissipation per unit mass ($cm^2 sec^{-3}$).

Ion exchange accompanied by chemical reaction (neutralization) can be used as a test reaction for film mass transfer studies.

Helfferich (24) developed in 1965 the theoretical basis for this process although very few results are available for testing it. We used the exchange Cl^-/OH^- accompanied by neutralization, in a batch reactor, for testing that theory. We recall the process overall reaction

$$ROH^- + Cl^- H^+ \rightarrow RCl^- + H_2O$$

The electroneutrality condition requires that:

$$c_{OH^-} + c_{Cl^-} = c_{H^+} \qquad |66|$$

Taking into account the dissociation equilibrium of water it results that $c_{OH^-} \leqslant 10^{-7}$ M; then for solutions with $c_{H^+} \gg 10^{-7}$M the ions OH^-

are consumed at the particle surface and we have diffusion of H^+ and Cl^- ions in a film of water. As a result fluxes are expressed as

$$\phi_{H^+} = \phi_{Cl^-} = -D \text{ grad } c_{Cl^-} \qquad |67|$$

where D is the interdiffusion coefficient:

$$D = 2D_{Cl^-} D_{H^+} / (D_{Cl^-} + D_{H^+}) = \text{const.} \qquad |68|$$

In fact the interdiffusion coefficient can be easily derived from:

$$\phi_{H^+} = - D_{H^+}(\text{grad } c_{H^+} + z_{H^+} \frac{c_{H^+} F}{RT} \text{ grad } V) \qquad |69a|$$

$$\phi_{Cl^-} = - D_{Cl^-}(\text{grad} c_{Cl^-} + z_{Cl^-} \frac{c_{Cl^-} F}{RT} \text{ grad } V) \qquad |69b|$$

We simply get (F grad V)/RT from Eq.(69a) and replace it in Eq.(69b) to get Eq.(68) taking into account that $\phi_{H^+} = \phi_{Cl^-}$ (electroneutrality condition) and $c_{H^+} = c_{Cl^-}$ (absence of electric current).

Assuming a linear concentration profile in the film we can derive solutions for the cases of infinite bath, finite bath with $c_{so}v_s > \bar{q}$ and finite bath with $c_{so}v_s < \bar{q}$, where c_{so} is the initial ion concentration in the reactor, v_s the volume of solution and $\bar{q}$ the available capacity of the resin.

Finite bath, $c_{so}v_s > \bar{q}$ (case 1)

$$c(t)/c_{so} = \exp(-\frac{v_r}{v_s} k_f a_p t) \qquad |70|$$

and

$$\ln(1 - \frac{\bar{q}}{c_{so}v_s}) = -\frac{v_r}{v_s} k_f a_p t \quad , \quad 0 < t < t_c \qquad |71a|$$

$$\bar{q}/(Q - q_o) = 1 \quad , \quad t > t_c \qquad |71b|$$

Finite bath, $c_{so}v_s < \bar{q}$ (case 2)

The solution is given by Eq.(70) but after $t = t_c$ there are no exchangeable ions in the solution and then $c_s = 0$.

Figure 14a shows pH versus time from which we calculated $1 - \bar{q}/c_{so}v_s$ as a function of time; from the slope of this line, plotted in Figure 14b we get k_f (case 2).

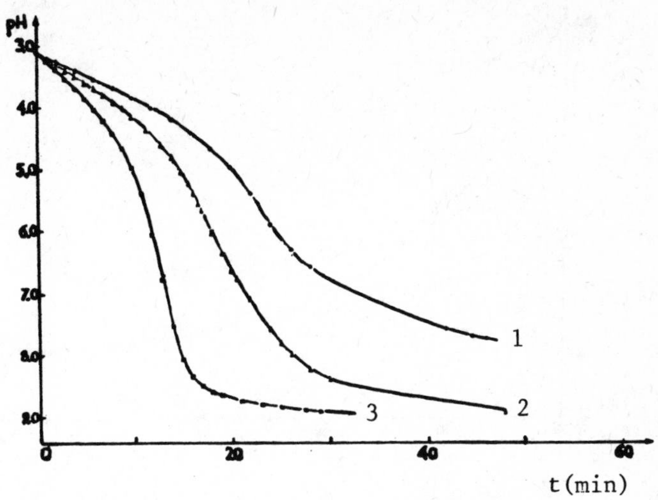

Figure 14a - pH versus time in batch reactor;exchange Cl^-/OH^- with neutralisation

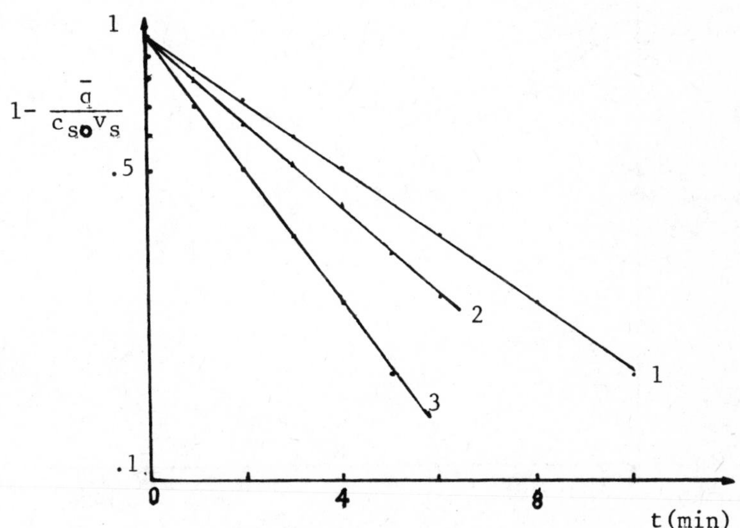

Figure 14b - Semilog plot of $1-\frac{\bar{q}}{c_{so}v_s}$ versus time

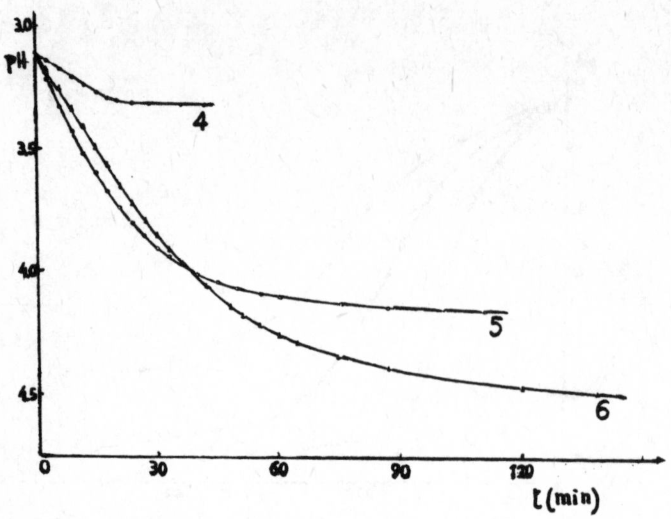

Figure 15a- pH versus time(excess of meq in solution)

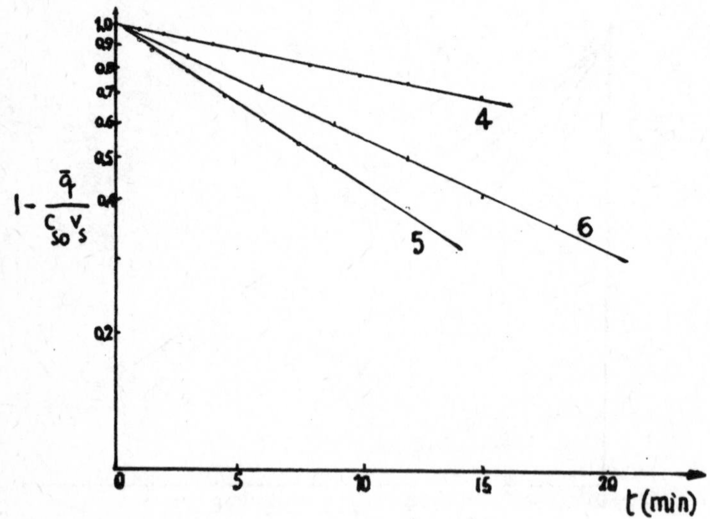

Figure 15b- $1-\frac{\bar{q}}{c_{so}v_s}$ versus time (semilog plot)

Figures 15 a and b are similar plots for case 1.

The information obtained from k_f values in batch systems can not be used in fixed bed calculations since hydrodynamics is very different;however it can be of interest to know ranges of concentration in which film mass transfer is the controlling mechanism.These values can probably be used also in multistage fluidised beds since each stage behaves as a perfectly mixed tank.

Film mass transfer coefficients in fixed bed processes can be found from experiments carried out in a differential column,inserted in a bed of inert material in order to keep hydrodynamics as close as possible from reality.

In Table III we present the experimental conditions and the results obtained in the runs of Figures 14 and 15.

TABLE III- Film mass transfer in batch systems

Run	v_s(ml)	v_r(ml)	tank H(cm)	tank T(cm)	Case	k_f(cm/sec)	Sh_p
1	700	2	8.1	10.5	2	1.1×10^{-2}	23.3
2	700	3	8.1	10.5	2	$.43\times10^{-2}$	19.7
3	700	5	8.1	10.5	2	$.86\times10^{-2}$	18.1
4	1400	1	11	12.7	1	$.65\times10^{-2}$	13.7
5	700	1	5.5	12.7	1	1.04×10^{-2}	15.2
6	1400	2	11	12.7	1	$.72\times10^{-2}$	21.9

d_p=650 μ ;c_{so}=7.8×10^{-4}N HCl ;D_s=4.5cm ;T=25C ;N=500 rpm

Resin was a Duolite A 102D (Diaprosim)

4.2 Diffusion Inside Particles

The question of transport phenomena inside particles is quite complex because we need to know how the ion exchanger structure is in order to develop realistic models.

Let us illustrate these ideas with models for adsorption in batch and continuous perfectly mixed adsorbers for two extreme cases: gel type resins (homogeneous diffusion model) and porous adsorbents ,monodisperse (pore diffusion model).

Table IV shows the model equations for Homogeneous Diffusion Model,while in Table V we present model equations for the Pore Diffusion Model applied to both systems:batch and CSTR adsorbers.

Now we will take one run for phenol adsorption in adsorbent resin Duolite ES861 (25).Experimental points for a batch operation

TABLE IV - HOMOGENEOUS DIFFUSION MODEL

Model equations:

Mass balance in the particle

$$\frac{\partial q(r,t)}{\partial t} = \frac{1}{r^2}\frac{\partial}{\partial r}\{ r^2 D_I \frac{\partial q(r,t)}{\partial t}\}$$

Average particle concentration

$$\bar{q}(t) = \frac{3}{r_p^3}\int_0^{r_p} q(r,t)r^2\, dr$$

Equilibrium at the interface

$$q(r_p,t) = f(c(t))$$

Boundary conditions

$r=0$, $\partial q(r,t)/\partial r=0$

$r=r_p$,

Batch $\varepsilon\{c_o - c(t)\} = (1-\varepsilon)\bar{q}(t)$

CSTR $c_o = c + \tau\, dc/dt + (1-\varepsilon)/\varepsilon\ \ \tau d\bar{q}(t)/dt$

Initial conditions

$r < r_p$, $q(r_p,0)=0$

$r=r_p$,

Batch $q(r_p,0)=f(c_o)$

CSTR $q(r_p,0)=f\{c(0)\}$

---

r_p -particle radius

TABLE V - PORE DIFFUSION MODEL :MODEL EQUATIONS

Mass balance in the particle

$$\frac{\partial}{\partial t}\{\chi c_p(r,t)+q(r,t)\} = \chi \frac{1}{r^2}\frac{\partial}{\partial r}\{ r^2 D_p \frac{\partial c_p(r,t)}{\partial r}\}$$

Average particle concentration

$$\overline{\chi c_p(r,t)+q(r,t)} = \frac{3}{r_p^3}\int_0^{r_p}\{\chi c_p(r,t)+q(r,t)\} r^2 dr$$

Equilibrium

$$q(r,t) = f'\{c_p(r,t)\}$$

with $c_p(r_p,t)=c(t)$

Boundary conditions

$r=0,\quad \partial c_p(r,t)/\partial t = \partial q(r,t)/\partial t = 0$

$r=r_p,$

Batch $\quad \varepsilon\{c_o-c(t)\} = (1-\varepsilon)\{\overline{\chi c_p(r,t)+q(r,t)}\}$

CSTR $\quad c_o = c + \tau dc/dt + (1-\varepsilon)/\varepsilon\ \tau \frac{d}{dt}\{\overline{\chi c_p(r,t)+q(r,t)}\}$

Initial conditions

$r<r_p,\quad c_p(r,0)=0$

$r=r_p,$

Batch $\quad c_p(r_p,0)=c_o$

CSTR $\quad c_p(r_p,0)=c(0)$

χ - particle porosity ; c_p -concentration in the pores

carried out in a basket adsorber are plotted in Figure 16 as c/c_o versus time. The adsorber volume is 470 ml and each basket is of dimensions 7.5 x 3 x 1 cm. Experimental conditions were: phenol concentration at t=0, c_o=87.3 mg/1, d_p=0.06 cm and adsorber porosity ε=0.968.

By optimization we obtain with the HDM a value $D_I=1.2\times10^{-11}\,m^2/sec$ while we get with the PDM a value $D_p=1.4\times10^{-9}\,m^2/sec$. Predicted curves are also shown in Figure 16.

It is obvious that if one carefully look at short times then the conclusion is that the models are not good enough to describe what is going on inside the solid.

This brings us to the need of more complete models to describe particle structure (26). This is particularly true for macroreticular resins in which a bead can be considered as an ensemble of microparticles with large pores between them. This suggests mechanisms of diffusion which should be association of the models described above either in series or in parallel.

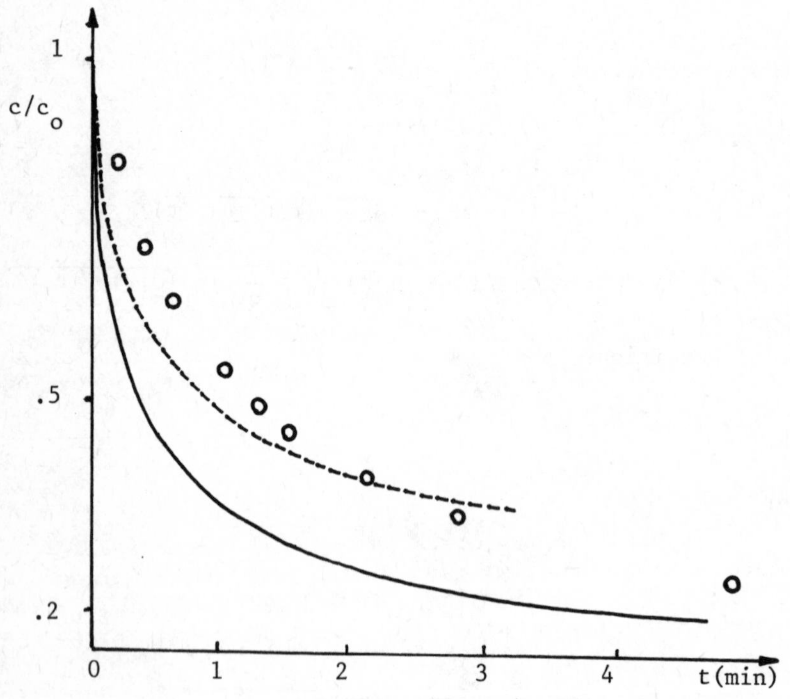

Figure 16- Comparison between experimental results and model predictions; —— homogeneous diffusion model and ---- pore diffusion model

In applications of macroreticular ion exchangers as catalysts the particle structure implies a redifinition of the catalyst effectiveness factor.In that case an overall catalyst effectiveness factor can be defined,for n^{th} order irreversible reactions as:

$$\eta_{ov}=\frac{3}{\phi_a^2}\left(\frac{dy_a}{dx_a}\right)_{x_a=1}=\{(1-\gamma)\bar{\eta}_i+\gamma\}\eta_a \qquad |72|$$

where

ϕ_a -Thiele modulus for the macroparticle
y_a -reactant concentration in macroparticle at radius x_a.
γ -fraction of active sites at the surface of microparticles

and η_i,η_a are the microeffectiveness and macroeffectiveness factors, respectively :

$$\eta_i=\frac{3}{\phi_i^2\, y_a^n}\left(\frac{dy_i}{dx_i}\right)_{\substack{x_a\\ x_i=1}} \qquad |73|$$

$$\eta_a=\frac{3}{\bar{M}^2}\left(\frac{dy_a}{dx_a}\right)_{x_a=1} \qquad |74|$$

with y_i -concentration in the microparticle at radius x_i,ϕ_i -Thiele modulus defined for the microparticle and $\bar{M}^2=\phi_a^2\{\gamma+(1-\gamma)\bar{\eta}_i\}$ .An important industrial application of this type of resins as catalysts is the production of MTBE.

It is obvious that for reaction order different from 1 numerical solutions should be required and is particularly important at this pointto be familiar with collocation methods.Such methods are appropriate to solve boundary value problems in PDEs and ODEs;let us consider the following ODE:

$$L\,u=g \qquad |75|$$

where L is a linear differential operator.The solution is approximated by a trial function v,which is a linear combination of basis functions,$\psi_1,\psi_2,\ldots,\psi_n$,i.e.,

$$u\simeq v=\sum_{j=1}^{n} c_j\psi_j \qquad |76|$$

The technique requires the determination of c_i such that Eq.(75) is satisfied exactly at n collocation points ,x_i=1,2,..,n.Then we get a system of n simultaneous linear equations whose coefficient matrix contains elements $a_{ij}= L\,\psi_j(x_i)$.If the basis functions are orthogonal polynomials the method is called orthogonal collocation.

Let us consider diffusion and first order reaction in a slab catalyst.

The model equations are:

$$\frac{d^2f}{dx^2} - \phi^2 f=0 \qquad |77a|$$

$$x=0, df/dx=0 \qquad |77b|$$

$$x=1, f=1 \qquad |77c|$$

or ,changing variables, $u=f-1$

$$\frac{d^2u}{dx^2} - u = 1 \qquad |78a|$$

$$x=0, du/dx=0 \qquad |78b|$$

$$x=1, u=0 \qquad |78c|$$

assuming that Thiele modulus is one.

Let us consider a basis function

$$\psi_j=(1-x^2)\ P_{j-1}(x^2)$$

and

$$v=(1-x^2)\sum_{j=1}^{n} c_j P_{j-1}(x^2)$$

Polynomials should satisfy the orthogonality condition, such as $P_o(x^2)=1; P_1(x^2)=1-5x^2$ with $x_1=0.447214$,$P_2(x^2)=1-14x^2+21x^4$ with roots $x_1=0.285232, x_2=0.765055$.

In the example above we have for one-point collocation, $v=c_1(1-x^2)$ and substituting in Eq.(78a) we get $c_1=-0.355$.Then the solution is:

$$f=1-0.355(1-x^2) \qquad |79|$$

Using now two-point collocation, $v=(1-x^2)\{c_1+c_2(1-5x^2)\}$; substituting in Eq(78a) we get for the collocation points $x_1=0.285232$ and $x_2=0.765055$ a system of equations:

$$-2.91865\ c_1 - 7.63123\ c_2 = 1$$

$$-2.4147\ c_1 + 23.9169\ c_2 = 1$$

Then $c_1=-0.35$,$c_2=0.0056$ and the solution is finally:

$$f=1+\{-0.35+0.0056(1-5x^2)\}(1-x^2) \quad |80|$$

Now let us compare one-point collocation, two-point collocation and analytical results, f=cosh x/(cosh 1):

x	Analytical solution	One-point collocation	Two-point collocation
0	.648	.645	.656
0.25	.668	.667	.675
0.50	.731	.734	.737
0.75	.839	.844	.843
1	1	1	1

5 APPLICATION TO WATER DEMINERALISATION

We will present in this section some results concerning the anionic step of water demineralisation and show how to apply part of the material reported here.

In general the feed of the anionic column is a mixture of acids (HCl, H_2SO_4..) and eventually a leakage from the cationic column(Na^+) can also be found.

What is needed to predict the useful capacity in cyclic regime is first of all a complete saturation curve and a complete regeneration curve. We start with the exchange Cl^-/OH^-; in Figure 17 we show different saturation fronts at different levels of column regeneration. Now from saturation fronts, at different flowrates, in column completely regenerated we get a curve of saturation factor, F_s as a function of the saturation potential, τ_s (Figure 18). The experimental conditions are shown in Tables VI and VII, respectively.

In Figure 18 we also show the reduction in the breakthrough point due to the increase of flowrate.

Similar curves should be obtained for the regeneration step. Figure 19 shows regeneration front in differently saturated columns. From the complete regeneration curve in a completely saturated bed we get F_r versus τ_r as shown in Figure 20. We should say that this curve is not very dependent on flowrate (during regeneration equilibrium conditions prevail).

Let us try a sample calculation; for reasonable practical conditions τ_r=1.51 and τ_s=0.745; from Figures 18 and 20 we read, respectively F_s=0.745 and F_r=0.84. Then Y_r=0.876 and the useful capacity is finally U=0.652 or 0.896 eq/liter of resin.

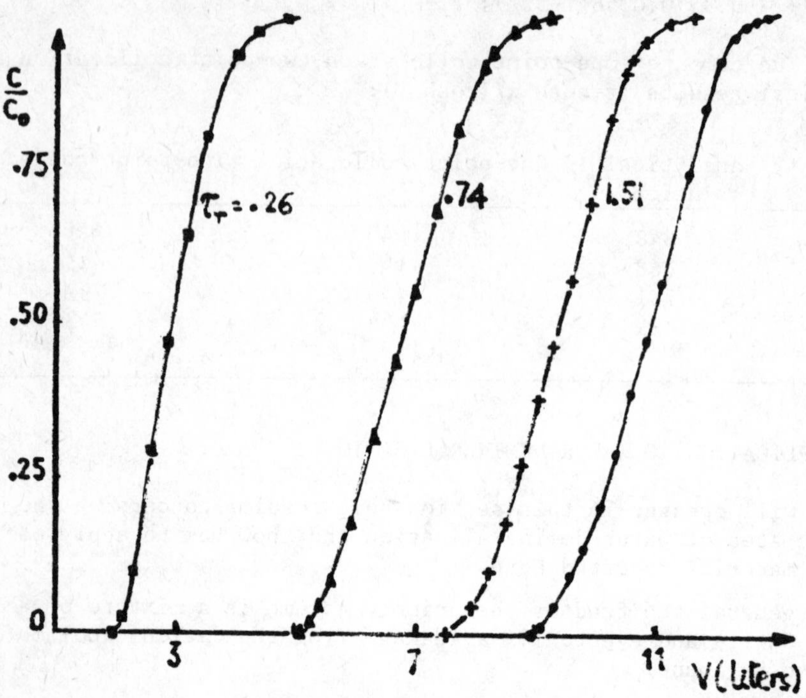

Figure 17- Exchange Cl^-/OH^- :saturation fronts at different regeneration levels
• saturation front after complete regeneration

TABLE VI - Experimental conditions for runs of Figure 17

Saturant : HCl 0.01N

Flowrate :4.76 ml/min

Regenerant: NaOH 1N

Column: L=15.5 cm, A=5.3 cm^2

Resin :Duolite A102D

Data from Diaprosim are presented in terms of useful capacity U in meq/ml of bed versus a certain regeneration potential, T_r, expressed as gr of NaOH 100%/liter of bed. Our conditions correspond to T_r=85 and then from Figure 21 we get U=0.88 eq/liter resin, which is in good agreement with our predictions.

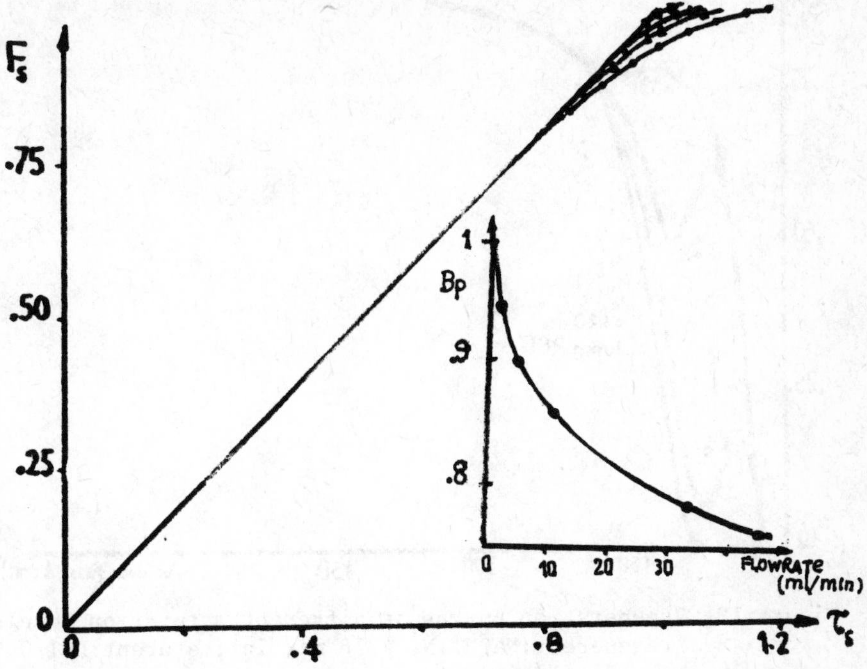

Figure 18- Exchange Cl^-/OH^- : F_s versus τ_s at different flowrates

TABLE VII - Experimental conditions for runs of Figure 18

Resin :Duolite A102D ,ROH^-

Saturant:HCl $1.5x10^{-2}N$

Flowrate:

- ■ 2.65 ml/min
- \+ 4.76 "
- ▲ 10.60 "
- ● 45.50 "

It should be said that our results were obtained in a laboratory column with L_1=15.5 cm;scale-up can be made if stationary front hypothesis is accepted.The useful capacity in large units,U_2 is rela-

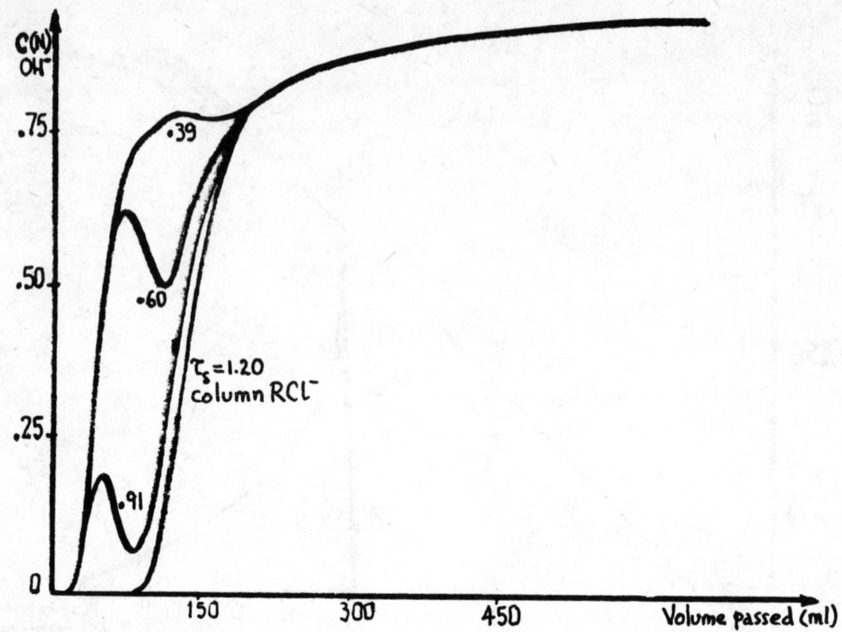

Figure 19- Regeneration curves at different saturation levels (regenerant:NaOH 1N, 4.76 ml/min ;saturant:HCl 0.01N at 4.76 ml/min)

ted with the useful capacity at the laboratory scale, U_1 by :

$$U_2 = U_1 \frac{L_1}{L_2} \frac{L_2 - Z_e/2}{L_1 - Z_e/2} \qquad |81|$$

What happens now if sulphate ions are present in the feed?We are faced with a ternary ion exchange,still accompanied by neutralization reaction.

The first remark is that the capacity of the anionic column increases with the concentration of sulphuric acid in the feed stream. The reasons for this can be found through several arguments:

a) Sulphuric acid in solution can be present in three forms: nondissociated acid,bisulphate ion HSO_4^- and sulphate ion $SO_4^=$ depending on the sulphuric acid concentration.In fact we have

$H_2SO_4 \rightleftarrows HSO_4^- + H^+$, $K_1 \simeq 10^3$

$HSO_4^- \rightleftarrows SO_4^= + H^+$, $K_2 \simeq 0.01$

At low acid concentrations, $SO_4^=$ is the main species while at

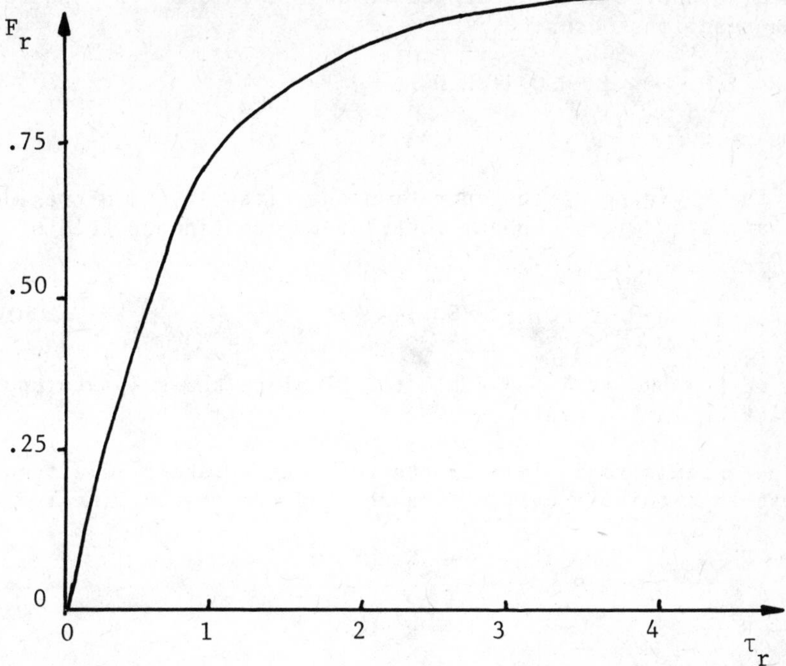

Figure 20 - Regeneration factor, F_r versus regeneration potential, τ_r

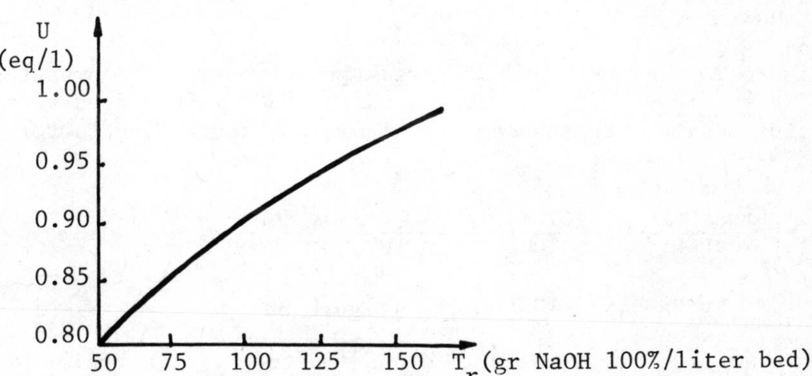

Figure 21- Useful capacity, U versus a regeneration potential, T_r (data from Diaprosim).

high concentrations of acid, bisulphate ion predominates.

b) when sulphuric acid is passed through a column ROH^- the following reactions occur:

$$ROH^- + H^+HSO_4^- \rightleftarrows RHSO_4^- + H_2O$$

$$2ROH^- + 2H^+ SO_4^= \rightleftarrows (R)_2SO_4^= + 2H_2O$$

At the begining of the operation the first layers of resin become saturated with sulphate ions; but the continuous feed of acid leads to:

$$R_2SO_4^= + 2\ HSO_4^- \underset{K_3}{\rightleftarrows} SO_4^= + 2RHSO_4^-$$

If we know K_3 it is possible to calculate the sulphate equivalent fraction in the resin.

c) the anionic resin in sulphate form can uptake acid without exchange; this is called site sharing phenomenon since:

$$\begin{matrix} -R^+ \\ -R^+ \end{matrix} \Big\rangle SO_4^= + H_2SO_4 \rightarrow \begin{matrix} -R^+\text{-}HSO_4^- \\ -R^+\text{-}HSO_4^- \end{matrix}$$

These points can explain why measured capacities increase with sulphuric acid concentration in the feed.

However the main factor to be considered is the influence of the ratio $Cl^-/SO_4^=$ in the feed. Experimental data were obtained using a total concentration $c_T=9.66\times10^{-2}N$ and a flowrate of 18 ml/min in all runs.

Figure 22 shows Cl^- and $SO_4^=$ breakthrough curves. The breakthrough point (in volume or time) increases as the $\%SO_4^=$ in the feed increases; so for design purposes we can calculate a correction factor, f_1 defined as

$$f_1 = \frac{\text{breakthrough volume for a } Cl^-/SO_4^= \text{ feed}}{\text{breakthrough volume for a 100\% } Cl^- \text{ feed}}$$

and plot f_1 versus $\%Cl^-$ in the feed (Figure 23)

Experiments were also carried out to study cyclic regime. Starting with a column saturated with sulphuric acid 0.023N we used NaOH 1N as regenerant and a feed $Cl^-/SO_4^= =50/50$ at total concentration 0.05N as saturant; the co-ion was H^+ (Figure 24)

Table VIII summarizes the experimental conditions used in the run.

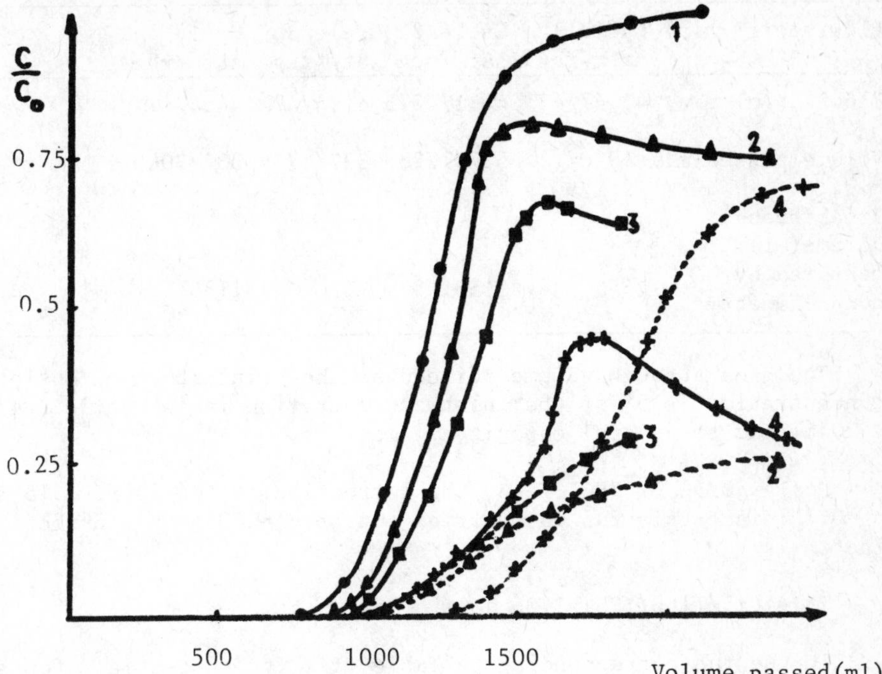

Figure 22 - Influence of the ratio $Cl^-/SO_4^=$ in the feed on the breakthrough curves

1- 100%Cl^- ,2-$Cl^-/SO_4^=$ =70/30 , 3- $Cl^-/SO_4^=$=50/50
4- $Cl^-/SO_4^=$ =25/75

——Cl^- ; ----- $SO_4^=$

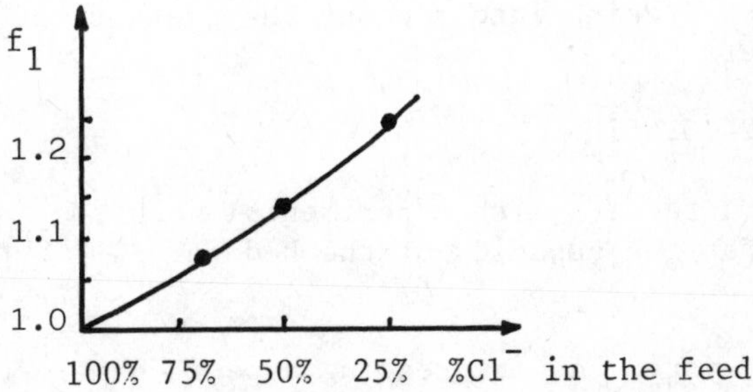

Figure 23- Correction factor,f_1 versus Cl^- % in the feed

TABLE VIII- Cyclic run:experimental conditions

Flowrate(ml/min)	Cycle 1 sat /reg	Cycle 2 sat /reg	Cycle 3 sat/ reg	Cycle 4 sat /reg
Flowrate(ml/min)	43.4/ 47	46.1/45.5	44.4/47	44.4/47
Volume passed(ml)	1910/ 1790	337/1728	337/1776	337/3000
Breakthrough volume(ml) measured by conductimetry	1455	1155	1170	1185

The breakthrough volume defined as the point at which outlet concentration is 5% of the inlet concentration is,in cyclic regime, 1420 ml and the useful capacity 71 meq.

For a 50/50 $Cl^-/SO_4^=$ feed, the correction factor is f_1=1.15 at c_T=0.1N;since this run was carried out at c_T=0.05N the correction factor is:

f_1^* =1.15/(150/138)=1.06

In the run corresponding to Table VIII ,the potentials for saturation and regeneration were

τ_s=0.633 ; τ_r=2.65

and then

F_s = 0.633 ; F_r=0.95

Finally U=0.62 ;taking into account the presence of sulphate ions in the feed we get:

$U^* = f_1^* \times 0.62 = 0.657$

which is in good agreement with experimental value,U_{exp}=0.645(we should mention that the capacity of the bed for 100% Cl^- feed is 112 meq).

This ternary system has interesting aspects namely the existence of inversions due to the shape of equilibrium isotherms.This can lead to practical implications on the regeneration efficiency,according to the ratio $Cl^-/SO_4^=$.

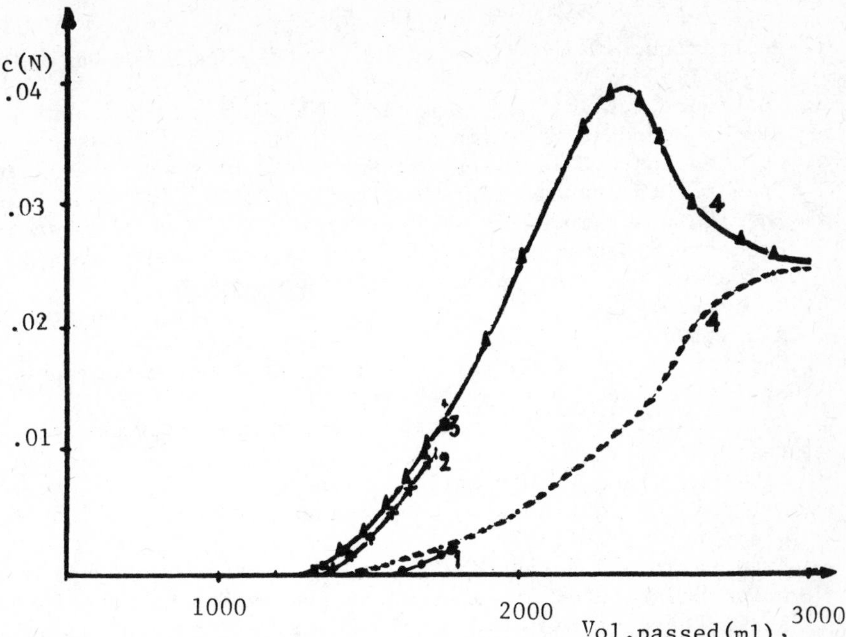

Figure 24- Ternary ion exchange $Cl^-/SO_4^=/OH^-$:cyclic regime. Feed $Cl^-/SO_4^=$ =50/50;cyclic regime is reached in 3-4 cycles. ——Cl^- ; ----$SO_4^=$

We can conclude by saying that dynamics of ion exchange can be understood if realistic models are developed;namely particle structure should be carefully considered.Models become complex ;however developments in numerical techniques make the solution possible in most cases.Computer aided design will then be more and more used in cyclic ion exchange processes.The applications of ion exchangers in catalysis also open interesting studies on catalyst effectiveness factors,as shown before.

ACKNOWLEDGEMENT

Anionic resin (Duolite A102D) and adsorbent resin (Duolite ES861) used in the experiments described here are produced by Diaprosim (Chauny-France).We gratefully acknowledge Diaprosim by providing us with free samples.

Some material presented here is part of the thesis by C.Costa and J.Loureiro ,now in preparation.

REFERENCES

1. Helfferich,F.G.The Promises of Ion Exchange,Chem.Eng.Prog. 73(1977)53-55
2. Rodrigues,A.E.Ion Exchange ,AIChEMI Modular Instruction, Series B,module B5.4 (in press)
3. Gold,H. and C.Calmon.Ion Exchange-Present Status,Needs and Trends.70th Annual AIChE Meeting,New York(1977)
4. Rodrigues,A.E.Modeling of Percolation Processes in Percolation Processes:Theory and Applications,edited by Rodrigues,A. and D.Tondeur (Sijthoff & Noordhoff,1981)
5. Tondeur,D.Théorie des colonnes d'échange d'ions,Thèse,Univ. Nancy(1969)
6. Helfferich,F. and G.Klein.Multicomponent Chromatography:Theory of Interference (Marcel Dekker,1970)
7. Rhee,H.K.Studies on the Theory of Chromatography,Univ.Minnesota(1968)
8. Schlogh,R. and F.Helfferich.Comment on the Significance of Diffusion Potentials in Ion Exchange Kinetics.The Journal of Chem.Physics 26(1957)5
9. Kataoka,T. and H.Yoshida.Effect of Electric Field and Equilibrium Relation on Breakthrough Curve in Ion Exchange Column: Resin Phase Diffusion Control.Journal of Chem.Eng.Japan 11 (1978)408-410
10. Van Brocklin,L. and M.David.Coupled Ionic Migration and Diffusion During Liquid-Phase Controlled Ion Exchange.Ind.Eng.Chem. Fund.11(1972)91
11. Michaels,A.Simplified Method of Interpreting Kinetic Data in Fixed Bed Ion Exchange.Ind.Eng.Chem.44(1952) 1922
12. Moison,R. and H.O'Hern.Ion Exchange Kinetics.Chem.Eng.Prog. Symp.Series.55(1959)71
13. Conder,J. and N.Fruitwala.Comparison of Plate Numbers and Column Lengths in Chromatography and Distillation.Chem.Eng. Science.36(1981)509-513
14. Valentin,P.Separation of Components by Gas Liquid Chromatography.Elf Solaize Research and Development Center(1978)
15. Thomas,H.Chromatography:a Problem in Kinetics.Ann.N.Y.Acad. Sci.49(1948)161
16. Rosen,J.Kinetics of a Fixed Bed System for Solid Diffusion into Spherical Particles.J.Chem.Phys.20(1952)387
17. Rodrigues,A.E.Application des Méthodes du Génie Chimique à l'Échange d'Ions.Thèse Univ.Nancy(1973)
18. Loureiro,J.,Costa,C. and Rodrigues,A.Propagation of Concentration Waves in Fixed Bed Adsorptive Reactors.Submitted to AIChEJournal(1982)
19. Safonov,M.A More Precise Criterion For The Formation of a Steady-State Sorption Front.Separation Science.6(1971)35

20. Dodds,J. and Tondeur,D.The Design of Cyclic Fixed Ion Exchange -Part I.Chem.Eng.Science.27(1972)1267-1281
21. Tondeur,D.Theorie des Colonnes d'Echange d'Ions.Chimie et Industrie-Génie Chimique.100(1968)1058-1067
22. Wesselingh,J.A. unpublished results
23. Levins,D. and Glastonbury,J.Particle Fluid Mass Transfer in a Stirred Vessel.The Trans.of Inst.Chem.Engrs.London.50(1972) 132
24. Helfferich,F.Ion Exchange Accompanied by Reactions. J.Phys. Chem.69(1965)1178
25. Costa,C. and Rodrigues,A. Adsorption of Phenol on Adsorbent Resins and Activated Carbon:Equilibrium and Kinetic Studies in Batch and Open Systems.in Adsorption at the Gas-Solid and Liquid-Solid Interfaces. ed.J.Rouquerol and K.Sing (Elsevier, 1982)
26. Weatherley,L. and Turner,J.Ion-Exchange Kinetics-Comparison Between a Macroporous and a Gel Resin.Trans.Instn.Chem.Engrs. 54(1976)89-94

SIMPLIFIED APPROACH TO DESIGN OF FIXED BED ADSORBERS

Roberto Passino

Istituto di Ricerca Sulle Acque
Consiglio Nazionale delle Ricerche
1,via Reno,00198 Roma,Italy

1.INTRODUCTION

A comprehensive treatment of mass transfer in ion exchange and adsorption processes,based,among others,on Thomas(1) and De Vault(2) theories,has been given by Hiester,Vermeulen and Klein(3) for continuous fixed bed and batch operations,under both favorable and unfavorable equilibrium conditions.Complete analytical solutions for real systems,however,were provided only when limiting boundary conditions are assumed (e.g.,constant separation factor,linear equilibrium,etc.).

To reduce extensive experimentation and boring calculation,use is often made of the simplified concept of the 'exchange zone',through which design data can be approximated fairly good,as shown by Michaels(4) and Lukchis(5).This latter approach,however,can only be applied to systems with favorable equilibrium.

In previous papers(6,7) we introduced a simple treatment which permits discontinuous adsorption and ion exchange processes with both favorable and unfavorable equilibrium to be approximated to counter-current,continuous ones,so that process performance and basic design data (HTU,NTU,mass transfer coefficients,etc.) are easily obtained.The method proved particularly useful for the design of pilot and large scale plants based on the DESULF process,where sulphates in sea water are ion-exchanged with chlorides by weak anion exchange resins to prevent $CaSO_4$ scale in evaporation plants.(6,7).

In this lecture the general characteristics of this method are illustrated and applied to the solution of some examples of general validity.

2.DESCRIPTION OF THE METHOD

Let us consider a simple fixed-bed liquid-solid mass transfer process such as binary ion exchange,i.e.,the exchange of ion B in solution by a resin (R) in A form, represented by the reaction

$$RA + B \rightleftarrows RB + A \tag{1}$$

Neglecting minor operations (backwash,rinse,etc.), essentially two operations alternate discontinuously in this process:resin exhaustion and regeneration.During exhaustion a solution containing the species B to be exchanged is fed at a certain flow rate to a column containing a given volume of exchanger in A form.This operation is discontinued when the capacity of the bed for that ion has been exhausted.(Usually this point,called the breakthrough point,corresponds to reaching a prefixed leakage, say 5 to 10%,of the exchanged species in the effluent). Regeneration is then performed:a concentrated solution of A species is fed at a certain flow rate to the column to restore it to its original condition,in preparation for the next cycle.

Column performance in these processes is commonly described in terms of effluent concentration histories (or breakthrough curves;see Fig.1),displaying the concentration of B in the effluent vs the volume V of fluid that has passed through the column (or vs time,if the flow rate is constant).The average composition of the effluent can be easily calculated by graphic integration of these curves.By reference to Fig.1A,integrating the breakthrough curve between the extremes 0 and V one obtains the area A_2,which is equal to the area A_1 that would have been obtained if an effluent with a constant average concentration X_2C had been exited from the bed throughout the run.

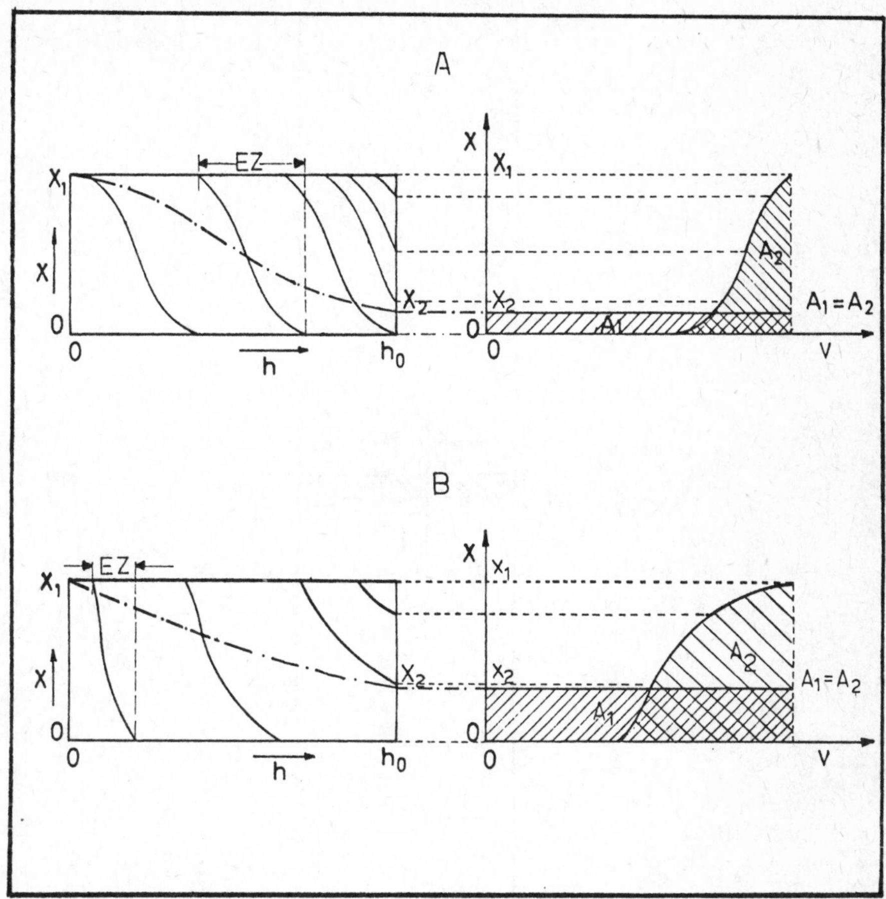

Fig.1. Schematic representation of concentration profiles inside the bed during a fixed bed ion exchange process (left) and corresponding breakthrough curve of the effluent(right).

A: fixed-bed process with favorable equilibrium

B: fixed-bed process with unfavorable exchange equilibrium

EZ: exchange zone.

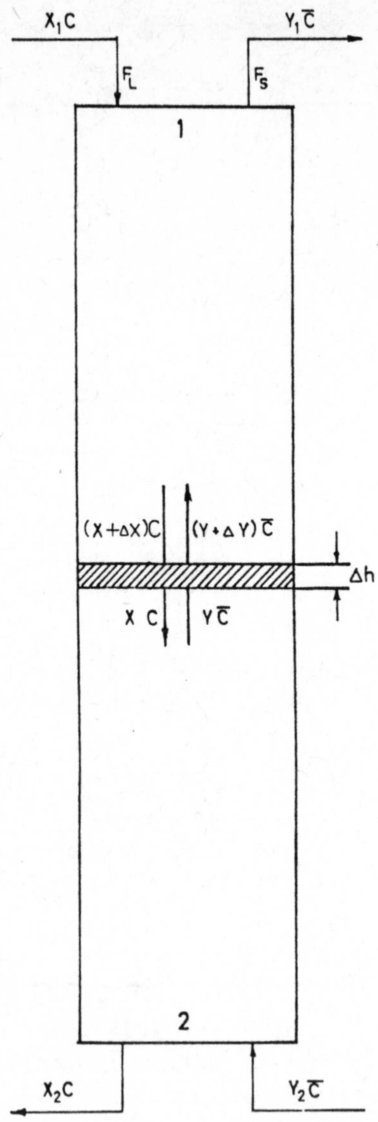

Fig.2.Schematic representation of continuous counter-current unit.

The average is given by

$$X_2 = \frac{\int_0^V X \, dV}{V} \qquad (2)$$

Let us now make the following assumptions:

1) in each operation the fixed phase moves countercurrently to the fluid phase at a relative flow rate given by the ratio between the bed volume and the operation time
2) the effluent through the operation has a constant composition X_2C corresponding to the average calculated by graphic integration of the breakthrough curve.

These assumptions correspond to an idealized process in which the solid phase, as soon as it becomes exhausted, is withdrawn continuously from one end of the contact apparatus, being replaced by fresh material at the opposite end. The second assumption corresponds to the use of a suitable holding tank, in which all the effluent is collected and mixed before being released.

With these assumptions the fixed bed process can be treated as a continuous countercurrent one, whatever the nature of the equilibrium involved, the concentrations at the extreme sections of the contactor being constant with time.

Assuming that composition changes in the intraparticle liquid are irrelevant, the following mass balance can be written for any finite portion of the contact bed (see Fig. 2)

$$VC\Delta X = v\bar{C}\Delta Y \qquad (3)$$

Introducing the (apparent) solid and fluid phase flow rates

$$F_S = v/t_o \quad \text{and} \quad F_L = V/t_o \qquad (4)$$

eq. 3 may be written as

$$\frac{Y_1 - Y}{X_1 - X} \left(\text{or } \frac{Y - Y_2}{X - X_2}\right) = \frac{CV}{\bar{C}v} = \frac{C\,F_L}{\bar{C}\,F_S} \qquad (5).$$

This represents the operating line (OL) of the proces considered,with slope proportional to the specific output V/v (i.e.,volume of solution treated per volume of bed), or to the ratio of flow rates,F_L/F_S,for a given ratio of total solution concentration,C, to total resin exchange capacity,$\bar{C}$.These quantities are usually constant during each ion-exchange operation,and so is the slope of the OL representing that operation.

The graphic description of the process greatly sim plifies the solution of mass transfer problems.First of all,process feasibility for any given set of conditions can be ascertaind by considering the position of the OL with respect to the equilibrium line(EL,or equilibrium isotherm),as driving forces are positive if the OL lies below the corresponding EL during exhaustion and above during regeneration.The optimization of operating variables is also greatly simplified.

3.APPLICATION OF THE OL METHOD TO WATER SOFTENING

For a better illustration of this method,a binary ion exchange such as water softening,(i.e.,Ca^{++} removal from a hard water by a fixed bed of a cation resin,regen erated by concentrated NaCl solution,depicted in Fig.3), may serve as an example.

The upper part of the diagram shows the fixed bed contactor containing a volume v of resin to which during the exhaustion (left side) a volume V_{ex} of a hard water (total solution concentration C_o) is fed at a flow rate $(F_L)_{ex}$ for an exhaustion time t_{ex},so that its average Ca^{++} equivalent fraction is reduced from X_{ex1} to X_{ex2}. Correspondingly,the average Ca^{++} equivalent fraction in the resin (total exchange capacity $\bar{C}$) increases from Y_{ex2} to Y_{ex1}.In the model,fresh resin is continuously supplied throughout the operation to the contactor from the left (opposite to the solution entrance) at a flow rate $(F_S)_{ex}$ $= v/t_o$.

Similar considerations apply during the regeneration operation (right side of Fig.3),performed with a volume V_{reg} of concentrated NaCl solution (total concentration $C_{reg} \gg C_o$),fed at a flow rate $(F_L)_{reg}$ for a time t_{reg}.

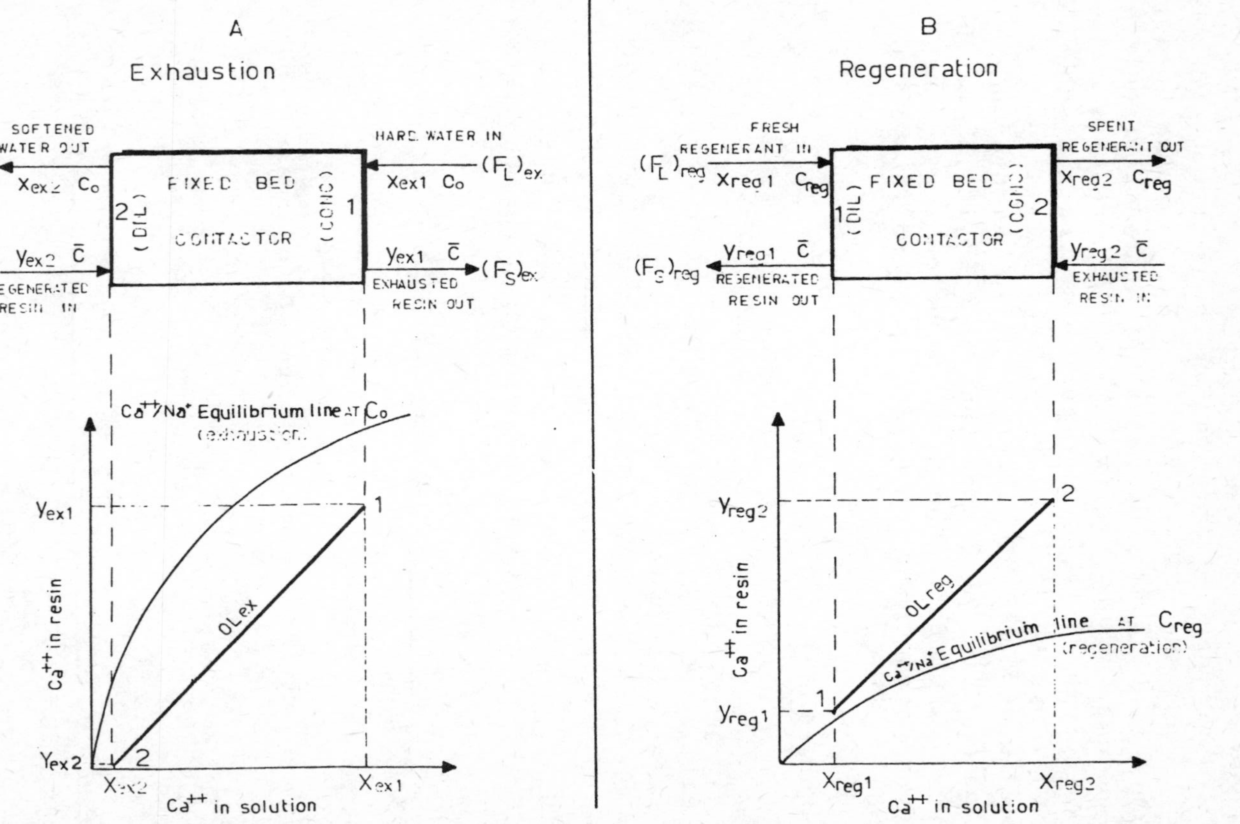

Fig.3.Representation of softening of a hard water on ideally continuous bed of strong cation resin in Na cycle.(DIL and CONC are contactor diluted and concentrated ends).

At the end of regeneration operation, the average Ca^{++} equivalent fraction has increased from X_{reg1} to X_{reg2} in solution and decreased from Y_{reg2} to Y_{reg1} in the resin phase.

The lower part of the diagram shows the OLs for these operations and the ELs for the Ca^{++}/Na^{+} exchange on a cation resin. EL, usually described by Henry's, Langmuir's or other equations of the type $Y^* = f(X)$, presents a peculiarity: typically, in heterovalent exchanges (such as Ca^{++}/Na^{+}) resin selectivity toward the polyvalent ion decreases with the ionic strenght of solution (8). Hence, the EL at C_{reg} is lower than that at C_o, lying above the corresponding OL for exhaustion, and below for regeneration.

Coordinates of the OL for these operations are easily obtained.

A. Exhaustion

$$X_{ex1} = X_o \qquad \text{(Ca in the feed)} \tag{6a}$$

$$Y_{ex1} = Y_{ex2} + (X_{ex1} - X_{ex2}) \frac{C_o V_{ex}}{\bar{C} v} \tag{6b}$$

(Ca in the exhausted resin)

$$X_{ex2} = \frac{\int_0^{V_{ex}} X dV}{V_{ex}} \qquad \text{(average Ca in the effluent)} \tag{6c}$$

$$Y_{ex2} \qquad \text{(Ca in the regenerated resin)} \tag{6d}$$

and slope

$$\left(\frac{\Delta Y}{\Delta X}\right)_{ex} = \frac{Y_{ex1} - Y_{ex2}}{X_{ex1} - X_{ex2}} = \frac{C_o V_{ex}}{\bar{C} \cdot v} = \frac{C_o}{\bar{C}} \left(\frac{F_L}{F_S}\right)_{ex} \tag{7}$$

The value of C_o,$\bar{C}$ and v is given,so that for any V_{ex} value the slope of OL is determined by eq.7.As for the coordinates,X_o is also given,X_{ex2} can be obtained by graphic integration from the breakthrough curve and Y_{ex2} is usually specified (regeneration procedures are set up normally to give a residual amount of Ca in the regenerated resin,Y_{ex2},below 2-5%,a value considered acceptable for most practical applications).The OL for exhaustion is thus completely defined.

B.Regeneration

Assuming that no chemical changes occur in the interval between exhaustion (or regeneration) and regeneration (or exhaustion),the coordinates of the end points of the OL for regeneration are

$$X_{reg1} \quad \text{(Ca in fresh regenerant)} \tag{8a}$$

$$Y_{reg1} = Y_{ex2} \quad \text{(Ca in the regenerated resin)} \tag{8b}$$

$$X_{reg2} = X_{reg1} + \frac{(Y_{reg2} - Y_{reg1})\bar{C}v}{C_{reg}V_{reg}} \tag{8c}$$

(Ca in the spent regenerant)

$$Y_{reg2} = Y_{ex1} \quad \text{(Ca in the exhausted resin)} \tag{8d}$$

while the slope is given by

$$\left(\frac{\Delta Y}{\Delta X}\right)_{reg} = \frac{Y_{reg2} - Y_{reg1}}{X_{reg2} - X_{reg1}} = \frac{C_{reg}V_{reg}}{\bar{C}\,v} = \frac{C_{reg}}{\bar{C}}\left(\frac{F_L}{F_S}\right)_{reg} \tag{9}$$

The OL for regeneration can accordingly be easily drawn.

Once each operation has been characterized graphically by the proper OL and the corresponding EL (this latter taken from the literature or determined experimentally), several familiar graphic procedures for design may be applied.More generally,the many established theories,such as those concerned with linear driving force kinetics,the number of transfer units (NTU),the height of transfer units (HTU),the bed height (h_z),as described in chemical engineering textbooks(9),may become more readily applicable to cyclic fixed bed operations,like those described, when used with the present method.Of course,allowance must be made for the simplifications involved.

4.GENERAL CASES OF APPLICATION TO THE DESULF PROCESS

The practical validity of the OL method has been demonstrated during the design of a 100 m^3/d pilot plant based on the DESULF process(7).The interested reader is referred to refs.(6 and 7) for demonstration of detailed calculation of basic design parameters (NTU,HTU,h_z and mass transfer coefficient) of that process by means of teh OL methods.By reference to that same process,here we will demonstrate how the Ol method may be used to solve problems of general occurrance in ion exchange practice, while Fig.4 depicts graphically the DESULF process

A.First problem

At a given flow rate,define the optimal value of specific output $(V/v)_{ex}$ in the exhaustion step.

As shown by Fig.5,different OLs are obtained depending on the selected V_{ex} value.By reference to sections B and C of Fig.5,it is clearly seen that the slope of the OL representing this operation,still confined between the two ELs,arises from its minimum value,(corresponding to arresting the exhaustion at the initial sulphate breakthrough,$X_{ex2} \cong 0.05\ X_{ex1}$;$V_{ex} = V^b$),to the highest value, (when resin almost complete saturation is attained,$X_{ex2} \cong 0.95\ X_{ex1}$;$V_{ex} = V^s$).

Between these extreme cases the practical solution was adopted as the one giving a sulphate leakage ($X_{ex2} = X_{lim}$) not exceeding $CaSO_4$ solubility in the evaporation plant.

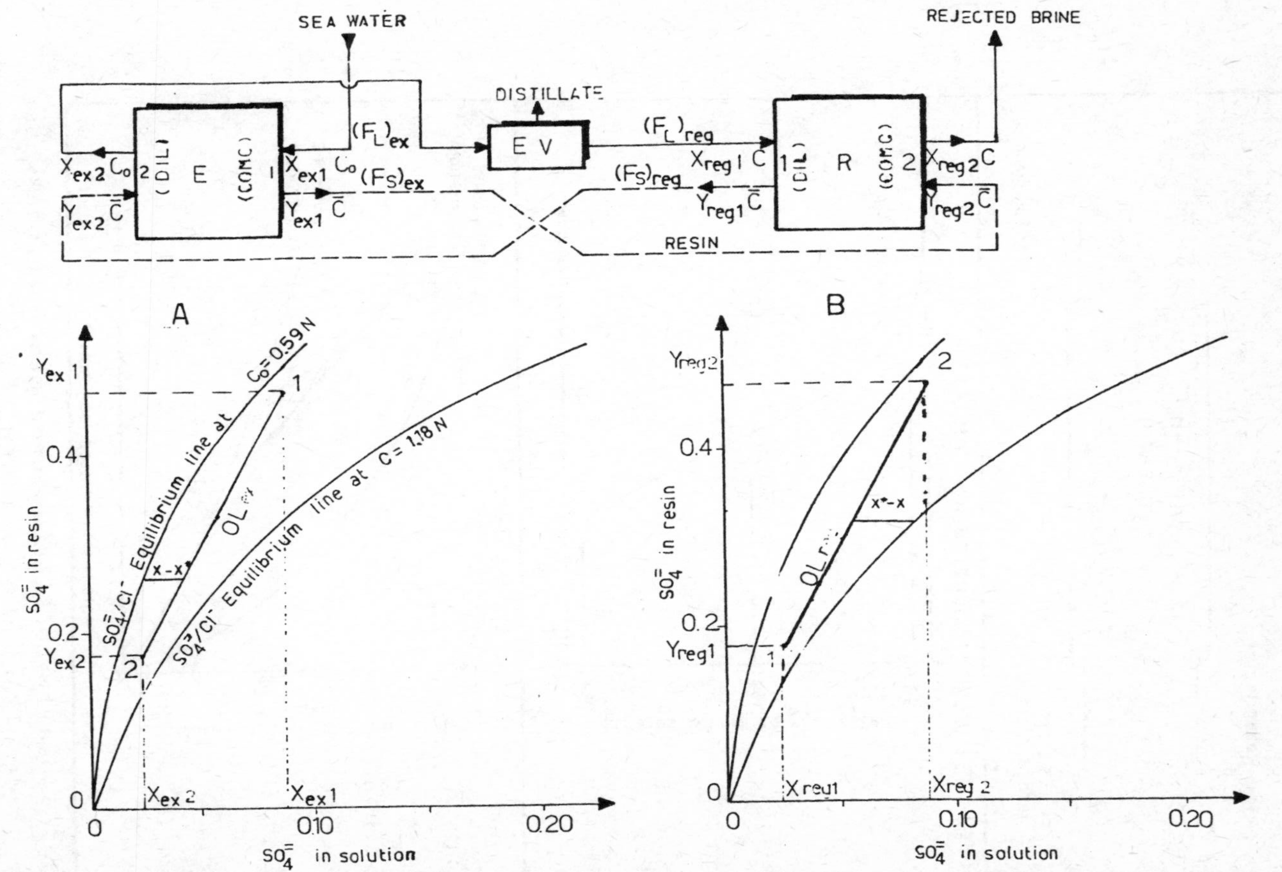

Fig.4.Representation of the combined ion exchange-evaporation DESULF process on ideally continuous resin bed.(E,exhaustion;R,regeneration;EV, evaporator)

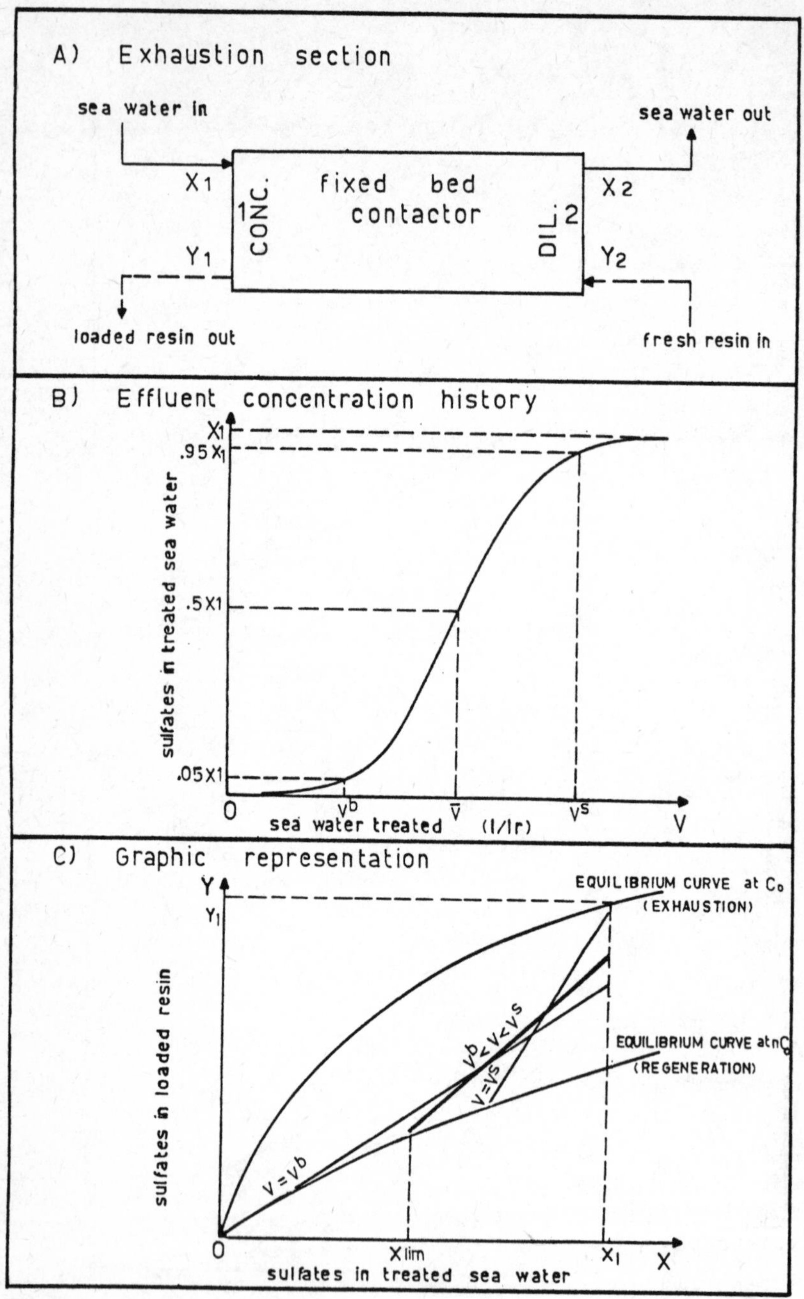

Fig.5.Ideally continuous treatment for sea-water desulfation on a fixed bed anion exchange resin Case of different throughputs.

B. Second problem

Optimize the exhaustion flow rate $(F_L)_{ex}$, still leading to resin almost complete saturation with sea water ($V_{ex} = V^s$).

As shown by Fig.6, now the corresponding OLs range from the vertical one, (corresponding to a shallow bed of resin fed at a very high solution flow rate, $(F_L)_{ex} \cong \infty$), to that passing through the origin (exhaustion performed in semi-equilibrium conditions, at a very low solution flow rate, $(F_L)_{ex} \cong 0; V_{ex} = \bar{V}$). Again, the maximum allowable flow rate should be that givin place to an average $X_{ex2} = X_{lim}$ compatible with $CaSO_4$ solubility in the operating conditions selected.

5. CONCLUSIONS

With the proposed method, cyclic exhaustion-regeneration operation of fixed-bed ion exchange processes under conditions of favorable and unfavorable equilibrium can be calculated. The method uses a model in which solid and fluid are viewed as contacting one another countercurrently and continuously at their respective average concentrations. Once the operating and the equilibrium lines for a given ion-exchange operation have been determined, the described method allows fixed bed performances over a wide range of operating conditions to be predicted and the basic kinetic parameters for bed design to be calculated with acceptable precision. This avoids tedious calculation and extensive laboratory experimentation usually required for unfavorable equilibrium processes (i.e., during regeneration of process with favorable exhaustion equilibrium) and provides the basic data for plant design with reasonable approximation.

The method has been successfully applied to seawater desulfation (DESULF process) by fixed bed ion exchange.

In addition to two major idealizations inherent in the model (use of average concentrations and continuous countercurrent movement of phases), the following simplifications were introduced: constancy of liquid phase flow rate and of total concentration of both phases, uniform saturation of the resin bed, uniform fluid distribution in the column, negligible concentration gradient in intra

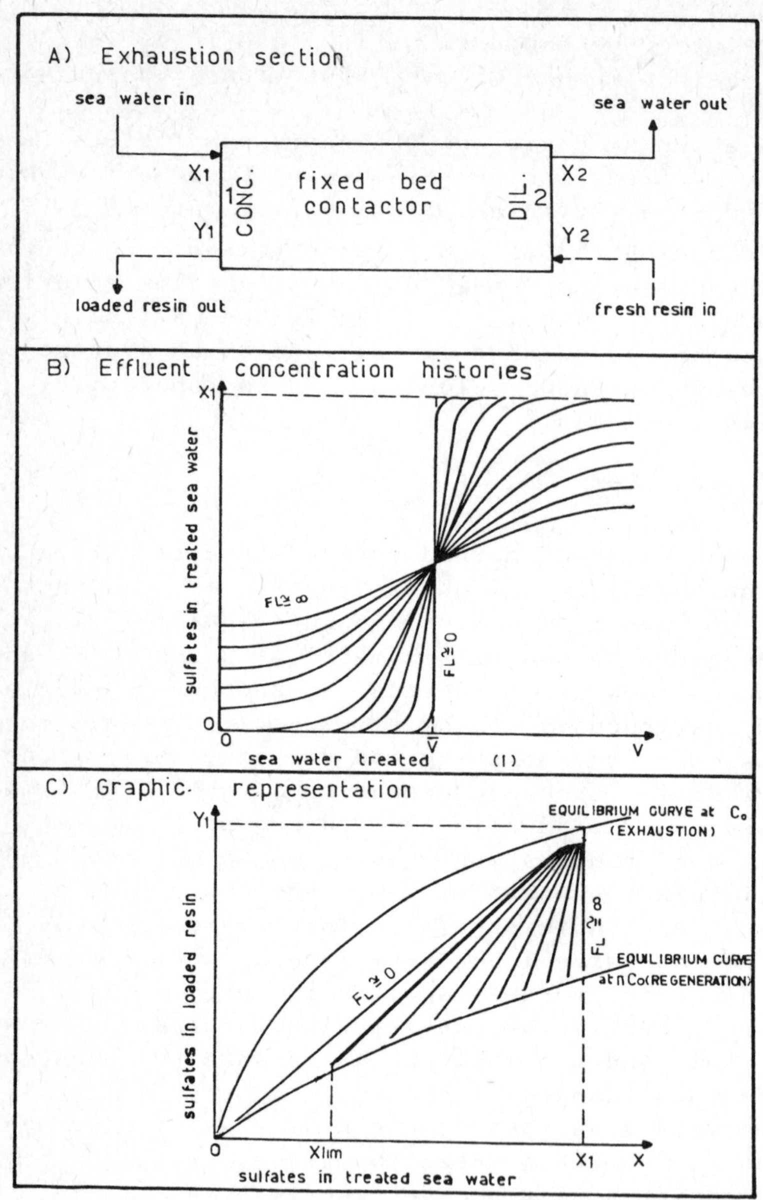

Fig.6. Ideally continuous treatment for sea-water desulfation on a fixed bed anion exchange resin. Case of different liquid flow rates.

particle liquid.

The method is expected to be of great utility for other fixed-bed mass transfer processes,such as multicomponent ion exchange or adsorption,provided the effect of the approximations is carefully accounted for.

REFERENCES

1.H.C.Thomas,J.Am.Chem.Soc.,66,1664(1944)
2.D.De Vault,J.Am.Chem.Soc.,65,532(1943)
3.N.K.Hiester,T.Vermeulen,G.Klein,Sect.16 in J.H.Perry's Chemical Engineers' Handbook,Int.Stud.Ed.,5th ed., McGraw-Hill,New York,N.Y.,1973
4.A.S.Michaels,Ind.Eng.Chem.,44,1922(1952)
5.G.M.Lukchis,Chem.Eng.,June 11(1973)111
6.G.Boari,L.Liberti,C.Merli,R.Passino,Env.Prot.Eng.6(1980) 251
7.L.Liberti,R.Passino,Ind.Eng.Chem.Proc.Des.Dev.,21,2 (1982)197
8.F.Helfferich,Ion Exchange,McGraw-Hill,New York,N.Y., 1962.(University Microfilms International:Ann Arbor, Mich.No.2003414)
9.J.H.Perry,Chemical Engineers' Handbook,Int.Stud.Ed., 5th ed.,McGraw-Hill,New York,N.Y.,1973,ch.16

LIST OF SYMBOLS

C = total solution concentration,equiv./l
$\bar{C}$ = total exchange capacity of resin,equiv./l_r
EL= equilibrium line (or equilibrium isotherm) describing a mass transfer process in equilibrium conditions at a given temperature
EZ= exchange zone
F_L= liquid phase flow rate,l/h
F_S= solid phase flow rate,l_r/h
h_z= height of the contact bed,cm

HTU = height of transfer unit, cm
n = concentration factor; i.e., ratio of regenerant to exhaustion solution concentration
OL = operating line describing specified process step
NTU = number of transfer units
t = time, h
t_o = operation time, h
V = volume of treated solution, l
$\bar{V}$ = 'equivalent volume', i.e., maximum solution volume that can be treated without leakage by one volume of resin in semi-equilibrium conditions, l
V^b = solution volume treated at breakthrough, l
V^s = solution volume treated at resin saturation, l
v = resin volume, l_r
X = equivalent fraction of the exchanged ion in solution
Y = equivalent fraction of the exchanged ion in resin
* = refers to equilibrium conditions
ex = refers to exhaustion
reg = refers to regeneration
o = refers to sea water conditions
1 = refers to the concentrated side of the contactor
2 = refers to the diluted side of the contactor
lim = indicates conditions of saturation with respect to $CaSO_4$ solubility

ION-EXCHANGE MEMBRANES

Patrick Meares

Chemistry Department, University of Aberdeen

Old Aberdeen AB92UE, Scotland

Introduction to membrane phenomena

The name "ion-exchange membranes" arose because the first such articles were made by preparing in sheet form the conventional organic ion-exchangers available 30 years ago. The purpose of these membranes is not to exchange ions but to transmit them in a controlled way. The name "selectively ion-permeable membranes" would indicate their function more clearly. From these origins the polymer chemistry of ion-exchange membranes has become more sophisticated and this article will consider some of these chemical developments. To understand them in context it is necessary to say something first about the physical chemistry of membranes and to outline some processes in which they are either in use already or under serious consideration.

An ion-exchange membrane can be thought of as a polymeric matrix, usually crosslinked, within which ions of one charge sign are confined by being bound covalently to the polymer. These membranes are thick in comparison with molecular dimensions and must therefore be electrically neutral overall, apart from effects confined to the electrical double layers inside and outside the membrane faces. The charge of the membrane bound ions must therefore be balanced by an equivalent number of oppositely charged ions. These are not covalently bound and, in a suitable matrix, they can diffuse freely inside the membrane. They can also cross its surfaces provided they are replaced by other ions sufficient to maintain electrical neutrality. These mobile ions are called counter-ions and the membrane is selectively permeable to them.

A flux of counter-ions through the membrane will occur from a region of high to one of low electrochemical potential of these ions. A difference in electrochemical potential $\Delta\mu_i$ may arise from differences in concentration, electric potential and pressure. It can be conveniently expressed by

$$\Delta\mu_i = RT\,(\ln a_i'' - \ln a_i') + z_i F\,(\psi'' - \psi') + V_i\,(p'' - p') \qquad (1)$$

where a_i, z_i and V_i are respectively the activity, charge number and molar volume of the ions i. ψ and p are electric potential and pressure; and constant temperature T is assumed to exist. If the superscript " refers to the upstream and ' to the downstream side of the membrane, positive forces give rise to positive fluxes. It is seen that this driving force is established by the thermodynamic conditions existing outside the membrane at each of its faces. The resulting flux density of i is controlled also by the concentration of the counter-ions in the membrane and by their mobility in the matrix.

A theory sufficient to explain in all essentials the interplay of these factors became available more than 10 years before the first membranes were developed. Two principles were involved: the Donnan membrane equilibrium (1) and the fixed charge theory of Teorell and Meyer and Sievers [TMS] (2). Under the requirements of the former it is assumed that the distribution of ions between the membrane and the solution at each interface is always at equilibrium even when there is a chemical potential gradient within the membrane leading to the passage of ion fluxes. The TMS theory assumes that the processes controlling the ion fluxes are confined entirely within the membrane phase.

The extension of this theory from the aqueous colloidal and biological systems for which it was conceived (3,4) to an organic polymeric matrix requires some consideration. The low thermal jumping frequencies of the molecular segments in polymers, even in elastomers, relative to the frequency of the molecular Brownian motion in liquids results in the diffusion coefficients of the counter-ions in the membrane matrix being relatively low. Furthermore, the low dielectric constant of a typical organic medium results in the counter-ions and fixed charges being only weakly dissociated. To overcome these barriers to free ion transport it is necessary for the membrane to swell spontaneously and imbibe water from contacting aqueous solutions. Water entering the membrane hydrates the ions and encourages the dissociation of the counter-ions from the fixed charges. Dissociation of the counter-ions creates an osmotic pressure within the membrane which causes further swelling of the otherwise hydrophobic matrix. The water absorbed plasticizes the polymer and if the swelling exceeds about 20% by volume it creates also essentially liquid-like pathways within the membrane.

The dissociated counter-ions have a relatively high mobility although they are constrained to follow tortuous pathways. Such high mobility is not exclusive to the counter-ions, it is shared by all reasonably small ions of either charge and by neutral molecules, including the water, present in the membrane. Selective permeability towards certain species rather than high permeability towards all results from the exclusion of some species, the relative retardation of others and the unhindered passage of the preferred species.

Ions of charge sign similar to that of the bound ions and opposite from the counter-ions are called co-ions. It is easy to see qualitatively why they are excluded from the membrane. The counter-ions may be imagined to be trying to leak out of the membrane into the surrounding aqueous solution in which their concentration is lower. Thus the membrane tends to acquire a net charge and a potential difference arises between it and the solution. The passage of an almost trivial number of ions across the interface is sufficient to create a considerable potential difference. If the fixed charges are anions the membrane becomes negatively charged and vice versa. As a result co-ions are repelled and counter-ions attracted by the membrane. The bound ions can be regarded as acting like the impermeant ions in a classical Donnan system. When a membrane is immersed in a solution of an electrolyte its activity in the membrane must become equal to that in the solution. Application of this thermodynamic principle leads to an expression for the ratio of co-ion and counter-ion concentrations in the membrane.

When equilibrium exists between the membrane and solution at an interface the chemical potential of each component that can diffuse between the membrane and the solution must be the same in both phases. For the counter-ions, subscript g, the requirement is

$$\mu_g^o + RT \ln a_g + z_g F \psi = \bar{\mu}_g^o + RT \ln \bar{a}_g + z_g F \bar{\psi} \tag{2}$$

where symbols with the superior bar refer to the membrane phase. Similarly for the co-ions, subscript n, the equilibrium condition is

$$\mu_n^o + RT \ln a_n + z_n F \psi = \bar{\mu}_n^o + RT \ln \bar{a}_n + z_n F \bar{\psi} \tag{3}$$

A relatively small and controversial term involving a pressure difference across the membrane/solution interface has been omitted because it cannot be measured directly. It can be thought of as having been included in the activity coefficient or in the standard chemical potential depending on the choice of standard state.

The distribution of "neutral electrolyte" is obtained by

adding $|z_n|$ times eqn.(2) and $|z_g|$ times eqn.(3) to give

$$\mu_{ng}^{o} + RT \ln a_n^{|z_g|} a_g^{|z_n|} = \bar{\mu}_{ng}^{o} + RT \ln \bar{a}_n^{|z_g|} \bar{a}_g^{|z_n|} \qquad (4)$$

thereby eliminating the Donnan potential difference $(\bar{\psi} - \psi)$.

If the same standard state, i.e. the hypothetical ideal one molal solution, is chosen for both phases then

$$a_n^{|z_g|} a_g^{|z_n|} = \bar{a}_n^{|z_g|} \bar{a}_g^{|z_n|} \qquad (5)$$

Electroneutrality in the membrane phase requires that

$$z_g \bar{m}_g + z_n \bar{m}_n + \omega \bar{M} = 0 \qquad (6)$$

where m stands for molality in the membrane, and ω and $\bar{M}$ are the charge number and molality of the ions covalently bound to the membrane.

To proceed further it is convenient to introduce ν_g the number of counter-ions and ν_n the number of co-ions per mole of neutral electrolyte, with $\nu = \nu_n + \nu_g$. For example if the counter-ions were cations then the salt would have the formula $M_{\nu_g} X_{\nu_n}$ whence

$$\nu_g = |z_n| \quad \text{and} \quad \nu_n = |z_g| \qquad (7)$$

For all practical cases ω is either +1 (anion exchange membrane) or -1 (cation exchange membrane) so that eqn.(6) becomes

$$\nu_n \bar{m}_g - \nu_g \bar{m}_n - M = 0 \qquad (8)$$

Introducing the mean molal activity coefficients $\gamma_{\pm}$ and $\bar{\gamma}_{\pm}$, eqn.(5) becomes

$$\bar{m}^{\nu_m}(\bar{m} + M/\nu_n \nu_g)^{\nu_g} = (\gamma_{\pm} m/\bar{\gamma}_{\pm})^{\nu} \qquad (9)$$

where the molality of salt in the solution is m and in the membrane it is $\bar{m}(=\bar{m}_n/\nu_n)$.

Glueckauf (5) has shown that over a wide range of concentration $\gamma_{\pm}/\bar{\gamma}_{\pm}$ is almost constant. It can be represented by α. Because the membrane bound ions are not uniformly distributed eqn.(9) needs a structure-dependent empirical factor β in order to represent the experimental facts. Hence, finally,

$$\bar{m}^{\nu_n} (\bar{m} + \bar{M}/\nu_n \nu_g)^{\nu_g} = (\alpha\, m)^{\beta \nu} \qquad (10)$$

may be used to explore the relationship between ion concentrations in the membrane and in the adjacent solution. It has been found that α is close to unity and β less than unity in several cases (6).

Qualitatively one can explore the effects of the operational variables on the relative concentrations of co-ions and counter-ions in the membrane by setting α and β equal to unity. In the case of a uni-univalent electrolyte ($\nu_g = \nu_n = 1$) one finds

$$\frac{\bar{m}_n}{\bar{m}_g} = \frac{\sqrt{\bar{M}^2 + 4\,m^2} - \bar{M}}{\sqrt{\bar{M}^2 + 4\,m^2} + \bar{M}} \tag{11}$$

Clearly the ratio of the ion concentrations is very small when the solution concentration is much less than the fixed charge concentration. This effect is often referred to as the Donnan exclusion of the co-ions. By considering other values of ν_g and ν_n and also variations in $m/\bar{M}$ a wider range of behaviour can be studied. In summary, the higher the valency of the co-ions the better they are excluded i.e. the greater the repulsion. The higher the valency of the counter-ions the less well the co-ions are excluded. The greater the ratio of the molal concentration of bound ions to the molal concentration of the solution the greater the Donnan exclusion.

Finally, for the Donnan potential difference $\Delta\psi_D$ $(= \bar{\psi} - \psi)$ which arises across the interface between membrane and solution, one may probably write with little error

$$\Delta\psi_D = \frac{RT}{z_n F} \ln \frac{m_n}{\bar{m}_n} \tag{12}$$

because the activity coefficient of the co-ions appears to behave normally in polyelectrolyte solutions although that of the counter-ions is depressed.

Selective permeability is influenced not only by the relative concentrations of co-ions and counter-ions but also by their mobilities. For a pair of ions of about equal hydrated size and of equal and opposite charge the co-ion mobility will be relatively greater than the counter-ion mobility because of the stronger frictional interaction of the latter with the polymeric matrix to which it is electrostatically attracted (7). As is found so often in membrane processes, the ratio of the mobilities tends to oppose the ratio of the concentrations created by the thermodynamic partitioning effect because the permeation flux can be regarded as the product of the applied force, the concentration and the mobility for each species.

It would be valuable if membranes were able to selectively transport one counter-ion from a mixture and thus to enable separate ions of like charge to be separated. Such possibilities are governed by the relative concentrations and mobilities in the membrane of the ions to be separated. The relative concentrations are determined by the influences already well known in the sphere of ion separations by ion-exchange resins. For ions of equal valence the larger and more polarizable are taken up preferentially and organic ones especially so because of the hydrophobic effect unless they are physically too large to enter the polymeric matrix. For counter-ions of unequal valence those of higher valence are preferred, very strongly so when the solutions are dilute. The counter-ions thus taken up in preference interact the more strongly with the bound ions and the membrane matrix. Hence they become relatively less mobile in the membrane than the less preferred ions. It is by no means universally true that membranes are selectively permeable to the counter-ions which are preferentially absorbed from a mixture. Often counter-ion selectivity leads to operational problems rather than to useful separations, especially in electrodialysis.

The permeability of membranes to water is also important. Ions driven through a membrane by an electric field move mainly in a locally aqueous environment and drag water molecules with them. Not only is their water of hydration transferred in this way but the viscous interactions that would lead to the electrophoretic effect in conduction in free solution lead to electro-osmotic flow of water in a membrane. The net flow is directed with the counter-ion current and can exceed 50 mole per Faraday. The electro-osmotic flow sets an upper limit on the efficiency of membrane processes that involve concentrating solutions.

In practice, the electro-osmotic flow is often far lower than this figure. The high ionic concentration in a membrane may restrict the degree of hydration of the ions to values below those operating in dilute solutions. The free water dragged by the ions is retarded by its viscous interaction with the membrane matrix. Thus the precise molecular structure and morphology of the polymeric network is very important in controlling electro-osmosis. Two membranes of similar ion and water contents and conductances may have very different water transference numbers t_w. Expressed as mole per Faraday t_w decreases as the concentration of the solutions increases. Expressed as mole per volt between the membrane faces it is not far from being independent of solution concentration. Under operational conditions values in the range 4 - 10 mol F^{-1} are commonly found.

Because of the Donnan exclusion of the co-ions, ion-exchange membranes show some degree of semipermeability. As a result some normal osmotic flow tends to occur across a membrane separating two different solutions. The effective osmotic pressure is usually far below the ideal thermodynamic value because the co-ion exclusion is incomplete and the reflection coefficient of the membrane is therefore less than unity. The phenomenon of osmosis is further complicated by the existence of the membrane diffusion potential which tends to generate an electro-osmotic water flow even in the absence of an electric current. Except in the special case of membranes prepared for use in hyperfiltration, the hydraulic permeabilities of ion-exchange membranes are normally low and the osmotic transport of water is, in the practical application of membranes, much less important than the electro-osmotic transport.

Some processes that use ion-exchange membranes

A wide range of uses has been suggested which exploit the properties of ion-exchange membranes. Table 1 lists representative examples of the main categories. Here it will be possible to discuss only a few.

The largest class of processes uses electrical energy to replace the free energy dissipated by molecular friction when substances move through a membrane. Such processes are known as electrodialysis. They can be used to demineralise as well as to increase the concentration of solutions, to recover a salt from a mixture, to adjust the ion content of a solution and to carry out oxidation and reduction reactions with separation of products. Other processes use concentration gradients and pressure gradients as driving forces. Each presents its own particular problems and, for optimum performance, requires specially developed membranes.

The desalination of brackish water was the first electrodialytic process to receive serious attention. It has now been in large scale use for about 25 years. Subsequently the means have been developed for the desalination of many other solutions. The scale of the operation differs greatly among these uses but the basic principles are always the same. In the complete process two kinds of membrane are required. Usually one kind is cation selective and the other anion selective although in some applications neutral membranes may replace the anion selective ones with advantage.

Several hundred membranes of area at least 1 m^2 are assembled in a horizontally alternating arrangement of the two kinds in a stack. The membranes about 0.2 mm thick are separated by spacers often less than 1 mm thick which provide room for the solution phases between the membranes. The spacers are designed so as to promote good mixing of the solutions which flow through the system. Turbulent flow would be desirable but is rarely achieved in practice

Table 1 Ion-exchange membrane processes

1. Electrical driving force:

   a) Desalination, Demineralization

Desalination	Demineralization
Brackish water	Whey
Sea water	Sugar
Sewage recycling	Blood serum

   b) Concentration of solutions

      Salt production from sea
      Nickel recovery from plating waste
      Radio-active waste treatment

   c) Exchange of ions

      Deacidification of fruit juice
      Baby food from cows' milk
      Double decompositions
        e.g. acid phosphate production

   d) Oxidation - reduction

      Chlor-alkali industry
      Adiponitrile from acrylonitrile
      Uranium refining $UO_2^{2+} \rightarrow UCl_4$

2. Concentration gradient driving force:

   a) Diffusion dialysis for acid recovery from pickling liquor.

   b) Solid electrolyte in chemical (e.g. Ag - Zn) cells and fuel cells.

   c) Ion specific electrodes.

3. Pressure driving force:

   a) Reverse osmosis

   b) Piezodialysis.

and the spacers induce either rapid flow along a smooth tortuous path or uniform sheet flow over an expanded and corrugated material (8).

An electric current is passed through the stack between electrodes in the solutions bathing the outsides of the two end membranes. The electrodes now used are frequently of metal e.g. platinised titanium. The electrode reactions are not involved in the separation process and care must be taken to isolate their products from the process solutions. The cathode compartment generates hydrogen gas and alkali; it must be continuously fed with sufficient acid to maintain neutrality. At the anode, oxygen and chlorine are generated which are harmful to the membranes and constant flushing is therefore required. The mechanism of the process can be understood by referring to Figure 1a which shows a very simplified arrangement. It is desirable to be able to reverse the polarity of the current passing through the stack so this has to be built with a plane of symmetry. It is terminated with anion selective membranes because these are more resistant to oxidation by the products of the electrode reactions.

When current is passed separation occurs because the transport numbers of the ions in the solution have different values in each type of membrane (1) and (2) in Figure 1a. Let the transport numbers of cations and anions be t_+' and t_-' respectively in membranes of type (1) and t_+'' and t_-'' in membranes of type (2). When one Faraday of electricity flows through the stack t_+'' equivalents of cations enter compartment (b) from (a) and t_+' are transferred from (b) to (c). At the same time t_-' equivalents of anions enter (b) from (c) and t_-'' go from (b) to (a). The overall result is that there is a gain of $(t_+'' - t_+')$ equivalents of electrolyte by (b) and a compensating loss from (c). Since membranes (1) are anion selective and (2) are cation selective, t_+' is small and t_+'' is large (i.e. close to unity) and the difference above is positive. Thus the solution in (b) becomes more concentrated and that in (c) more dilute as current is passed.

Compartment (d) behaves similarly to (b) and (a) similarly to (c), that is assuming effects due to the proximity of the electrodes in our simple diagram may be ignored. In a large stack of many membranes and compartments alternate streams become diluted and intermediate ones concentrated. Reversing the current simply reverses the roles of the compartments. The greater the difference $(t_+'' - t_-'')$ the more salt is transferred per Faraday, i.e. the greater the current efficiency. It is a maximum when t_+' is zero and t_+'' is unity. Where the concentrated stream is to be rejected, as in desalination, its volume is minimised by driving its concentration as high as possible. Inevitably this leads to some drop in current efficiency due to the co-ion current which increases as the Donnan exclusion is progressively overwhelmed.

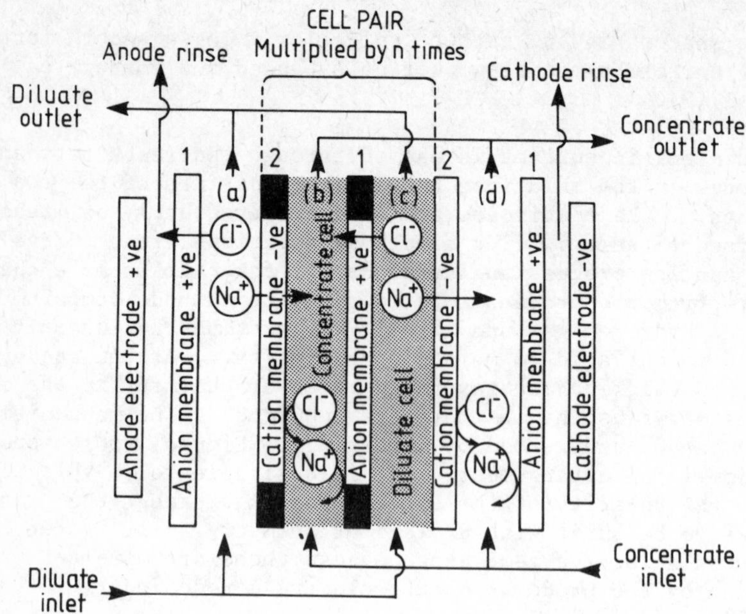

Fig. 1a Schematic representation of an electrodialysis cell for brackish water desalination. See text for explanation.

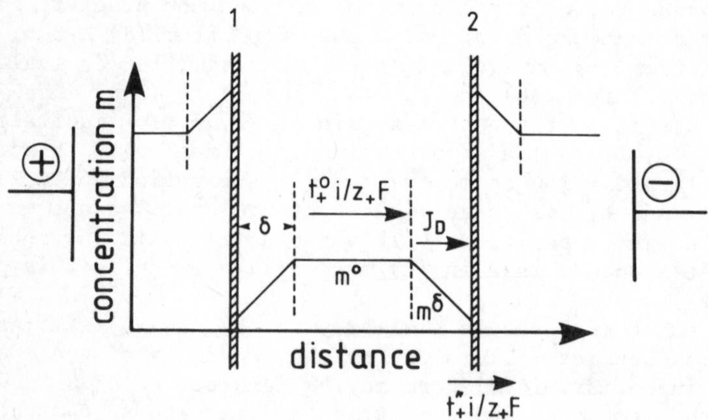

Fig. 1b Concentration profiles and cation fluxes in a cell pair undergoing concentration polarization. Analogous arguments yield the, oppositely directed, anion fluxes.

Just as water distillation processes are overshadowed by the problems of scale formation, so electrodialysis is dominated by polarisation and membrane poisoning. The need to control these factors has had a great influence on stack design. Other important problems arise because of the difficulty of constructing a leak-free stack of perhaps 1,000 soft membranes between which an electrolyte is circulated at entry pressures of 2 atmospheres or greater. Because so many membranes are in series the potential difference applied between the electrodes may well exceed 1000 volts and leaking electrolyte provides short circuit paths and causes serious electrochemical corrosion of all metal parts.

Further difficulties arise from the design of the manifolds which distribute the diluate and concentrate streams to the compartments. The flows must be rapid to minimise polarisation, they must be in parallel in all diluate and in all concentrate compartments to minimise power consumption in pumping and they must be co-current to minimise pressure differences across the membranes. Countercurrent flow might be more desirable from the electrochemical standpoint but is mechanically too difficult to arrange. The manifolds which distribute and collect these streams constitute leak paths for an electric current bypassing the membranes because they interconnect compartments at the high and low potential ends of the stack. This problem is most serious in the concentrate manifolds because the solution in these has a relatively high conductance.

The conditions in a desalination plant during operation are very non-uniform. Feed water enters at, typically, 2000-3000 ppm dissolved salt after some pretreatment stage. After flowing through the plant, which will consist of several stacks in parallel and series arrangement chosen to given the desired quantity and quality of product, the diluate output will contain 300-500 ppm solute whereas the concentrate, usually 10-15% of the feed in volume, has a concentration 6-8 times that of the feed. These concentration changes develop progressively along the process streams and, because the conductances are approximately proportional to concentration, the electrochemical conditions vary greatly over the surfaces of the membranes. This inevitable non-uniformity in operating conditions aggravates the problem of controlling polarisation at the membrane/solution interfaces which is related to a combination of cell and membrane parameters.

The current i in the diluate stream is carried by all the ions in proportion to their transport numbers in that solution. Across the interface between the solution and the cation-selective membranes the current is carried largely by cations; across the anion-selective membrane interfaces it is carried by anions. Considering the cation-selective membrane as an example, the difference between the electrical flux in the solution up to the interface and that away from it into the membrane $(t_+'' - t_+^o)\ i/z_+F$, where t_+^o is the transport number of the cations in the solution, has to be compensated by a mass flux in the solution because the total fluxes on the two sides of the interface must be equal. The rapid circulation of the solution over carefully designed inter-membrane spacers provides a convective contribution to this mass flux but the agitation is poor close to the membrane surface and the flux there must be supplemented by diffusion. For the purposes of an approximate treatment the poor agitation can be represented by an equivalent layer of stationary solution of thickness δ. If the concentration at the interface falls to a value m^δ, which is less than the bulk value m^o, a diffusion flux $J_D = D(m^o - m^\delta)/\delta$ is generated where D is the diffusion coefficient of the salt in the solution. The concentration profile which develops across the stack as a result of this effect is represented by the dotted line in Figure 1b.

The principle of continuity requires that

$$D(m^o - m^\delta)/\delta = (t_+'' - t_+^o) i/z_+F \tag{13}$$

whence

$$m^\delta = m^o - (t_+'' - t_+^o) i\delta/z_+FD \tag{14}$$

It can be seen that a critical current i_c exists at which m^δ becomes 0. Furthermore i_c decreases as m^o decreases and also as δ increases. Even in a well designed stack δ is rarely less than 10 μm and it is larger than this in the stagnant areas. A similar depletion in concentration occurs at the anion-selective membrane. On the concentrate side of the membranes a local increase in concentration arises but this is not important except in extreme cases where it may lead to supersaturation and the formation of scale.

This depletion of concentration at the membranes in the diluate stream is called polarisation. It increases the electrical resistance of the stack and places a limit on the quality and rate of production from a stack of given size. For good economic results electrodialysis cells must be operated close to their critical currents. It must however be noted that the concentrations close to the membrane change due to progressive desalination along the flow path and the flow rates over the membrane surfaces may vary due to the imperfect flow patterns thus leading to local variations in the stationary layer thickness δ. Consequently the critical current density varies from place to place and the whole stack does not polarise suddenly at a unique value of the current.

Certain electrochemical events occur as the critical condition is approached which make it desirable to limit the voltage across the stack so that the critical condition is not exceeded anywhere. In practice 1-2 volts per membrane is usually a satisfactory working condition. Among the reasons why polarisation should be avoided is that, as the concentration at the membrane surfaces falls, a level is reached at which the specific conductance of the salt approaches that of the H_3O^+ and OH^- ions from the water dissociation. These ions then begin to carry part of the current resulting in a loss of product and of electrical efficiency. In practice the critical situation usually arises first at the anion-selective membrane. As a result OH^- ions flow from the diluate which becomes acid and, at the same time, the concentrate becomes alkaline. If there are divalent ions, especially Ca^{++} and Mg^{++}, in the concentrate and this is allowed to become alkaline, scale may be deposited on the membranes rather than form as a precipitate at the concentrate side of the membranes. This leads to a great increase in electrical resistance and in power consumption. If the streams are interchanged and the polarity reversed the scale is mostly redissolved. This is one reason for designing plants symmetrically so as to permit polarity reversal.

The phenomena of membrane fouling and poisoning should also be mentioned. The former is due mainly to small quantities of organic ions, usually anions, in the feed water. These are taken up by the anion selective membranes and adsorbed onto the polymeric organic matrix. Such ions are almost immobile in the membrane but neutralise its charges. As a result its resistance is increased and its selectivity destroyed, effects which are to a large extent irreversible. The cation-selective membranes can also be fouled by organic ions but, more seriously, they can be poisoned by ter- and quadrivalent metal ions which are strongly preferred by the ion-exchange material. They are so strongly held by the fixed charges that the properties of the membranes may be reversed from being cation to becoming anion selective. If necessary the feed water must be pretreated to remove such poisoning ions but these additional treatments add greatly to the running cost of the desalination process.

In addition to its major use in the desalination of raw water, electrodialysis can be applied as a desalting step in sewage recycling. Its use for the treatment of waste liquors particularly from the paper industry has been explored with a view to recovering some chemicals in reusable form as well as for reducing the discharge of pollutants (9). In the food industry the use of electrodialysis for demineralisation is already important. Examples are in sugar refining and in cheese whey processing (10). In such organic-containing media fouling of the anion-selective membranes creates serious difficulties. They may be replaced instead by neutral membranes of regenerated cellulose which are resistant to fouling. The transport numbers in wet cellulose membranes are not very different from those in aqueous solution hence the electrical efficiency of the cation-neutral membrane process is lower than that of conventional electrodialysis and it is referred to under the name transport depletion. The additional operating cost may however be justified by the elimination of membrane fouling.

The electrical power required to transfer a given amount of salt is determined by the quantity of electricity which has to be passed and by the electrical resistance of the stack. It is clear that everything must be done to keep that resistance low. For this reason the membrane spacers are as thin as is compatible with providing a good flow path between the membranes and the resistance of the membranes themselves must be low. A good membrane must therefore have an almost ideal ion selectivity and a high conductivity in a variety of electrolytes over a range of concentrations. In addition it must be thin, strong, dimensionally stable, resistant to chemical and thermal degradation and have a working life of at least two years. The manufacture of such membranes is described below as well as discussing some of the modifications in membrane design that can help to reduce problems due to polarization, fouling and poisoning.

Manufacture of ion-exchange membranes

A variety of methods has been developed, especially in Japan and U.S.A., to meet the requirements of electrodialysis and this has led to the commercial production of membranes of various chemical structures. Representative examples of the most important types will be considered here.

Polycondensates

The earliest membranes were polycondensates of formaldehyde or urea with phenol and either a phenol sulphonic acid or an amino-phenol. In a typical example dimethyl o-hydroxy-tolylamine (1 part) and phenol (4 parts) are condensed in aqueous solution with formaldehyde to produce an anion exchange membrane. The following exemplifies the preparation of a polycondensate cation membrane. Sodium metabisulphite, formalin and phenol are dissolved in 15% w/w sodium hydroxide and reacted overnight at 80°C. The quantities are chosen to give mainly $HOC_6H_4CH_2SO_3Na$. This solution is mixed with more phenol, paraformaldehyde and water and reacted under reflux to produce a solution of a linear polymer. The amounts of the reagents control the ion concentration and water content of the final membrane. This liquid resin is poured and held between spaced glass plates before curing at 94°C for about seven hours to form the solid membrane. Careful control of all reaction variables is essential to produce membranes with reproducible properties.

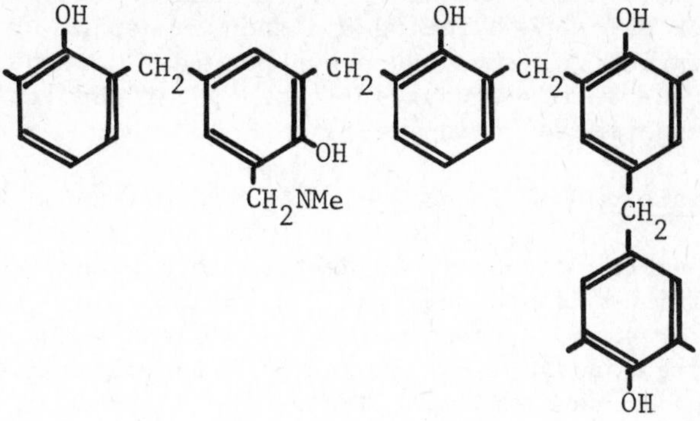

Representative anion exchange membrane unit

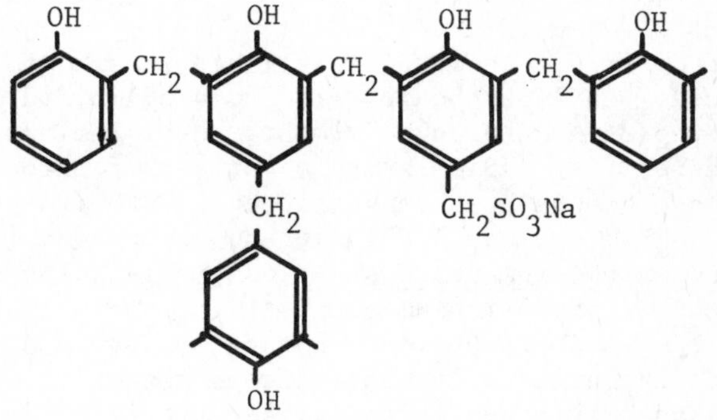

Representative cation exchange membrane unit

Membranes of this type having been prepared in the presence of 50 - 80% water by volume could not be dried and rewetted reversibly. This is a great disadvantage in practical terms. Their exchange capacity had to be kept low, about 1 m fixed ions was a typical value, otherwise the osmotic pressure of the counter-ions made the swollen molecular network fracture readily into small particles. Additional strength was achieved by forming the membrane on a woven mesh of an inert fibre such as nylon or glass but fragility, especially to sudden stresses, remained. These condensation networks have been shown by narrow angle X-ray and light scattering studies to consist of tightly crosslinked regions interconnected by looser material which causes the mechanical weakness and reduces also the Donnan exclusion of co-ions and hence the electrical performance of the membrane.

Condensation polymers were soon abandoned in favour of addition polymers. However polyelectrolytes with a high content of ionizable groups are soluble in water unless crosslinked. Attempts to follow the chemistry of the highly successful polystyrene + divinyl benzene-based ion-exchange beads by sulphonating or chloromethylating and then aminating sheets of crosslinked styrene + divinyl benzene copolymers were not at first successful. The sheets always shattered after chemical substitution and immersion in water.

Unusual materials

Exotic chemical methods were devized to overcome some of these difficulties and produce polystyrene type ion-exchange membranes. For example, a mixture of styrene and the propyl ester of p-styrene sulphonic acid can be copolymerized with divinyl benzene in sheet form. The ester can subsequently be hydrolyzed to the sulphonic acid and a cation exchange membrane of reasonably good properties obtained. Such complex methods are however not of commercial interest because of their high costs.

It is not difficult to make crosslinked sheets of methacrylic acid, vinyl sulphonic acid and vinyl pyridine but either their properties are poor, or their uses limited or the monomers too expensive. Similarly, a variety of anionic and cationic exchange groups has been explored for their potential use in membranes. They include phosphate and arsenate anions and phosphonium and sulphonium cations. Today only sulphonate, carboxylate, quaternary ammonium and pyridinium are of practical importance. The majority of ion-exchange membranes has the fixed ion attached to aromatic nuclei but this is not exclusively so. The principal types will now be described.

Heterogeneous membranes

The wide range of ion-exchange resins in production with various types of ion content, crosslinking, swelling and exchange capacity has been exploited by Rohm and Haas, among other companies. By reducing the beads to a fine powder, 1 - 10 μm particle size, mixing this with a non-ionic linear polymer and sheeting on a roller mill at temperatures up to about 150°C a number of useful heterogeneous membranes has been produced. Polyethylene, butadiene + styrene and vinyl chloride + vinyl acetate copolymers are typical of the binder polymers that have been used in amounts from 25 - 75% by weight (11).

Heterogeneous membranes have reasonably good mechanical strength and dimensional stability and they played an important part in enabling electrodialysis to become adopted in the 1950's. Although such membranes are still in use today their electrical resistance is high and the swelling and shrinking of the ion-exchange particles causes them to de-bond from the binder thus allowing aqueous solution to creep along the particle boundaries. This reduces the current efficiency in desalination.

Homogeneous addition-polymer membranes

The problem of activating sheets of crosslinked polystyrene by the methods standard in ion-exchange resin technology (12) without disintegration was solved by mixing the monomers with 30 - 60% by volume diethyl benzene and polymerizing on a supporting cloth between glass plates, usually with benzoyl peroxide as the initiator. The diethyl benzene, a compatible solvent for polystyrene, produces a homogeneous gel which is then further swollen in ethylene dichloride before reaction either with concentrated sulphuric acid at $50^{o}C$ for 18 hours or with chlorosulphonic acid. The supporting cloth may be glass, polypropylene, or polyvinyl chloride + polyacrylonitrile copolymer. Improved bonding of the resin to the cloth is obtained if the ethylene dichloride is substituted by soaking first in a polar solvent such as methanol or dioxane before sulphonation.

Anion exchange membranes may be prepared from similar polystyrene + divinyl benzene boards by chloromethylation of the styrene with chlormethyl methyl ether and aluminium chloride following by treatment with a tertiary amine while swollen in dioxane. In this method, used for example by Ionics Inc. (13), the membranes have to be stored wet.

Another route to preventing distruption of crosslinked polystyrene on activation is to copolymerize it with a diene monomer such as butadiene or isoprene. If a random copolymer is used rupture still occurs on sulphonation but it is reported that Asahi Chemical Co. (14) forms its membranes from a block copolymer of styrene and butadiene. Slices are cut from a large mass of this copolymer with wood veneer making machinery. The uniformity of the network structure in the interior of the mass of polymer is so good that the sheets can be chemically activated by the usual reactions and still remain sufficiently robust to be used in electrodialytic cells without the need for use of a supporting mesh. This enables the electrical resistance of the membranes to be kept extremely low because reinforcing cloth always interferes with the electric current flow.

Graft and snake-cage membranes

Instead of using polyethylene as a binder for powdered resin, polyethylene films can be swollen in styrene + divinyl benzene and then exposed to γ-radiation to induce polymerization. Subsequent chemical attack is used to introduce ionogenic groups onto the styrene residues. This type of membrane, introduced by A.M.F. Company (15) and manufactured now by Raipore is usually described as a graft copolymer but the direct evidence of grafting of the polystyrene onto the polyethylene is rather limited and it may well be that the styrene forms a network interpenetrating the semi-crystalline mass of polyethylene. Such membranes are micro-heterogeneous at the molecular level with the polyelectrolyte material concentrated into regions a few tens of nm in diameter embedded in the amorphous regions of the polyethylene (16).

A chemically and mechanically more stable membrane of this type has been made similarly by using polyvinylidene fluoride film instead of polyethylene (17). In this case photodegradation and extraction experiments have proved that grafting of the styrene to the polyvinylidene fluoride does not take place and that a snake-cage structure is formed from which the polystyrene sulphonic acid cannot be leached out even if no divinyl benzene is included in the polymerizing mixture.

Sulphonated polyethylene membranes

It has been discovered that polyethylene alone can be sulphochlorinated on treatment with a mixture of chlorine and sulphur dioxide or directly sulphonated with a complex of sulphur trioxide and triethyl phosphate. Hydrolysis of the products then produces a polyethylene membrane with chemically attached sulphonate groups (18). Chemical attack on the polyethylene occurs only at the chain folds on the surface of the crystallites. Further treatment of the membrane with diethylamino-propylamine forms a sulphonamide with free diethylamino-groups. These can then be quaternized with methyl bromide to produce anion exchange sites. Such cation-exchange and anion-exchange membranes prepared directly from polyethylene are cheap and tough. They are manufactured in Isreal and are claimed to function satisfactorily in electrodialysis plants (19).

Membrane manufacture by the paste method

One more method of membrane manufacture remains to be described. This is the paste method used by Tokuyama Soda Company. In this method a paste is prepared by mixing finely powdered polyvinyl chloride, styrene or 2-methyl 5-vinyl pyridine, divinyl benzene, benzoyl peroxide and ioctyl phthalate. The polyvinyl chloride powder becomes swollen by the monomers and plasticizer and the paste is then coated onto a loosely woven polyvinyl chloride cloth before being heated between cellophane sheets to polymerize the monomers. A cation exchange membrane is made from the resulting membrane by sulphonating the styrene while the anion-exchange membrane is obtained by quaternizing the pyridine residues. Membranes prepared in this way have been shown to have a micro-heterogeneous structure with interconnecting pockets of ion-exchange material about 50 nm diameter occupying about 50% of the total volume (20). In view of the ease with which polyvinyl chloride takes part in chain transfer processes, it is possible that some grafting of polymer to the surfaces of the polyvinyl chloride particles occurs. The membranes prepared as described have very good strength and electrical properties.

PROBLEMS IN ELECTRODIALYSIS

Membranes of acceptable strength, chemical stability and electrical properties have been successfully produced by all of the above methods. Their exchange capacities lie typically in the range 1 - 3 equiv. kg^{-1} i.e. slightly less than that of many particulate resins. Their water contents are higher than those of particulate resins; usually they lie in the range 20 - 50%. Thicknesses range around 0.15 - 0.30 mm and electrical resistances are commonly 2 - 8 ohm cm^{-2} in dilute salt solutions.

The chemistry and properties of a new class of perfluorinated ionomer membranes will be discussed later. First the use of these more conventional ion-exchange membranes in electrodialytic processes is discussed so as to highlight some difficulties and the means used to overcome them. The difficulties to be considered are polarization, fouling and poisoning.

The mechanism of concentration polarization and the possibility of pH changes being caused by it were described earlier. The details of the phenomenon are influenced by several membrane properties listed below:-

(i) the relative exchange selectivity of the membrane towards hydrogen and the other cations present or towards hyroxyl and the other anions present,

(ii) the mobilities of hydrogen and hydroxyl ions in the membrane relative to those of the other cations and anions,

(iii) the effect of the large electric field in the interfacial layer at the surface of the membrane on the dissociation equilibria or rate of dissociation of the water molecules,

(iv) the orientation of the water quadrupoles at the membrane surface as determined in part by the chemical nature of the matrix,

(v) the roughness or molecular diffuseness of the membrane/ solution interface.

The problem of polarization is further complicated because at sufficiently high currents other mechanisms can arise which inject the wanted ions into the membrane and it is important therefore to be able to suppress the water splitting process up to the point where these other mechanisms take over. Usually pH changes appear first in the neighbourhood of the anion exchange membrane. This is thought to be because the water molecules in the electrical double layer there are arranged with their oxygen atoms pointing towards the membrane.

In an ingenious development, Israeli workers (21) have introduced a membrane spacer formed by knitting cation-exchange fibres and anion-exchange fibres onto a nylon mesh in such a way that the cation exchange material is exposed on one side of the spacer and the anion-exchange material on the other side as shown in Figure 2. The ion-exchange fibres are made by chemical modification of polyethylene monofilaments using the reactions described earlier (19). These spacers serve the dual purpose of promoting good liquid flow in the cell compartments and because the cation-exchange fibres contact the cation permeable membrane and the anion-exchange fibres contact the anion permeable membrane they act to increase the area for ion transfer from solution into the membranes and particularly extend this area outside of the range of the concentration depleted layer. Figure 3 shows that in this way the limiting current density, indicated by the knee in the current/voltage plots, is at least doubled and the demands for high pumping speeds and consequent power loss in the diluate flow are reduced.

Natural raw waters frequently contain organic anions such as humic acids. Furthermore, increasing amounts of synthetic cationic and anionic surfactants are finding their way into the environment. Such ions are strongly taken up and held by ion-exchange membranes. They become absorbed onto the polymeric matrix so that they can travel only slowly once inside the membranes (22). The accumulation of such ions in the membranes reduces their effective capacity for small inorganic ions and so their electrical resistance increases. Such surfactant ions may even be accumulated in excess of the exchange capacity so that the anionic membrane becomes effectively cationic and vice versa.

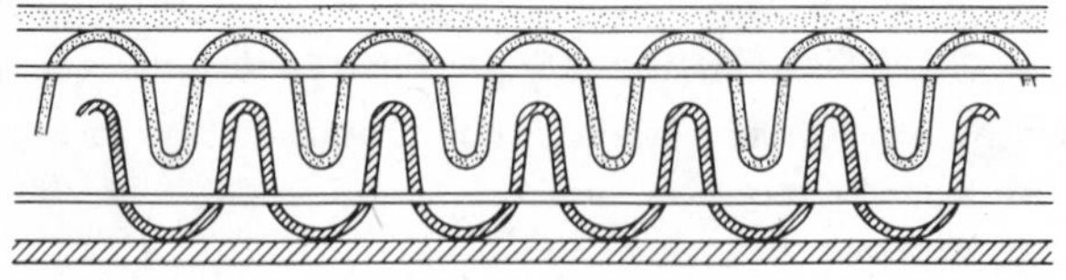

Fig. 2 Membranes separated by a spacer made of ion exchanging fibres. The anion exchange fibres contact the anion permeable membrane and vice versa.

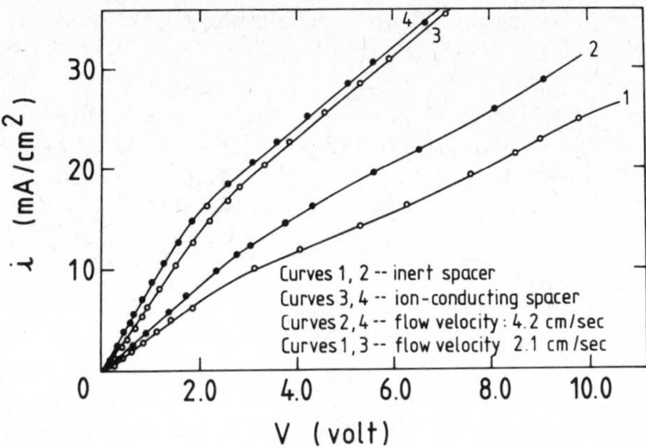

Fig. 3 Current density versus cell voltage showing the increase in limiting current (knee in the curves) resulting from the use of an ion-conducting spacer and the reduced dependence on diluate flow velocity.

Furthermore, since the ions are taken up at one face of the membrane it may become essentially bipolar during the fouling process. This greatly exacerbates the problem of water splitting which takes place inside the membrane at the junction between the fouled and the non-fouled material. As a result serious pH changes occur in the liquid streams and there is increasing electrical power loss.

Several means have been found to modify anion and cation-exchange membranes to protect them against organic fouling. An early method was to include t-amyl alcohol in the mixture of styrene, divinyl benzene and other solvents used in membrane manufacture by several methods (23). t-amyl alcohol, being a non-solvent for the polymer, led to the formation of a macroporous membrane. Membranes made in this way were less susceptible to fouling but their selectivity as between anions and cations was also impaired and their water permeability increased. By using instead ethyl or methyl alcohol as non-solvent it proved possible to restrict the macroporous region to the surface layers of the membrane (24). In this way the electrical properties of the membranes were preserved but the resistance to fouling was found to be only temporary.

More recently it has been found that when membranes are dipped into a solution of a polyelectrolyte, the polyions of which would act as counter-ions in the membrane, a strongly adsorbed layer of these polyions forms on the membrane surface and protects it against organic fouling during electrodialysis (25). For example, an anion-exchange membrane can be protected by immersion in a solution of sodium polystyrene sulphonate and a cation-exchange membrane can be protected by using polyethylene imine.

The adsorbed layer of polyions is long lasting but is not truly permanent. By lightly sulphonating the base membrane by treatment with sulphuric acid for a controlled period before the normal process of chloromethylation and amination of the body of the membrane to form an ion-exchange membrane, permanent protection against fouling is achieved because the sulphonate ions are covalently bound to some of the aromatic nuclei close to the membrane surfaces (26). Provided the sulphonated layer is kept thin, the electrical properties of the membranes as ion-selective conductors are not impaired by this treatment (27). Figure 4 illustrates how effective such antifouling measures can be in delaying the voltage rise which is caused by fouling. The mechanism of protection has been claimed to be the electrostatic repulsion of the foulant ions by the treated surface layer through which they could pass only as co-ions. This explanation does not seem to be wholly satisfactory because the foulant ions and the small anions to which the membrane is selectively permeable are usually both univalent and so must experience an equal repulsion. Possibly the inner boundary of the surface treated layer, where anionic and cationic fixed charges co-exist, is less swollen than the rest of the membrane because no osmotically active counter-ions are required there to achieve electro-neutrality. Although such a region would still be hydrophilic and permeable to small ions organic ions might be excluded from it both by their greater size and by their lower polarity.

In support of this explanation it may be noted that when an anion-exchange membrane is being prepared by the normal route of chloromethylation followed by amination, if the chloromethylated polymer is suface oxidised with potassium permanganate prior to amination and quaternization the surface of the membrane is made more hydrophilic and this has been found to protect it against organic fouling. Evidently hydrophilicity alone enables inorganic ions to be preferred over organic ones.

Although comparable means have been found for protecting the cation permeable membranes against fouling the problem is usually less serious with these. The development of fouling resistant membranes has contributed not only to the development of electrodialysis for desalination but has also made possible the electrodialytic demineralization of aqueous organic products such as cheese whey and blood serum which would seriously foul conventional membranes.

The poisoning of cation-exchange membranes by the entry of ions of high valence such as ferric,which are among the common ions

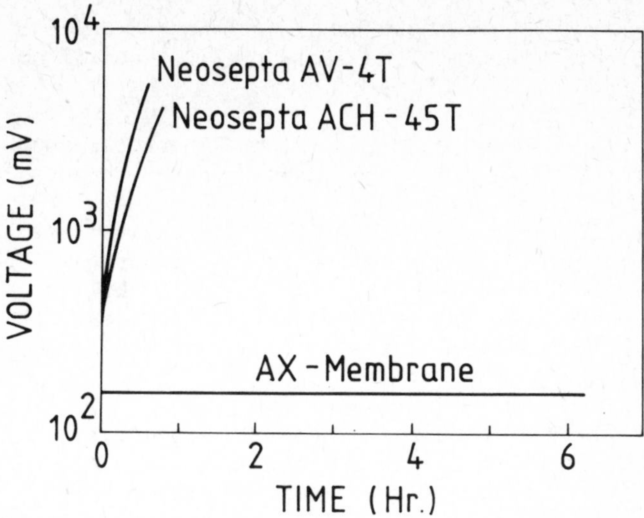

Fig. 4 Comparison of the voltage/time curves, which indicate fouling, of anion-exchange membranes AV-4T and ACH-45T made by Tokuyama Soda Co. Ltd., with that of the surface-sulphonated membrane AX which is otherwise similar in properties to AV-4T and ACH-45T. The measurements were made at 25°C and 2 mA cm^{-2} by electrodialysing 0.05M NaCl containing 100 ppm sodium dodecylbenzene sulphonate.

strongly preferred by cation exchangers, occurs right through the body of the membrane. Once inside the membrane these ions are so strongly held that they are essentially immobile. Their effect is to cause the normal counter-ions, such as sodium, to be excluded and there is a great increase in the electrical resistance of the membrane (28,29). At the present time there is no full protection against poisoning. Surface treatments merely delay its onset. Th poisoning ions should be removed by pre-treatment of the raw water.

Concentrating solutions by electrodialysis

By changing the balance between the flows in an electrodialysis cell the major function can become the concentration of one stream. Thus, for example, concentrated brine can be produced from sea water. This process is important in Japan which lacks salt deposits for its chlor-alkali industry (30). The main practical difficulty is that sea water contains besides sodium and chloride, calcium, magnesium and sulphate in substantial amounts. Conventional membranes are preferentially permeable to these bivalent ions and calcium sulphate would soon deposit in the concentrate stream if normal membranes were used.

By using cation-exchange membranes which have been surface treated with polyethylene imine and ion-exchange membranes surface treated with polystyrene sulphonic or polymethacrylic acids, rather as in the production of fouling resistant membranes, preferential permeability to univalent ions is imparted to the membranes. Figure 5 shows how the relative permeability of a cation permeable membrane to bivalent cations declines with increasing time of treatment of the membrane with 10 ppm polyethylene imine solution (31).

By using such treated membranes the electrodialytic concentration of sea water can be brought about from 0.5 M to about 3.0 M sodium chloride with substantial exclusion of bivalent ions as is shown by the data in Table 2. In order to offset the high concentrations of the solutions the membranes must be highly crosslinked and have a high ion-exchange capacity otherwise the transport numbers of the counter-ions fall well below unity and the current efficiency is impaired due to the transport of co-ions. Tight crosslinking of the membranes also helps to reduce the electro-osmotic transfer of water from diluate compartments to concentrate compartments. The upper limit of the concentration of the product in this process is determined by the amount of water transferred with the ions; in most cases 9 - 10 mol F^{-1} through each membrane.

Table 2 Electrodialytic concentration of sea water

Ion	Concentration (equiv l^{-1}) in		
	sea water feed	electrodialysed brine	solar evaporation brine
Na^+	0.443	3.030	2.335
K^+	0.010	0.105	0.050
Ca^{2+}	0.020	0.075	0.082
Mg^{2+}	0.108	0.199	0.553
Cl^-	0.523	3.402	2.745
SO_4^{2-}	0.054	0.008	0.275

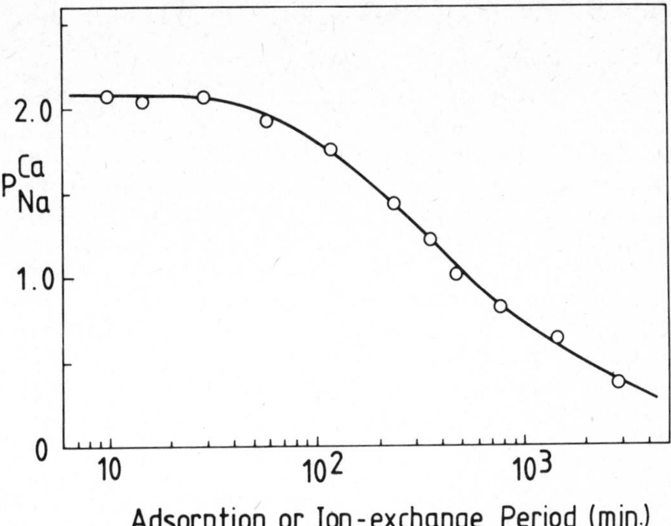

Fig. 5 Change in the relative permeability of Neosepta CL-25T cation exchange membrane to calcium and sodium ions P_{Na}^{Ca} $(=t_{Ca}C_{Na}/t_{Na}C_{Ca})$ with increasing time of exposure of the membrane to 10 ppm polyethylene imine (Everamine 210T) prior to use in electrodialysis.

Although the desalination of brackish water and the concentration of sea water together account for by far the greatest use of ion exchange membranes in electrodialysis, other applications are also important. Among well known examples is the demineralization of cows' milk. The salt content of cows' milk is more than three times that of human milk and must be reduced before it can be used as baby food. This has been successfully achieved by electrodialysis at fairly low current densities so as avoid pH changes due to polarization. A larger scale use which presents related problems is the demineralization of cheese whey. The presence of high concentrations of organic matter in the treated liquids causes problems with membrane fouling and in some plants the anion membranes are replaced by neutral, non-selective membranes. The organic solutes can form complexes with inorganic ions thus altering their transport numbers while pH changes can seriously damage by denaturation the valuable proteins in the whey. Thus the treatment of such complex liquids by electrodialysis introduces challenging problems. Salts are frequently by-products in the manufacture of organic chemicals such as amino-acids and have been successfully removed by electrodialysis in specially designed plants.

MEMBRANE CELLS IN THE CHLOR-ALKALI INDUSTRY

One of the principal electrochemical processes of the chemical industry is the manufacture of sodium hydroxide and chlorine by the electrolysis of brine. Annual production throughout the world amounts to several million tons. Traditionally the electrolysis has been carried out in mercury amalgam cells from which very pure sodium hydroxide can be recovered at very high concentrations. The risk of pollution of the environment by mercury has resulted in legislation that will soon demand the replacement of mercury cells in several countries. In an alternative process the anode and cathode are separated by an inert porous asbestos diaphragm. The alkali produced in these diaphragm cells is of low concentration and is heavily contaminated with chloride. Furthermore, the current efficiency of this process is low and the purity of the chlorine difficult to maintain.

If a cation-exchange membrane permeable to sodium and impermeable to hydroxide and chloride ions could be set up between the electrodes this would permit the construction of an efficient and non-polluting electrolytic cell. The main problem is to find a membrane which maintains its chemical integrity and its electrochemical properties during continuous exposure to 17 molal caustic soda at almost 100°C in an atmosphere of chlorine. Conventional ion-exchange membranes deteriorate very rapidly in such hostile conditions. In 1958 a membrane made from a dispersion of powdered styrene + divinyl benzene sulphonic acid resin in poly-(tetrafluoroethylene) or, better, Kel-F $[\!-\!(CF_2-CFCl)\!-\!]_n$ was described as being suitable for this process but it was never manufactured (32).

Recently Ionics Inc. has described an improved heterogeneous membrane in which the ion-exchanger is made by reacting polyphenylene sulphide with oleum (33). Polyphenylene sulphide $\left[\!-\!C_6H_4\!-\!S\!-\!\right]_n$ is a commercially available polymer and has excellent resistance to degradation and atmospheric oxidation up to temperatures above 400°C. When treated with 10% sulphur trioxide in sulphuric acid the sulphide links are oxidised to sulphone and, with difficulty over a period of several hours, about one quarter of the phenylene rings become sulphonated. At the same time some sulphone crosslinks are introduced giving a hard infusible resin which can be represented by

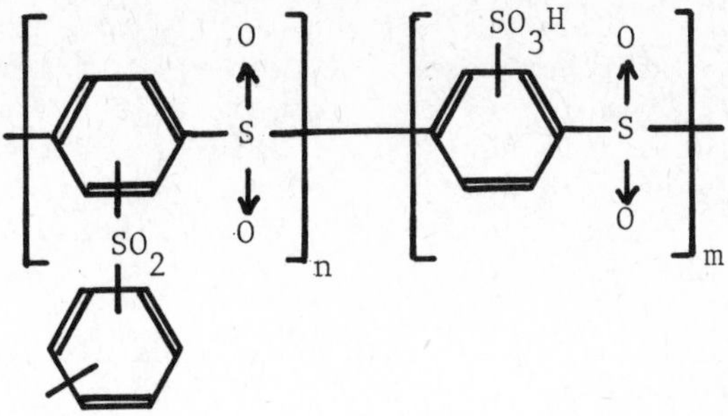

polysulphone sulphonic acid

The finely powdered resin is mixed with a similar weight of polyvinylidene fluoride, or one of its copolymers with CF_2CFCl, swollen to a paste with acetone or other suitable solvent, spread on a glass plate either with or without a backing cloth and dried in an oven. The resulting membrane is very resistant to oxidation and can be used as cell separator in the manufacture of hypochlorite by the anodic oxidation of brine. It is cheap to manufacture compared with the fluorocarbon membranes now to be described but it is not adequate to withstand the extreme conditions met in alkali manufacture.

Nafion membranes

The basic process of chlorine and alkali manufacture in a membrane cell is very easily understood. A concentrated solution of sodium chloride is fed continuously to an anode chamber separated from a cathode chamber by a cation exchange membrane. Water is supplied to the cathode compartment to the extent necessary to supplement that arriving electro-osmotically along with the sodium ions through the membrane. Chloride ions are discharged at the anode and chlorine is evolved. Water is split at the cathode, hydrogen is evolved and sodium hydroxide remains in the catholyte and is run off continuously. The product should contain up to 40% caustic by weight, free from chloride ions. To ensure that the chlorine produced is free from oxygen the pH of the anolyte must not rise. In order to be economically competitive with mercury cells the current efficiency of the caustic production must be very high, preferably greater than 95%.

The foregoing three conditions demand excellent co-ion exclusion by the membrane. The rigour of this demand can be judged from the fact that not only is the catholyte highly concentrated but the anolyte feed is usually 3.5 M sodium chloride. Furthermore the required product concentration can be achieved only by ensuring that the electro-osmotic and osmotic transfer of water through the membrane does not exceed about 4 mol F^{-1}.

There are two further essential requirements for the process to be economically viable: a very low electric resistance in the membrane and solutions is necessary to minimize electric power consumption and very high operating current densities are needed, usually 1 - 3 kA m^{-2}, in order to make the maximum use of the capital cost of the plant.

The first real indication that a membrane adequate to meet these demands might be devized came in 1968 when E.I. Dupont de Nemours & Co. produced a perfluorinated ion-exchange resin, developed a few years earlier (34), in sheet form (35). In essence the material was a copolymer of tetrafluoroethylene and the sulphonyl fluoride of a perfluorovinyl polyether which forms a sulphonate on hydrolysis. Originally called XR, this material is now known as Nafion. Its chemistry is summarized below.

A perfluoro-olefin reacts with sulphur trioxide to form a cyclic sultone which on rearrangement and reaction with hexafluoropropene ethoxide gives a compound of the general formula $FSO_2CF_2O[CF(CF_3)CF_2O]_nCF(CF_3)COF$. Treatment with sodium carbonate removes COF_2 and leaves a vinyl ether which can be copolymerized with tetrafluoroethylene to a high molecule weight polymer

$$FSO_2CF_2O[CF(CF_3)CF_2O]_n \underset{|}{CF} - [CF_2 - CF_2]_m -$$

Control of m controls the equivalent weight of the resin and control of n influences the mechanical and morphological properties of the matrix. It is melt processed into a membrane and the sulphonyl fluoride FSO_2 - groups are then hydrolyzed to sulphonic acid or sodium sulphonate.

Properties of Nafion

Although Nafion withstands the severe thermal and chemical resistance demanded by the chloro-alkali process, in its original form it possessed properties which adversely affected its economic viability. In particular the electrically driven passage of OH^- co-ions from catholyte to anolyte at caustic concentrations above about 25% by weight was unacceptably high. This resulted in reduced current efficiency, to the formation of chlorate and hypochlorite in the anode compartment and to the contamination of the chlorine with oxygen unless acid dosing was used to control pH.

The economic importance of the chlor-alkali industry is so great that far more research effort has been devoted already to improving these perfluorinated cation-exchange membranes than to any other class of ion-exchange membranes over the past thirty years

A typical Nafion membrane originally contained about ten $-C_2F_4-$ links for every sulphonated perfluoro-polyether link giving an equivalent weight of about 1300. This was controllable and materials were made with molecular weights ranging from below 1000 up to 1500. The relatively large average distances between the ionic groups in this polymer taken together with the hydrophobic nature of the perfluorinated intermediate sections puts Nafion into the class of ionic polymers called ionomers.

In water these swell and the ionic groups become hydrated and dissociated but the amount of water taken up is too small to convert the mass into a fairly homogeneous hydrogel. Conductances are normally low in such materials but at a sufficiently low equivalent weight the polymer segment mobility is sufficient to allow some limited clustering of the ionic groups. This process is favourable because the configurational restrictions it places on the polymer chains are more than offset by the reduction in electrostatic free energy which results. The clusters absorb more water and, at the equivalent weights chosen for Nafion, they touch and interconnect giving rise to an internal network of conducting channels in the polymer through which the cations and some water can be transported electrically.

The internal structure of Nafion has been intensively studied by many spectroscopic and thermodynamic techniques and a fairly complete understanding of the factors which determine this structure is now emerging (36).

Improvement of Nafion

Partly through intuition and partly through an extended understanding various means have been found to improve the performance of Nafion. Although high water absorption lowers the electrical resistance it permits co-ion uptake and hence allows a considerable current of hydroxyl ions to pass. By forming the membrane on a Teflon woven fabric swelling was restricted and the current efficiency somewhat improved. In most ion-exchange resins and membranes the lower the equivalent weight the higher the fixed ion concentration and the better the Donnan exclusion of co-ions. In the case of ionomers, owing in part to their low dielectric constant, the reverse holds. A lower equivalent weight encourages better fixed ion clustering and hence higher swelling. As a result the molal concentration of fixed charges in the conducting aqueous regions goes down with the equivalent weight decreasing. Thus material of high equivalent weight excludes the transport of hydroxyl ions up to a higher caustic concentration but its high electrical resistance constitutes a serious disadvantage.

The melt-processable thermoplastic nature of Nafion enabled a compromise to be achieved by bonding together two layers one of high and one of low equivalent weight. The membrane is mounted with the high equivalent weight side exposed to the catholyte so as to restrict the entry of the hydroxyl ions. The low equivalent weight material imparts high conductance to the composite membrane which results in high current efficiency being achieved with relatively good voltage/current characteristics.

It is necessary to restrict not only hydroxyl transport from catholyte to anolyte but also water transport from anolyte to catholyte. 40% w/w sodium hydroxide solution contains only 3.3 mole water per mole sodium hydroxide. Allowing that 1 mole of water is split at the cathode per Faraday, a product consisting of 40% caustic soda can be reached in the cell provided no more than 4.3 molecules of water are transferred with each sodium ion. If the equivalent weight of the polymer is large the ion clusters, which are the regions where most of the absorbed water accumulates, are well separated. In consequence the electro-osmotic transfer of water is impeded but at the cost of a higher electrical resistance and power dissipation. It has been proposed to impregnate the surface of the membrane facing the cathode compartment with a monomer such as p-vinyl benzyl alcohol and then to polymerize this (37). Closing the surface structure of the membrane in this way reduces both the hydroxyl and the water transport.

One of the most effective methods of improving the current efficiency of Nafion membranes has been to react one surface of the material at the sulphonyl fluoride stage with an aliphatic amine or diamine for a few minutes. The effect of this treatment is to create a layer, usually 1 - 100 μm thick, of N-monosubstituted sulphonamido groups. The remainder of the membrane is then hydrolyzed to give sulphonate exchange sites in the normal way.

The surface layer of e.g. $-SO_2NHCH_3$ groups behaves as weakly acidic and forms cation exchange sites which absorb relatively little water. These treated membranes when arranged with the sulphonamido layer facing the cathode are found to resist very well the passage of hydroxyl ions but there is only a slight increase in overall membrane resistance as the price to be paid for this improvement in current efficiency.

Carboxylic membranes

The next stage and the current stage of membrane development came with the realization that at the high pH of a chlor-alkali cell carboxylic groups are as effective as sulphonic groups in providing cation exchange sites. The Japanese membrane companies have been especially prominent in developing membranes containing carboxyl groups. Thus the Asahi Chemical Co. (39) have modified Nafion membranes while at the sulphonyl fluoride stage by reduction first to the sulphinic acid and then by further and stronger reductions to a carboxylic acid:-

$$-OCF_2CF_2SO_2F \rightarrow -OCF_2CF_2SO_2H \rightarrow -OCF_2COOH$$

Either the process was carried out as a surface treatment of the membrane or a laminate was made of a carboxylic and a sulphonic sheet and used in the electrolysis cell with the carboxylic groups at the cathodic side.

Most recently Asahi Chemical Co., Asahi Glass Co. and Tokuyama Soda Co. have each developed carboxylic perfluorocarbon membranes. These membranes are copolymers or terpolymers containing carboxyl and sometimes also sulphonate groups. By adjusting the monomer ratios the glass temperature and water content of the membrane can be adjusted so as to give optimum performance. Many monomers and precursors of the ionic groups have been described. In most membranes an oxygen atom is used as a hinge attachment for the side groups on the perfluoroethylene chain which contributes to the elasticity and to a desirably low glass temperature of the final copolymer.

Using these new membranes, about 100 μm thick, alkali containing up to 50% caustic by weight can be produced with 95% current efficiency. Excellent reviews of this new work can be found in a recently published book (36). It is likely that within the next ten years a major part of the world's alkali will come to be produced in membrane cells.

COMPOSITE MEMBRANES

Several references have been made above to the use of membranes with graded or layered structures. Examples are the use of two films of different equivalent weights bonded together for use in chlor-alkali cells and the use of an adsorbed layer of polyanions on a cation-exchange membrane in a cell for the recovery of salt by the concentration of sea water. In both of these cases the purpose of the second layer was to improve the performance of the membrane in an essentially single-layer process. Composite membranes can carry out also processes which could not be performed by simple uniform membranes. Two examples will be discussed briefly in this article. The first concerns a composite membrane made of layers arranged in series. Such membranes find application in electrically driven processes. In the second example the different membrane elements are arranged in a mosaic for parallel flows. Such membranes are used in a process driven by an applied pressure and will be described later.

Bipolar membranes

When a cation permeable and an anion permeable membrane are bonded together face to face the result is a bipolar membrane. They were first made by heat-bonding a pair of membranes that contained polyethylene as a matrix binder. Such membranes were prepared with a thick cation exchange layer and a thin anion exchange layer and were proposed for use in electrodialytic desalination (40). The beneficial effects claimed for such membranes were the prevention of the precipitation of hydroxides in the intermembrane spaces of the cell by maintaining the concentrate stream acid and also to simplify the control of organic fouling by frequent reversal of the current because, with such membranes, the concentrate and diluate streams do not become interchanged when the current is reversed. In fact the use of bipolar membranes in normal electrodialysis was never developed far partly because the membranes prepared by pressure bonding at 150°C tended to develop a highly resistive layer at their internal interface.

More recent patents of bipolar membranes describe the controlled functionalisation of styreneised polyethylene sheet by attacking first from one side with a sulphonating agent and then by chloro-methylating and aminating the other side (41). By judging correctly the conditions and times of attack the depth of penetration of the reagents into the film can be finely controlled so as to produce a membrane of low electrical resistance that can operate at current densities up to 1.5kA m^2 or more. It has been found that if the outer surfaces of the membrane are eroded by sanding, solvent wiping etc. before functionalisation, a surface skin of polystyrene formed during manufacture but not held within the matrix is removed and the chemical reactions then proceed more uniformly giving a better finished product.

Bipolar membranes formed in this way to have essentially electrically equivalent anion and cation permeable layers function as water splitters at potentials greater than about 0.8 volt. They generate H^+ and OH^- ions when electrodialysis is carried out with the cation exchanging side of the bipolar membrane facing the cathode. Such membranes can be used to carry out in a simple and economic way the reaction

$$MX + H_2O \rightarrow HX + MOH$$

with continuous separation of the products and without the need for acid or alkali dosing.

Research on the use of bipolar films is a rapidly growing area of membrane technology at the present time.

PRESSURE DRIVEN PROCESSES

It was pointed out at the beginning that a flux through a membrane is driven by the negative gradient of the generalised chemical potential of the species concerned. This driving force can include a gradient of pressure. Thus, in addition to the electrically driven desalination process, two other possibilities exist. They are hyperfiltration or reverse osmosis, in which water is driven through a membrane under pressure and the salt is retained, and piezodialysis in which salt is driven through the membrane and water, or at least a more dilute solution, is retained.

Hyperfiltration

The desalination of water by hyperfiltration through non-ionic membranes such as cellulose acetate is now widely used. Table 3 shows a typical set of working thermodynamic conditions in, say, sea water desalination. It can be seen that the driving force per mole of salt is greater than that per mole of water and that the separation of salt and water is hindered rather than aided by the relative thermodynamic forces on the two components. The mobilities of small inorganic ions and of water molecules differ but not sufficiently to offset alone this unfavourable thermodynamic factor. Good separation of salt and water therefore depends on having a large difference in their relative concentrations in the membrane. In the case of cellulose acetate, the ratio of the concentration of salt and water in the membrane is about 5% of the ratio in the contacting solution. In this case also the diffusion coefficient of water in the membrane is about 100 times that of salt. This is an unusually favourable factor and it is these two items taken together that give cellulose acetate its especially good powers of desalination by hyperfiltration.

It has often been suggested that the Donnan exclusion of the co-ions by conventional ion-exchange membranes would provide the concentration ratio between salt and water necessary to enable such membranes to act efficiently in desalination by reverse osmosis provided that the membrane could be made thin enough to pass an adequate flux under pressures up to 10 MPa and strong enough to withstand such pressures.

Experience so far with ion-exchange membranes in hyperfiltration has not been encouraging. In concentrated solutions the Donnan exclusion is only a weak effect and in dilute solutions it falls short of ideal thermodynamic expectations due to the slightly non-uniform distribution of the fixed charges on the molecular scale in

Table 3 Hyperfiltration of Sea Water at 25°C

Mole fraction of sodium chloride in brine feed = 0.01 i.e. 0.55 m

Mole fraction of sodium chloride in desalinated product = 0.0001 i.e. 325 ppm

Applied pressure $\Delta p = 10^7$ Pa (i.e. ∿ 100 atm)

Thermodynamic driving force on the water $\Delta\mu_1$, assuming ideality, given by

$$\Delta\mu_1 = \bar{V}_1\Delta p + 2RT \ln (0.99/0.9999) = 130 \text{ J mol}^{-1}$$

Thermodynamic driving force on the salt $\Delta\mu_2$ given by

$$\Delta\mu_2 = \bar{V}_2\Delta p + RT \ln (0.01/0.0001) = 11710 \text{ J mol}^{-1}$$

$$\Delta\mu_2/\Delta\mu_1 = 90 \ (\bar{V}_1 = 18 \text{ cm}^3, \bar{V}_2 \simeq 30 \text{ cm}^3)$$

It may be noted that as the driving pressure Δp is increased, $\Delta\mu_1$ is increased almost in proportion to Δp because the small improvement in product quality that results, i.e. by raising the mole fraction of water in the product above 0.9999, has no significant effect on the osmotic term:- 2 RT ln (0.99/mole fraction of water in the product).

Both terms that contribute to $\Delta\mu_2$ increase when Δp is increased because a further fall in the low concentration of salt in the product has a large effect on RT ln (0.01/mole fraction of salt in the product). This term normally greatly exceeds $\bar{V}_2 \Delta p$, which is even ignored in some simplified treatments.

No matter what driving pressure is used $\Delta\mu_2 > \Delta\mu_1$ because $\bar{V}_2 > \bar{V}_1$ and the osmotic term in $\Delta\mu_1$ is negative while that in $\Delta\mu_2$ is positive.

the membrane matrix (42). Furthermore, in a hyperfiltration cell in which the flux is large enough to be useful there will be concentration polarization at the upstream face. This increases still further the concentration of co-ions carried into the membrane. Thus one finds the Donnan exclusion is largely swamped.

The diffusion coefficient of water in ion-exchange membranes is typically about ten times that of the ions and there is strong coupling between the ion and solvent fluxes (43). This is another factor leading to poor salt rejection by such membranes. One of the few systems that has so far shown real promise in this connection,and is comparable in performance with cellulose acetate while being chemically and mechanically superior,is sulphonated polysulphone (44). In this material the water uptake is only about 12% with an ion-exchange capacity of about 0.9 mequiv per gram. The rejection of salt by sulphonated polysulphone membranes is comparable with that of cellulose acetate while their permeability to water is a little better than that of cellulose acetate. It is the low dielectric constant of the polymer containing only 12% of water that is a more important cause of salt rejection than the Donnan exclusion of the co-ions.

Piezodialysis

An alternative possibility is to exploit the high ion permeability of ion-exchange membranes to permit the salt to pass through under pressure leaving behind the desalinated product water. Such a process would benefit from low frictional dissipation of energy because the minor component was passed through the membrane. Also, a minor flaw in the membrane causes only a small loss of product but not a deterioration in its quality (45).

When solution is forced through an ion-exchange membrane under pressure coupling of the flows of ions and water causes a streaming potential to develop which prevents the counter-ions leaving the membrane. If a membrane were made which was a mosaic of anion and cation permeable patches there would develop on its faces a mosaic of opposite charges which would continuously discharge by ion currents flowing in the solutions at either side of the membrane (46). If the membrane mosaic is sufficiently fine grained the resistance of the current paths will be small and the overall effect would be to force out of the membrane a mixture of ions and water equivalent in concentration to that swelling the membrane i.e. characteristically 10 molal at least. This would be far more concentrated than the feed solution and so the latter would become diluted.

This process, called piezodialysis, is attractive and promising in prospect. Despite considerable research the development of suitable mosaic membranes has proved to be very difficult. Of the many ideas that have been tried one has shown substantial promise. This is the latex-polyelectrolyte membrane (47). A latex containing 60% solids of equal amounts of polystyrene and polybutadiene is prepared and mixed with sodium polystyrene sulphonate. When cast into a film the polystyrene and sodium polystyrene sulphonate separate into small islands embedded in polybutadiene. Chloromethylation and amination can then be used to convert the polystyrene islands into anion exchanger patches after first crosslinking the film with aluminium chloride. In small scale demonstration modules such membranes have shown up to 4-fold enrichment of the concentration of the permeate from 0.01 M salt solution under 10 MPa driving pressure. Further development of this process awaits the invention of stronger membranes.

REFERENCES

1. Helfferich, F. Ion Exchange (New York, McGraw-Hill, 1962) 134-143.
2. Teorell, T. Transport Phenomena in Membranes. Disc. Faraday Soc. 21 (1956) 9-26.
3. Teorell, T. An attempt to formulate a quantitative theory of membrane permeability. Proc. Soc. Exptl. Biol. Med. 33 (1935) 282-285.
4. Meyer, K.H. and J.-F. Sievers. La perméabilité des membranes I Théorie de la perméabilité ionique. Helv. Chim. Acta 19 (1936) 649-664.
5. Glueckauf, E. and R.E. Watts. The Donnan law and its application to ion exchanger polymers. Proc. Royal Soc. A 268 (1962) 339-349.
6. Meares, P. The Permeability of Charged Membranes in Ussing, H.H. and N.A. Thorn (eds) Transport Mechanisms in Epithelia (Copenhagen, Munksgaard, 1973) 51-67.
7. Meares, P. Transport in Ion-exchange Polymers in Crank, J. and G.S. Park (eds) Diffusion in Polymers (London, Academic Press, 1968) Ch. 10.
8. Solt, G.S. Electrodialysis in Meares, P. (ed) Membrane Separation Processes (Amsterdam, Elsevier, 1976) Ch. 6.
9. Ahlgren, R.M. Electromembrane Processes for Recovery of Constituents from Pulping Liquors in Lacey, R.E. and S. Loeb (eds) Industrial Processing with Membranes (New York, Wiley-Interscience, 1972) Ch. 5.
10. Ahlgren, R.M. Electromembrane Processing of Cheese Whey in Lacey, R.E. and S. Loeb (eds) Industrial Processing with Membranes (New York, Wiley-Interscience, 1972) Ch. 4.
11. Bodamer, G.W. Rohm and Haas Co. Philadelphia, Pa. U.S. Patent (1954) 2 681 320.
12. Helfferich, F. Ion Exchange (New York, McGraw-Hill, 1962) Ch. 3.

13. Eisenman, J.L., E.T. Roach and A. Scieszko. Ionics Inc., Watertown, Mass. U.S. Patent (1972) 3 607 706.
14. Tsunoda, Y., M. Seko, M. Watanabe, A. Ehara and T. Misumi. Asahi Chemical Industry Co. Japanese Patents (1957) 4142-4146.
15. Chen, W.K. American Machine and Foundry Co. U.S. Patent (1966) 3 247 133.
16. Mizutani, Y. and M. Nishimura. Studies on Ion-Exchange Membranes II Heterogeneity in Ion-Exchange Membranes. J. Applied Polym. Sci. 14 (1970) 1847-1856.
17. Hodgdon, R.B. and J.R. Boyack. Study of Swelling in Two New Ion-Exchange Membranes. J. Polymer Sci. A 3 (1965) 1463-1472.
18. De Korosy, F. Israel Patent (1970) 26 598.
19. Korngold, E. Present State of Technological Development of Permselective Polyethylene Membranes at the Negev Institute for Arid Zone Research. Adv. Desalin., Proc. Natl. Symp. Desalin. 7th (1970) 23-28.
20. Mizutani, Y., R. Yamane, H. Ihara and H. Motomura. Studies of Ion Exchange Membranes XVI The Preparation of Ion Exchange Membranes by the "Paste Method". Bull. Chem. Soc. Japan 36 (1963) 361-366.
21. Kedem, O. Reduction of Polarization in Electrodialysis by Ion-conducting spacers. Desalination 16 (1975) 105-118.
22. Sata, T., R. Izuo, Y. Mizutani and R. Yamane. Transport Properties of Ion-Exchange Membranes in the Presence of Surface Active Agents. J. Coll. Interf. Sci. 40 (1972) 317-328.
23. Kusumoto, K., H. Ihara and Y. Mizutani. Preparation of Macroreticular Anion Exchange Membrane and its Behaviour Relating to Organic Fouling. J. Appl. Polym. Sci. 20 (1976) 3207-3213.
24. Hogdon, R.B. Ionics Incorporated, Watertown, Mass. U.S. Patent (1975) 3 926 864.
25. Kusumoto, K., T. Sata and Y. Mizutani. Modification of Anion-Exchange Membranes with Polystyrenesulphonic Acid. Polymer J. 8 (1976) 225 - 226.
26. Kusumoto, K., T. Sata and Y. Mizutani. New Anion-Exchange Membrane Resistant to Organic Fouling. Proc. Int. Symp. Fresh Water from the Sea 4th. 3 (1973) 111-118.
27. Kusumoto, K. and Y. Mizutani. New Anion-Exchange Membrane Resistant to Organic Fouling. Desalination 17 (1975) 111-120.
28. De Korosy, F. and E. Zeigerson. Breakthrough of Poisoning Multivalent Ions across a Permselective Membrane during Electrodialysis. J. Physical Chem. 71 (1967) 3706-3709.
29. De Korosy, F. and E. Zeigerson. Interaction of Permselective Membranes and their Counterions. Desalination 5 (1968) 185-199.
30. Yamane, R., M. Ichikawa, Y. Mizutani and Y. Onoue. Concentrated Brine Production from Sea Water by Electrodialysis Using Ion-Exchange Membranes. Ind. Engng. Chem. Proc. Des. and Dev. 8 (1969) 159-165.
31. Sata, T. Modification of Properties of Ion-Exchange Membranes II. J. Coll. Interf. Sci. 44 (1973) 393-406.

32. Bodamer, G.W. Rohm and Haas Co. Philadelphia Pa. U.S. Patent (1958) 2 827 426.
33. Hodgdon, R.B. Ionics Inc. Watertown, Mass. U.S. Patent (1978) 4 110 265.
34. Gibbs, H.H. and R.N. Griffin. E.I. du Pont de Nemours, Wilmington, Del. U.S. Patent (1962) 3 041 317.
35. Wolfe, W.R. E.I. du Pont de Nemours, Wilmington, Del. British Patent (1970) 1 184 321.
36. Eisenberg, A. and H.L. Yeager. Perfluorinated Ionomer Membranes (Washington D.C., ACS Symp. Ser. 180, 1982).
37. Kiyota, T., S. Asami and A. Shimizu. Toyo Soda Mfg. Co. Ltd. Jap. Patent (1978) 78 60388.
38. Reswick, P.R. and W.G. Grot. E.I. du Pont de Nemours, Wilmington, Del. U.S. Patent (1978) 4 085 071.
39. Seko, M., Y. Yamakoshi, H. Miyauchi, M. Fukumoto, K. Kimoto, I. Watanabe, T. Hane and S. Tsushima. Asahi Chemical Co., Osaka. British Patent (1978) 1 523 047.
40. Leitz, F.B. Ionics Inc. Watertown, Mass. U.S. Patent (1971) 3 562 139.
41. Lee, L.T.C., G.J. Dege and K.-J. Liu. Allied Chemical Corp. Somerville, N.J. U.S. Patent (1977) 4 057 481.
42. Glueckauf, E. A new approach to ion-exchange polymers. Proc. Royal Soc. A 268 (1962) 350-370.
43. Foley, T., J. Klinowski and P. Meares. Differential conductance coefficients in a cation-exchange membrane. Proc. Royal Soc. A 336 (1974) 327-354.
44. Brousse, Cl., R. Chapurlat and J.P. Quentin. New Membranes for Reverse Osmosis I. Characteristics of the Base Polymer: Sulphonated Polysulphones. Desalination 18 (1976) 137-153.
45. Leitz, F.B. Piezodialysis in Meares, P. (ed) Membrane Separation Processes (Amsterdam, Elsevier, 1976) Ch. 7.
46. Gardner, C.R., J.N. Weinstein and S.R. Caplan. Transport Properties of Charge-Mosaic Membranes III Piezodialysis. Desalination 12 (1973) 19-33.
47. Shorr, J. and F.B. Leitz. Development of Membranes and Resins for Piezodialysis. Desalination 14 (1974) 11-20.

Acknowledgements

The author gratefully acknowledges permission to reproduce diagrams as follows:

Fig. 1a from G.S. Solt, Electrodialysis, Ch. 6 in Membrane Separation Processes, Elsevier, Amsterdam 1976.

Fig. 2 from O. Kedem, Desalination, 16 (1975) 105-118.

Fig. 3 from O. Kedem and Y. Maoz, Desalination, 19 (1976) 465-470.

Fig. 4 from K. Kusumoto and Y. Mizutani, Desalination, 17 (1975) 111-120.

Fig. 5 from T. Sata, J. Colloid and Interface Sci., 44 (1973)393-406.

THE NERNST-PLANCK EQUATION IN THERMODYNAMIC TRANSPORT THEORIES

G. Dickel

Institute of Physical Chemistry, University of Munich

1. INTRODUCTION

The first investigations of Boyd (1) concerning the kinetics of ion exchange are based on a solution of the second Fick's law. Thereby the boundary conditions were taken into account by the well known method developed by Fourier in his famous book "Theorie Analytique De La Chaleur" (1822) (2).

The application of this method is restricted to constant values of the diffusion coefficient. Since Boyd had investigated the exchange of isotopes, this condition is valid a priori in this case. Applying this method to the exchange of different ions, a restriction of small concentration intervals is necessary, where a mean value of the diffusion coefficient can be assumed. This restriction, however, can be set aside, if the concentration dependence of the diffusion coefficient is known.

According to Helfferich (3) the interdiffusion coefficient resulting from the Nernst-Planck eqn. is given by

$$D_{12} = D_1 D_2 (c_1+c_2)/(D_1 c_1+D_2 c_2) . \tag{1}$$

where D_1 and D_2 are the individual diffusion coefficients of univalent ions 1 and 2 and c_1 and c_2 the concentrations. Using this coefficient, the second Fick's law is given by

$$dc/dt = - \Delta(D_{12} c), \quad (\Delta = \text{Laplace operator}). \tag{2}$$

Taking a constant mean value of D_{12} and expanding the right hand side in a Taylor series we obtain

$$\Delta c \sim \bar{c} - c_o. \tag{3}$$

This means, that Δc is a measure of the deviation of c_o in P_o from the mean value $\bar{c}$ in the neighbourhood of P_o. Two cases are possible:

1. Regular case (Helfferich): Regarding the limiting case $dc/dt=0$, we have $\Delta(D_{12}c)=0$. In this case $c_o \rightarrow \bar{c}$ and we get the Laplace eqn.

$$\Delta c = 0 \,. \tag{4}$$

2. Singular case: In the case of sinks and/or sources a permanent deviation from eqn.(4), given by

$$\Delta c = \Psi(c) \tag{5}$$

results. Ψ is the Green's function or the source function of the traps (fixed ions).

1.1 Examples of Green's function:

1. A thread-shaped ion exchanger in a highly diluted solution of an electrolyte of concentration $\bar{c}$ (see fig.1). If c_o is the concentration of the counterions, eqn(5) is valid if $c_o \gg \bar{c}$. This is a consequence of eqn(3).

2. The electric potential inside and outside of charges. If U is the electric potential, we have:
Outside of charges : $\Delta U = 0$ (Laplace equation).
Inside of charges : $\Delta U = - 4\pi\varrho$ (Poisson equation).
The importance of the Poisson equation follows from the fact, that this equation, together with the Laplace eqn. gave rise to the developement of the potential theory. Thereby the introduction of the Green's function is necessary to solve boundary value problems occuring in this connection.

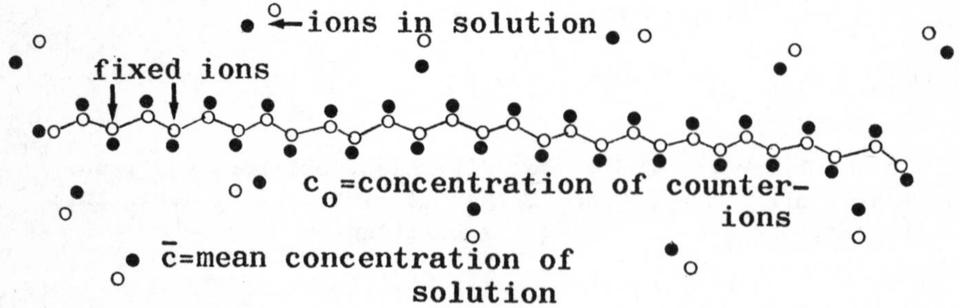

Fig. 1 A thread-shaped ion exchanger in a diluted solution of concentration $\bar{c}$

In the following the field theory should be introduced to solve complicated transport problems in membranes and in ion exchangers. Thereby the regular case defined by eqn(4) will control the stationary transport in membranes, whilst the singular case, defined by eqn(5), must be taken into account to obtain a general inpretation of the much more complicated ion exchange kinetic.

2. APPLICATION OF ONSAGER's PRINCIPLE OF LEAST DISSIPATION OF ENERGY TO MEMBRANE TRANSPORT.

As Gyarmati showed(4), the application of Onsager's principle of least dissipation of energy is an appropriate method for introducing the field theory as a base of thermodynamical transport processes (5,6). For this sake we regard a membrane separating two solutions I and II of different concentrations and consider dn_i molecules of the sort i moving in a volume qdx of a membrane with the mean velocity v_i along the x-axis. If v_L is the velocity of the solvent, according to the frictional model(7,8) the force

$$dK_i = f_i(v_i - v_L)dn_i \tag{6}$$

must be applied, if f_i is the frictional coefficient between the particle i and the solvent. Having N sorts of particles covering the path ds, the energy

$$d^2E = \sum^N dK_i ds_i \tag{7}$$

is necessary to maintain this state of motion in a volume qdx. Introducing instead of v_i the number of particles $\dot{n}_i$ entering or leaving the membrane in unit time

$$\dot{n}_i/q = v_i c_i \, , \qquad (\dot{n}_i = dn_i/dt) \tag{8}$$

and regarding

$$ds_i = (v_i - v_L)dt$$

and

$$dn_i = c_i q dx$$

equation (7) can be written, after integration ,

$$E = \int_o^{\Delta x}\int_o^{\Delta t} dL \, dt \, , \tag{7a}$$

where

$$dL = \sum^N (f_i/c_i)(\dot{n}_i/q - c_i\dot{n}_L/c_L q)^2 \, qdx \tag{9}$$

is the Lagrangian in the strip qdx. Because E is the energy of dissipation, we can apply Onsager's principle of least dissipation of energy (9)

$$\delta_t \delta_x E = \delta_t \delta_x \int_0^{\Delta x}\int_0^{\Delta t} dL\ dt = 0. \tag{10}$$

This variational problem means to find out the extremals $\dot{n}_i$ (extr.) which furnish the minimum value of the integral (10). According to Weierstrass (10) we have

$$\dot{n}_i(\text{extr.}) = J_i(\,t,x,n_i)\ , \tag{11}$$

where J_i is the slope at the point (t,x) of the unique extremal of the field of extremals passing through the point (t,n).

3. FIELD OF EXTREMALS

Assuming that we have found these extremals J_i, we obtain the field of tangents represented in fig. 2. Therefrom we can go over to Hamilton's field theory by introducing the canonical variables, defined by

$$\pi_i qdx = \left[dL_{\dot{n}_i} \right]_{\dot{n}_i = J_i}\ , \qquad (i = 1,2,3,\ \ldots\ N,L)\ . \tag{12}$$

The left hand side expressions represent the derivatives of dL (given by eqn. (9)) with respect to $\dot{n}_i$.

In fig. 3 Hamilton's transformation will be illustrated and compared with the Lagrangian representation.

4. HILBERT's INDEPENDENCE THEOREM

Consider a strip qdx , resp. dE instead of E in eqn (7a), the Lagrangian representation is given by the line integral $dE = \int dL\ dt$. According to Hilbert, this line integral which depends on the path, can be

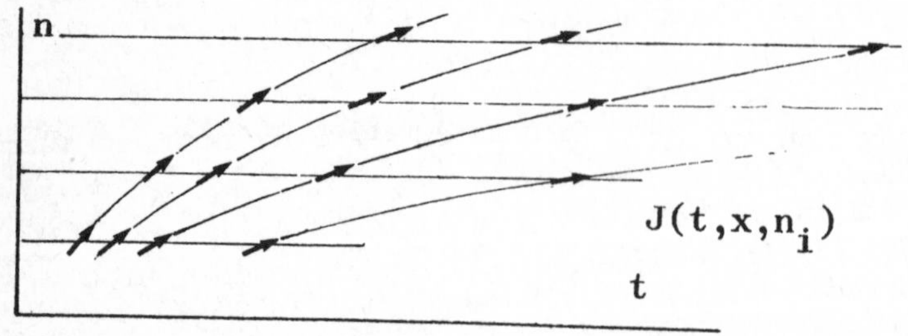

Fig. 2 Field of extremals (tangents)

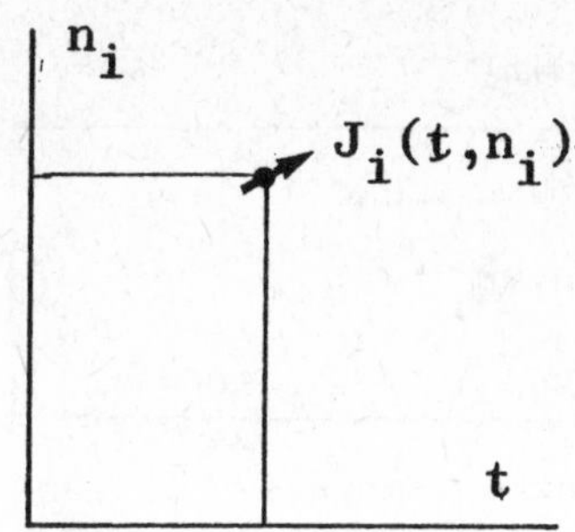

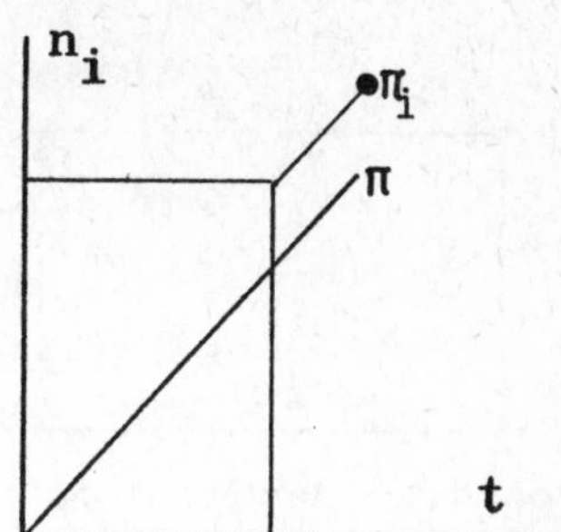

Fig. 3 Lagrangian representation

In the Lagrangian representation the state of a particle is given by the point (t,n_i) and the slope J_i at this point.

Hamiltonian representation

In the Hamiltonian representation the Lagrangian line elements in the R_{N+1} are replaced by the 2N+1 canonical variables t, n_i, π_i.

transformed into an integral, which is independent of the path, if the canonical variables are taken into account. This transformation is given by Hilbert's theorem of independence (11)

$$\int_0^{\Delta t} dL(\dot{n}_1,..\dot{n}_L)\,dt \geqq \int_0^{\Delta t} [dL(J_1,..J_L)+\sum^{N,L}(\dot{n}_i-qJ_i)\ \pi_i q dx]\,dt\ , \tag{13}$$

where the left hand side represents the Lagrangian line integral and the right hand side the Beltrami-Hilbert's independent integral. The sign of equality is valid only, if the integration on the left hand side is performed along the integral curves of the differential equation

$$\bar{\dot{n}}_i = J_i(t,x,n_i). \tag{14}$$

These integral curves are represented in fig. 2 for a fixed value of x.

5. SOLUTION OF THE VARIATIONAL PROBLEM

The solution of the double integral in eqn. (10) involves the following steps, (5,6).

1.) Following Hilbert, we conceive the variable x as a parameter and go over to the variation with respect to time.

2.) to fulfil the boundary conditions, we consider the membrane as an open system (fluxes through the boundaries), whilst the system membrane+solution must be taken as a closed system (no fluxes through the boundaries), (see Fig. 4).

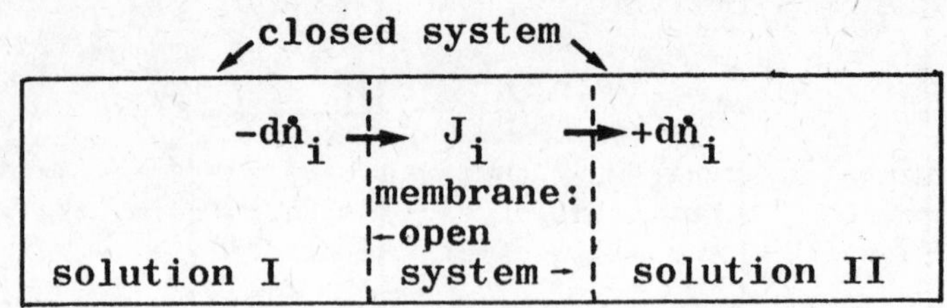

Fig.4 Boundary condition of the system membrane-solution

3.) With the help of the transformation of Routh, these conditions can be taken into account and we obtain, applying the Euler-Lagrange method, the condition for the canonical variable

$$dL_{\dot{n}_i} = \pi_i q dx , \qquad (i=1,2,..N,L) \tag{15}$$

in the strip dx at a fixed position x (parameter).
4.) According to Hilbert, the application of the theorem of independence yields the minimum condition

$$1/2\, \pi_i q dx = - d\mu_i , \qquad (i=1,2..N,L) \tag{16}$$

with respect to the position. μ_i are potential functions given by physical problems. Introducing eqn.(16) into the independent integral, we obtain the free enthalpy

$$\Delta G = \int_0^{\Delta x}\int_0^{\Delta t} \sum J_i \mu_i' dx\, dt = \sum^{N,L} n_i \Delta\mu_i , \qquad (\mu_i' = d\mu_i/dx). \tag{17}$$

5.) To fix the system of reference the following must be considered (11) a.) If a particle i should be the system of reference

$$J_i^* = \text{const.} \tag{18}$$

must be valid. The asterisk marks the system of reference. This means, that the particle i can assume any arbitrary but constant value of J_i. b.) Postulating an inertial system, the sum of all forces must vanish.

6.) To avoid the preferential position of the solvent in the Lagrangian eqn(9), we apply Euler's theorem of homogeneous functions to eqn(9) and obtain

$$dL = 1/2 \sum^{N,L} \dot{n}_i \, dL_{\dot{n}_i} . \tag{19}$$

In this representation all N+1 particles incl. the solvent are of equal relevance. Replacing in the right hand sum of N+1 terms, N arbitrary terms by their canonical variables, we obtain, considering eqn(15), the N+1 Routhian functions

$$dR_k = 1/2\ \dot{n}_k dL_{\dot{n}_k} - \sum^{N,L,(-k)} n_i d\mu_i \ , \quad (k=1,2..N,L) \tag{20}$$

which represent Lagrangians with respect to k and Hamiltonians with respect to the remaining particles. Introducing the Routhians for dL in eqn.(10), the calculus of variations yields

$$1/2\ dL_{\dot{n}_i} - 1/4\ \dot{n}_L\ dL_{\dot{n}_L \dot{n}_i} = -\ d\tilde{\mu}_i \quad (i=1,2,..N) \tag{21}$$

$$1/2\ dL_{\dot{n}_L} - 1/4 \sum^{N} \dot{n}_i dL_{\dot{n}_i \dot{n}_L} = -\ d\mu_L\ . \tag{22}$$

$\tilde{\mu}_i$ means the electrochemical potential of an ion i and μ_L the chemical potential of the solvent. If the solvent is the system of reference for a particle i, a variation must be excluded and we get

$$dL_{\dot{n}_i \dot{n}_L}{}^{*} = 0. \tag{23}$$

Multiplying all eqn. (21-22) by c_i we obtain specific forces. According to postulate 5b we get

$$1/2 \sum^{N,L} c_i dL_{\dot{n}_i} - 1/4\ \dot{n}_L \sum^{C} c_i dL_{\dot{n}_i \dot{n}_L} - 1/4\ c_L \sum^{C} \dot{n}_i dL_{\dot{n}_i \dot{n}_L} = = -\sum^{N} c_i\ d\tilde{\mu}_i - c_L d\mu_L\ . \tag{24}$$

Since the solute is the system of reference for the Donnan-ions, from eqn.(23) follows that the summation of the second and third term is restricted to the counterions, marked by C. As concerns the first term, this vanishes according to Euler's theorem because the summand is of zero degree in c_i (see eqn.9). This sum vanishes also, if the frictional action between all particles is taken into account, and not only the action between solvent and solute.

The second and third term vanish according to eqn.(23), if the solvent is the system of reference. In this case eqn.(24) yields the GIBBS-DUHEM equation

$$\sum^{N,L} c_i d\mu_i^o = 0 \ , \tag{25}$$

if the condition of electroneutrality is considered.

Going over to an ion exchanger, we must add to the system of equations(21) the constraint acting on the fixed ion

$$-\ 1/2\ dL_{\dot{n}_F^*} + 1/4\ \dot{n}_L\ dL_{\dot{n}_F^* \dot{n}_L} = d\tilde{\mu}_F\ , \tag{26}$$

in order to fulfil postulate 5a. Therefrom follows, as concerns the right hand side of eqn(24)

$$\sum^{N} c_i d\mu_i^o - c_L d\mu_L^o = - c_F z_C F d\varphi - dP. \qquad (27)$$

z_C means the charge number of univalent counterions. Beneath eqn(25) the relation $\mu_L = \mu_L^o + V_L dP$ was taken into account. Therefore eqn(24) yields regarding eqn.(9) and (27) in the case of an ion-exchanger mem= brane

$$1/2\sum^{C} f_i(J_i + J_L c_i/c_L)\,dx = - c_F z_C F d\varphi - dP . \qquad (28)$$

Eqn.(28) includes no chemical forces, but mechanical forces(frictional forces and pressure gradient) and electrical forces, only.

Thermodynamical transport systems which include no chemical potentials but mechanical forces only, have been dicovered by Prigogine(13) and called mechanical equilibria. In eqn(28) the electric potential forms together with the mechanical forces an electromechanical equilibrium(12). Such an equilibrium was first postulated by G.Schmid(14) in his investigations concerning the electrokinetic effects.

6. EXAMPLES OF ELECTRO - MECHANICAL EQUILIBRIA

The following examples arise from the mechanical equilibrium eqn(28) and/or from the flux relations(21-22). Using eqn(9) the flux relations can be written

$$f_i(J_i - J_L c_i/c_L)\,dx + \alpha f_i J_L (c_i/c_L)\,dx = - c_i d\tilde{\mu}_i =$$

$$= - c_i(d\mu_i^o + V_i dP + z_i F d\varphi), \quad (i = k, 2, \ldots N), \qquad (29)$$

$$- \sum^{N} f_i(J_i - J_L c_i/c_L)\,dx + \alpha \sum^{C} f_i J_i\,dx = - c_L d\mu_L - dP. \qquad (30)$$

We emphasize, that in these relations which are used in the following, the condition of the inertial system is not taken into account.

In eqn(29-30) $\alpha = 0$, if the solvent is the system of reference for a particle i, but $\alpha = 1/2$ if it is not the case. This means that the second terms on the left hand side in eqn(29-30) vanishes, if a solution is under consideration. In the case of a membrane $\alpha = 1/2$ for the counterions but $\alpha = 0$ for the co-ions, because the solvent is the system of reference for the latter but not for the former. This is a consequence of eqn(23).

6.1 Flux Equations in Solutions

In a solution, where no mechanical equilibrium is possible, eqn(29-30) yield, eliminating the arbitrary value J_L, the familiar system of equations used by Meares, Thain and Dawson(15), given by

$$\sum^{k} R_{ik} J_k = K_i , \qquad (i=1,2,\ldots N). \qquad (31)$$

In our case we have (6)

$$R_{ii} = f_i(f_i - \sum c_k f_k/c_i), \quad R_{ik} = f_i f_k, \tag{32}$$

and

$$K_i = -\sum f_k c_k \text{ grad } \tilde{\mu}_i - f_i\, c_L \text{grad } \mu_L \tag{33}$$

The values of $\tilde{\mu}_i$ and μ_L must be related with the help of the familiar boundary conditions to the chemical potentials of the adjoining solutions. The Onsager relations are fulfilled, in eqn(31). This is a consequence of the variational principle.

6.2 Osmotic Pressure in Membranes

To obtain the equation for the osmotic pressure in a membrane, we set $J_L = 0$ in the equation of the electromechanical equilibrium and solve for dP. We obtain, replacing the fluxes J_i with the help of the flux relations (29-30) by grad $\tilde{\mu}_i$ and considering eqn(25)

$$dP = -1/2(c_F - c_D)Fd\varphi - 1/2(c_L d\mu_L^o) + \sum c_D d\mu_D^o). \tag{34}$$

Regarding the boundary condition membrane-solution

$$c_L d\mu_L^o + dP_Q = c_L d\mu_L^s, \tag{35}$$

where the suffix s stands for the adjacent solution, we obtain for an ideal solution

$$dP = -1/2[(c_F - c_D)Fd\varphi + RT(\sum dc_i^s - \sum dc_D)] + dP_Q \quad . \tag{36}$$

P_Q is the swelling pressure.

A similar equation was derived by Schlögl(16) from phenomenological considerations in his theory of osmotic pressure in permeable membranes

$$dP = -c_F Fd\varphi + RT(\sum dc_i^s - \sum dc_D). \tag{37}$$

Besides the factor 1/2 the essential difference between eqn(36) and (37) results from the coefficient $c_F - c_D$ in eqn(36) and c_F in eqn(37). The swelling pressure was considered by Schlögl allready. The experimentum crucis for the electric term is the isotonic osmosis, where the second and third term on the right hand side of eqn(35) and (37) vanishes. In this case eqn(36) yields

$$dP = -1/2\ (c_F - c_D)Fd\varphi \quad , \tag{36a}$$

whilst Schlögl's equation turns into Schmid's equation(14).

By applying a solution of HCL to the one, and an isotonic solution of LiCl, or a mixture of HCl+LiCl to the other boundary of an ion-exchanger membrane (see fig.5) we obtain strong fluxes of water(17). The choice of the isotonic system HCl/HCl+LiCl results

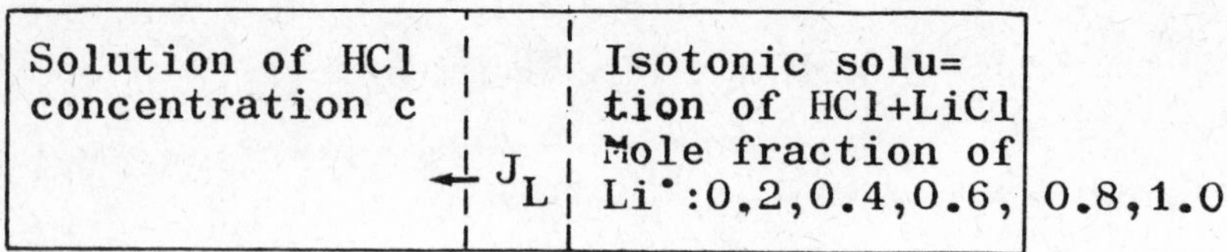

Fig.5. Schema of isotonic osmosis

from the fact that the activity coefficient of a HCl-LiCl solution is independent of the mole fraction of the mixture(18). Other systems e.g. the system HCl/HCl+NaCl shows an umimportant deviation(19).

In fig.6 the flux of water and the membrane potential as a function of the concentration of the adjacent solution is represented (20). In agreement with eqn(36), the flux of water vanishes,if $c_D = c_F$ but not the electric potential as postulated by formerly theories.

For better understanding the difference between eqn(36) and (37) we rearrange the electric term and get, regarding $c_F + c_D = c_C$.

$$dP = - 1/2\ (c_F - c_D) F d\varphi \ = \ - 1/2\ c_C F\ d\varphi \ + c_D F\ d\varphi\ . \qquad (36a)$$

Comparing the right hand side of eqn(36a) with Schmid's relation $dP = - c_F F d\varphi$ (14), we state, that the influence of the Donnan-ions was not taken into account; furthermore, the reaction between the fixed ions and the counterions, which yields the factor 1/2 ,was neglected.

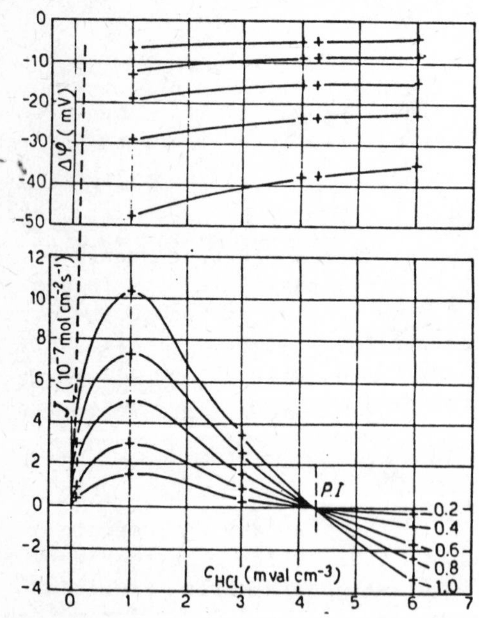

Fig.6.Membrane potential and flux of water in the isotonic system HCl/ HCl+LiCl Parameter:Mole fraction of LiCl in the solution. Abscissa:Concentration of HCl in solution.

To obtain a complete examination of isotonic osmosis we must proceed from the J_L-c representation to the dP-dφ representation. The necessary electric field follows from the mechanical equilibrium. From eqn(28) follows if J_L=0.

$$c_F F d\varphi = - 1/2 \sum^{C} f_i J_i - dP . \tag{38}$$

With the help of the flux eqn.(29-30) we replace dP by the ion fluxes and then all ion fluxes by the chemical potentials of the adjacent solution, using

$$\mu_i = \mu_i^s . \tag{39}$$

Furthermore the condition of electroneutrality

$$\sum J_i = 0 \tag{40}$$

and the equilibrium condition

$$c_1^s \, c_2 / c_2^s c_1 = K \approx 1 \tag{41}$$

must be regarded. After integration we obtain(21)

$$\Delta\varphi = - \frac{RT}{F} \ln \frac{c_2''/f_2 + c_D/f_D + c_1''/f_1}{c_2'/f_2 + c_D/f_D + c_1'/f_1} . \tag{42}$$

Setting $1/f_i = P_i$, we obtain Goldman's equation as used by Hodgkin and Katz in the investigations of nervous excitation(22).Measurements of the isotonic electric potential have been performed in the isotonic system HCl / HCl+LiCl. To investigate the influence of the membrane on the ion mobility all measurements were performed with and without membrane. These are represented in fig.7

As shown (21), the differences between the transmembrane potentials and the corresponding diffusion potentials result in the first place from the influence of the fixed ions and secondly from a change in the ratio of the ion mobilities.

Having investigated the isotonic transmembrane potential,the perfect evaluation of the equation of the electroosmotic pressure in isotonic osmosis is possible (20).In fig.8 the electroosmotic

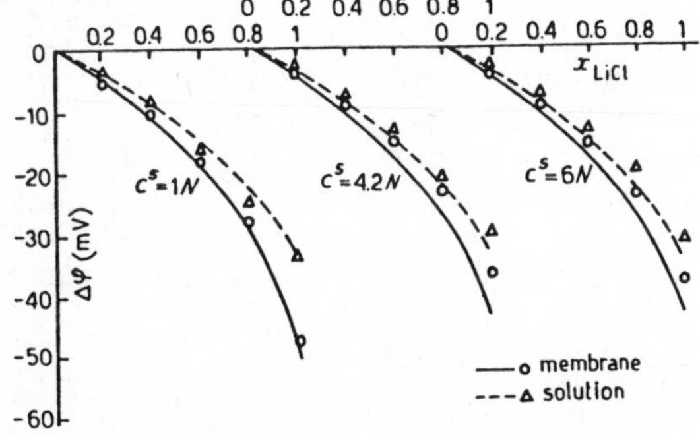

Fig.7.Transmembrane potential and the corresponding diffusion potential(in absence of the membrane) in the isotonic system LiCl+HCl/HCl vs. mole fraction of Li in solution. Parameter:concentration c^s of the HCl-solution.

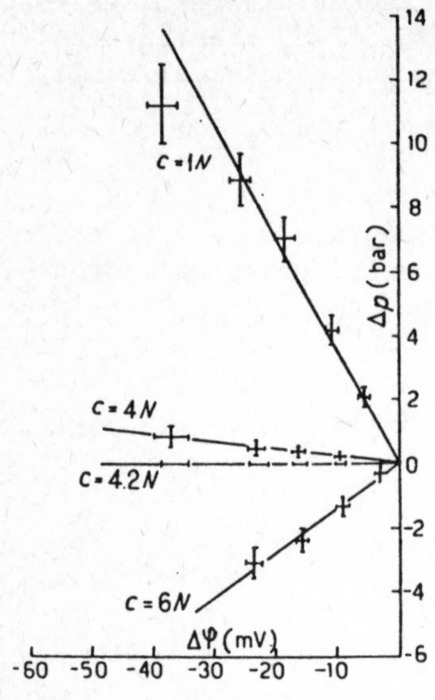

Fig.8. Electroosmotic pressure Δp vs. membrane potential $\Delta\varphi$ in the isotonic system LiCl+HCl/HCl. Parameter: concentration of HCl in the adjoining solution.

pressure is plotted v.s. the electric potential whereby the slope is given by the term $1/2\ (c_F - c_D)$.

6.3 Restrained Inversion of Electroosmosis

The equation of the electroosmotic pressure resulting from eqn(36a) involves no statement as concerns the origin of the field. Therefore the question arises whether an external field can produce the same effect as the transmembrane potential (diffusion potential in isotonic osmosis. Helfferich(23) says: "It is irrelevant whether the electric potential gradient is caused by an external voltage source or by a diffusion process within the system; the individual ion has no means of knowing the cause of the electric potential gradient".

Indeed, up to the point of inversion, internal as well as external potential brings about the same flux of water(24). Whilst the internal field gives rise to the inversion in the point $c_F = c_D$, in the case of an external field the flux of water vanishes and remains zero, if the concentration increases. This is the result of a fundamental theorem of potential theory: A potential assumes its minimum and maximum value at the boundaries. The theorem of contradiction explains this. Applying to the one side of a membrane 0 Volt and to the other 20 Volt, according to this theorem we have $0 < \varphi < 20$ inside the membrane. Taking in mind that a flux of water, originating from the electric current, gives rise to a streaming potential, this streaming potential reduces the initial potential up to the point of inversion. If in this point an inversion would take place, the streaming potential would raise the potential difference to a value higher than 20 Volt. This is impossible.

In the case of isotonic osmosis, however, no electric potential is applied to the boundaries but an electrochemical potential. Guggenheim says: "The conception of splitting the electrochemical potential $\tilde{\mu}_i$ into the sum of a chemical term μ_i and an electrical term $z_i\varphi$ has no physical significance", (25).

For example the Donnan potential represents a raising of the electric potential but this is no rise according to the potential theory, because the boundary conditions which represent the continuity are given by $\tilde{\mu}^I = \tilde{\mu}^{II}$ and not by $\varphi^I = \varphi^{II}$. The postulate that the electrochemical potential assumes its minimum and maximum value at the boundaries is valid indeed. The difficulty to understand this paradox, results from the fact that potentials have been introduced in the thermodynamics, without introducing the potential theory.

6.4 General Boundary Condition in Electromechanical Equilibrium

As mentioned above, the restriction of eqn(29-30) to a solution yields the system of eqn(31-33) which fulfills the Onsager's relations a priori. Analogously, we obtain in the case of binary isotonic osmosis a system of the form(31), which yields the following relation of the flux of water(5)

$$J_L = F(c_i, f_i)(c_F - c_D) \text{ grad } \mu_1^s \,. \tag{43}$$

In eqn(43), $F(c_i f_i)$ is a function of the frictional coefficients f_i and the concentrations c_i. In agreement with the experiments we state that the flux of water vanishes, if $c_D = c_F$, independent of the value of grad μ_1^s.

Though eqn(43) follows from a system of equations which fulfills the Onsager relations, and though the vanishing of J_L in the point $c_D = c_F$ is in accordance with the experiments, this equation is not correct. This is a consequence of the fact that the boundary conditions, necessary for the application of the Euler-Lagrange method (see fig.(4)) are not fulfilled. According to this theory, a closed system, where no fluxes pass the boundaries, must be taken into account. A device, which fulfills these conditions can be obtained with the help of the following steps. In the first step we introduce instead J_i the new variable

$$I = \sum z_i J_i \,. \tag{44}$$

Together with the condition $J_L = 0$ and $I = 0$ we obtain a closed system. The second step involves the transformation of Routh which is based on the idea, to find out a closed system in cases where $J_L \neq 0$ and $I \neq 0$. The Routh concept, which plays a fundamental rôle in the calculus of variations means the compensation of all fluxes by introducing suitable potentials. In our case these are given by(26)

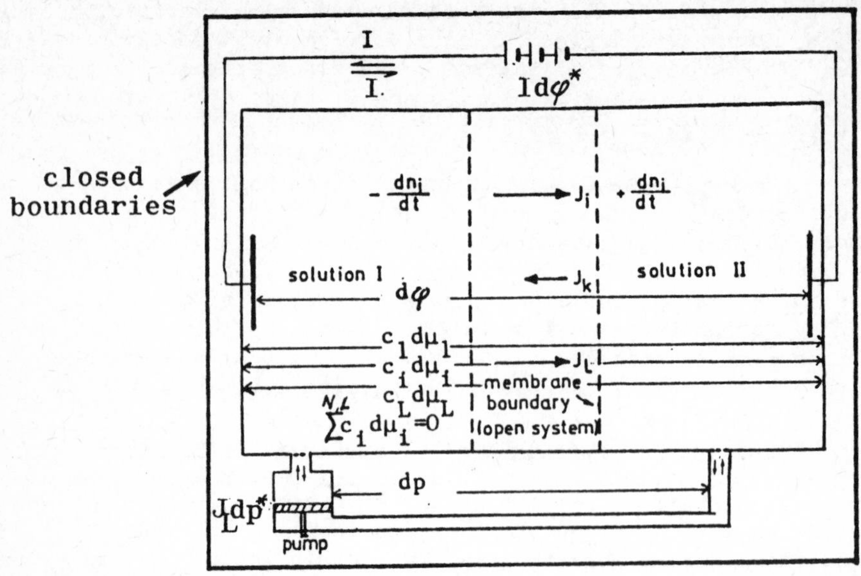

Fig.9 "Opening" of the osmotic cell. The outside boundary involves beneath the open osmotic cell, a pump and a reversible power source, yielding the energies $J_L dp^*$ and $Id\varphi^*$.

$$dp^* = dp - dp_c \,, \tag{45}$$

$$d\varphi^* = d\varphi - d\varphi_c \,. \tag{46}$$

In eqn(45-46) dp and $d\varphi$ represent the actual values of the hydrodynamic pressure and the transmembrane potential. dp_c is equal to dp, if $J_L = 0 = d\varphi^*$. Analogously we have $d\varphi_c = d\varphi$, if $I = 0 = dp^*$.

Such a system is represented in fig.9, where the external boundaries furnish a closed system. The inside system however, given by the originally closed system, is transformed this way to an open system.

Performing the transformation of Routh, calculus of variations yields

$$J_L/c_L = - L_{11} dp^*/dx - L_{12} F d\varphi^*/dx \tag{47}$$

$$I/F = - L_{21} dp^*/dx - L_{22} F d\varphi^*/dx \tag{48}$$

where

$$L_{11} = 2/\sum^{C} f_i c_i \,, \quad L_{12} = L_{21} = (c_F - c_D)/\sum^{C} f_i c_i$$

$$L_{22} = (c_F - c_D)^2/2\sum^{C} f_i c_i + \sum^{N} c_i/f_i \,. \tag{49}$$

dp^* and $d\varphi^*$ represent differences relative to the compensated state. It must be emphasized that these equations are valid for an arbitrary number of solved particles.

An example should demonstrate the application of these equations. From eqn(47) follows, if $d\varphi^*$ can be neglected

$$J_L = - (2c_L / \sum^{C} c_i f_i)\ dp^*/dx\ . \tag{50}$$

Contrary to this relation, from eqn(30) follows, if $d\mu^o_L = 0$ and $J_i = 0$

$$J_L = - (c_L / \sum^{N} c_i f_i) dp/dx\ . \tag{51}$$

Investigations(26) have confirmed the validity of eqn(50).

The difference between eqn(50) and (51) follows from the fact that eqn(50) involves the validity of the electromechanical equilibrium, but not eqn(51).

6.5 Physical Interpretation of the Electromechanical Equilibrium

A characteristic feature of the electromechanical equilibrium is the vanishing of the chemical potentials in eqn(28) and (47-49). According to the Gibb-Duhem eqn(25), the chemical potentials form a compensated system. But not only the chemical potentials from such a system, but also the frictional forces, whose sum, given by $\sum c_i\ dL_{n_i}$, vanishes according to Euler's theorem of homogeneous functions. This means, that the frictional force of the ions, acting on the solvent, is opposite to the frictional force of the solvent acting on the ions. The remaining forces, whose sum vanish, form the electromechanical equilibrium eqn(28)

For a better understanding of this equilibrium we illustrate it with a mechanical example. The total load of the abutments of a truss bridge is independent of the stress in the framework and is equal to the weight of the bridge. The forces in the framework compensate each other and have no influence on the load of the abutements. Therefore these forces represent internal forces. This does not mean, that these forces can be neglected. On the contrary, these are necessary and important for the constructor of the bridge.

Analogously, we conceive the chemical potentials and the frictional forces as internal forces. These forces have no immediate influence on the electromechanical equilibrium but an indirect one, as the electric potentials are functions of the chemical potentials. In the absence of chemical potentials eqn(49-51) present the familiar eqns. of the electro-kinetic effects, where an external electric field must be applied.

The electromechanical equilibrium represents the action of a constraint given by eqn(25) which must be taken into account, otherwise a theorem of Gyarmati(4) must be considered. This states:"If homogeneous linear relationships exist between the fluxes as well as the forces, the phenomenological coefficients are not uniquely defined, and the reciprocal relations are not necessarily fulfilled. This means, to avoid eqns. of the form

$$J_1 = L_{11} X_1 + L_{12} X_2 + L_{13} X_3$$

$$J_2 = L_{21} X_1 + L_{22} X_2 + L_{23} X_3 \qquad (52)$$

$$J_2 = L_{31} X_1 + L_{32} X_2 + L_{33} X_3$$

if X_i are no independent variables.

An example should explain the restriction of the system of eqn. (52). Applying this equation to the diffusion of a binary system, involving the components 1 and 2, we take the gradients of the chemical potentials $d\mu_1/dx$ and $d\mu_2/dx$ as the forces X_1 and X_2. Going from this system to a ternary system by adding the potential $d\mu_3$, this addition means a disturbed superposition of the forces, because the mere addition of $d\mu_3$ gives rise to a strong variation of $d\mu_1$ and $d\mu_2$. Therefore a separation of J_i into partial fluxes, resulting from each force X_i, is meaningless in this case.

Quite an other situation occurs if we introduce $dX_3 = dT/dx$ in eqn(52). According to the Gibbs-Duhem equation, the system(52) yields the well known relations for the thermal diffusion and the diffusion thermoeffect(27). In this case we have an undisturbed superposition of two forces which are independent of each other. Moreover the latter effects are an object of the Kinetic Theory of Gases and have been predicted in this connection (28). As concerns the former effect, called ternary diffusion, the interaction of force can be obtained with the help of the Gibbs-Duhem equation. In this case we come back to Gyarmati's statement.

As concerns the application of the system(52) to transport problems in membranes, the choice of the forces $d\mu_i/dx$, $d\varphi/dx$ and dp/dx and the fluxes J_i, J_L and I brings about the same problems. However, there is a more important aspect which must be taken into account. According to thermodynamics we have to regard the boundary conditions of the system membrane-solution, given by the equality of the electrochemical potentials. According to Guggenheim "the conception of splitting the electrochemical potential of an ion into the sum of a chemical term μ_i and an electrical term φ has no physical significance". Therefore the choice of μ_i and φ as individual forces is in contradiction to Guggenheim's theorem.

Underlying this statement, Gyarmati's theorem has been the

motive to look for flux equations which are in accordance with these postulations. Unfortunately a discussion concerning the eqn(47-48) and equations resulting from eqn(52) is not possible as the basic effect, the isotonic osmosis was not taken into account in previous theories.

6.6. Ion Fluxes in Membranes

From the electromechanical equilibrium follows, if $J_L=0$, for the case of two ions

$$1/2(f_1-f_2)J_1 dx = -c_F F d\varphi - dp . \tag{53}$$

From the flux, eqn.(29), of the ions we get

$$-\mathrm{grad}\varphi = [(RT/F)(1/f_1-1/f_2)/c_1/f_1+c_2/f_2)]\,\mathrm{grad}\ c_1 \tag{54}$$

and from the flux, eqn.(3o), of the solvent, if $J_D=0$,

$$dp = 1/2(f_1-f_2)J_1 . \tag{55}$$

Introducing eqn.(54-55) in (53) and considering $D=RT/f$ and $c_1+c_2=c_F$ we obtain Helfferich's equation

$$J_1 = -[(D_1D_2)(c_1+c_2)/(D_1c_1+D_2c_2)]\ \mathrm{grad}\ c_1. \tag{56}$$

Therefore Helfferich's equation is in agreement with the general field theory.

7. APPLICATION OF ONSAGER'S PRINCIPLE OF LEAST DISSIPATION OF ENERGY TO ION EXCHANGE

Up to now we have applied Onsager's principle of least dissipation of energy to the transport of particles through membranes. Going over to ion exchange, this principle postulates the application of the statement of energy balance to this process. To understand the difference between membrane transport and ion exchange, we will first set up the energy balance of the membrane transport. This balance is given by the dissipation energy represented by the integral in eqn(10). Restricting it to the differential state we have

$$dE = dL = TdS . \tag{57}$$

Generally the dissipation function dE/T is taken as the entropy production dS(15). Replacing dL by eqn(19) and introducing for $dL_{\dot n_i}$ the values resulting from eqn(21-22), we obtain after integration

$$d\dot S = -q\sum^{N} J_i(d\mu_i - 1/4\ qJ_L dL_{\dot n_i \dot n_L} - qJ_L(d\mu_L - 1/4\sum^{N} qJ_i dL_{\dot n_i \dot n_L} . \tag{58}$$

In the case of a compensated state ($J_L=0$), the terms involving $dL_{\dot n_i \dot n_L}$ vanishes and we obtain the familiar expressions of entropy production used in stationary thermodynamics.

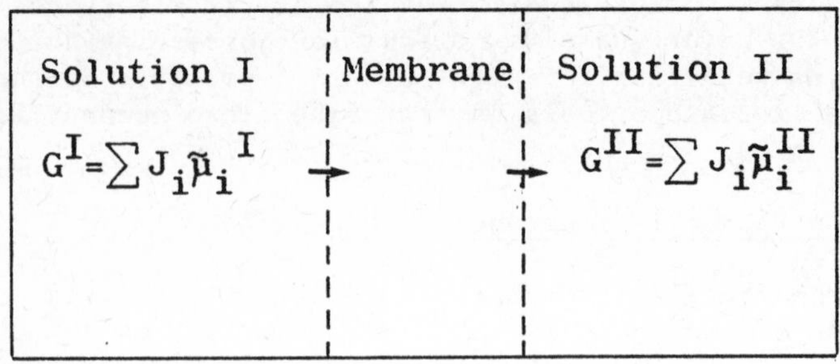

Fig. 10 Balance of energy

In fig. 10 the balance of energy in a membrane is illustrated. We see that this balance is determined by two boundary conditions. Whilst on boundary I the free energy $J_i\tilde{\mu}_i^{I}$ passes into the membrane, the energy $J_i\tilde{\mu}_i^{II}$ comes forth. The difference is the entropy production given by eqn(58).

Rolling up the membrane and putting this cylinder in a solution, we have only one boundary condition. The same is valid in the case of an ion exchanger bead. Thereby the only transport process is an ion exchange where the second Fick's law must be applied. As this law is represented by a differential equation of second order, two independent boundary conditions are necessary, but thermodynamics yield only the familiar boundary condition at the surface of the ion exchanger bead. With this problem we have to deal at some lenght in the following.

The solution of this problem is given in Hamilton's "Theory of Systems of Rays"(29). The following figure(11) should explain this. On the left hand side a system of rays arises from a source of light S. Whilst the rays represent the extremals, the potential lines are given by the "eiconals". Replacing the rays of light by ion fluxes we get the object under consideration. Regarding the geometrical element E, bounded by two extremals and two eiconals I and II (potential lines), in the steady state we have $\Delta c_i = 0$. This results from the fact that all ions passing on the one side in the element E, comes forth on the other. Regarding the source and rounding it with a circle, within this circle, however, $\Delta c_i \neq 0$ because any particle either enters or leaves the geometrical boundary if ion exchange takes place. This is possible only, if we replace the source of Hamilton's model by a source-sink model with a constant number of fixed ions.

The most ingenius features of Hamilton's field-theory, however, are the boundary conditions. Whilst the boundary conditions of our element E are given by the electrochemical potentials of the adjoining potential lines according to

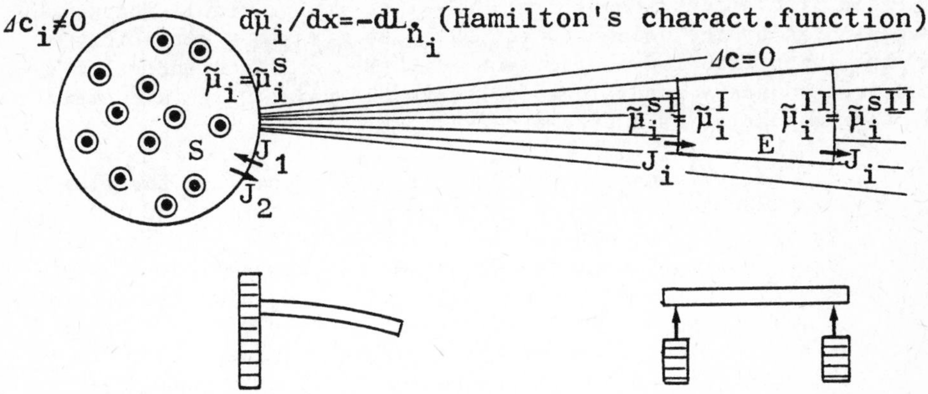

Fig. 11. Free boundary conditions

$$[\mu_i^s = \mu_i]^I \; , \; [\mu_i^s = \mu_i]^{II} \; , \tag{59}$$

The boundary conditions of the source S are given by Hamilton's characteristic functions which involves, beneath the equilibrium condition $\mu_i^s = \mu_i^o$ at the boundary 0 of the ion exchanger bead, functions of the derivations $[d\mu_i/dx]^o$ at this boundary.

Hamilton's conditions which arise from a variation of endpoint (boundary), yield free boundary conditions. An example should explain this designation. We regard a beam resting on two abutments (s. fig. 11). In this case the boundary conditions correspond to the conditions (59). Removing, however, one abutment, a stable state is possible only, if we clamp the one end of the beam to the resting abutment. In this case the boundary conditions at the remaining abutment are given by

$$y = \text{const}, \; dy/dx = 0, \; d^2y/dx^2 = f(p(x), W), \tag{60}$$

whilst no boundary condition rules the positions of the other "free end". In eqn(60) x is the length of the beam, y its displacement (bend), p(x) the load and W the resistance to bend.

7.1 Ion Exchange Kinetics and Energy of Reaction

Returning to the ion exchange, Hamilton's theory yields together with eqn(21) in the case where the energy of reaction $E_r = 0$, the differential equation for the singular case

$$J_i = -[D_{12}]^0 \, dc_i/dx \tag{61}$$

$[D_{12}]^o$ is the value of Helfferich's diffusion coefficient at the surface of the ion exchanger bead. Whilst in the regular case, represented by the volume E, eqn(59) give rise to a variation of the interdiffusion coefficient with the local concentration (3), the free boundary conditions (60) take the value of this coefficient at the boundaries as a parameter.

For a better understanding the difference between the fluxes through a membrane and an ion exchanger, two facts must be considered. First, owing to the electroneutrality, counterions together with co-ions must pass the membrane, whilst in ion exchange, co-ions are not necessary. Therefore the mechanism of ion exchange is different from the ion transport through membranes. This difference results from the exchange mechanism of the sink/source model. Secondly, Onsager's principle of least dissipation of energy presupposes the consideration of all sources of energy taking part in the transport process. This means, that the energy of reaction connected with an exchange process must be taken into account in addition to the energy passing through the boundaries (see Fig. 11). Regarding, as the energy is a scalar, the fluxes, however, being vectors, a combination of these quantities is not possible. Therefore the fixed ions, where the ion exchange takes place, are excluded by circles from the space, where diffusion takes place. This way, the fixed ions are conceived as singularities representing sources of the reaction energy. This energy must be taken into account besides the energy passing the boundaries of the ion exchanger beads.

The calculation yields the differential equation(30)

$$J_1 = - [D_{12}]^o[1 + (c_1^o c_2^o/c_F)(dE_r/RT\, dc)]dc_1/dx \qquad (62)$$

In addition to Helfferich's diffusion coefficient, there is a coefficient in eqn(63) involving the heat of reaction.

The following statements result from eqn(63).
1.) Because the value of the apparent diffusion coefficient depends on the concentration at the boundary only, the process of diffusion proceeds with a constant diffusion coefficient dependent on the end concentration only.
2.) Taking into account the fact that dE_r changes sign if we go from the forward to the backward reaction and vice versa, eqn(63) yields a branch for the forward and a branch for the backward reaction. The difference between the apparent diffusion coefficients of the backward and the forward reaction should be designated the "reaction jump". Its value follows from the reaction term $(c_1^o c_2^o/c_F)(E_r/RTdc)$ in eqn(62). In the range where c_1 or c_2 tends to zero, $c_1 c_2/c_F$ becames small and the reaction term can be neglected.

Fig. 12 and 13 show that these conditions are fulfilled. Attention should be paid to the fact that the diffusion coefficient depends

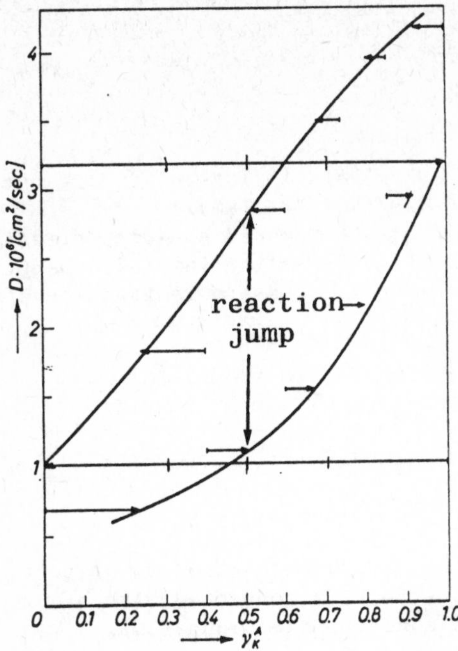

Fig. 12. Interdiffusion coefficient of RH⇄RK
Arrowhead: end-concentr.
Crossline: initial-conc.
Abscissa: mole fraction of $K^{\bullet}$.

on the end concentration(arrowhead) and is independent of the initial concentration(crossline). Concerning the reaction jump, we have used values of the reaction heat found with Levatit KS (31,32). Together with values obtained of the reaction jump in fig. 12 and 13, the values of the term $1 \pm (c_1^o c_2^o / c_F)(|dE_r|/RTdc)$ can be found.

Table 1. Values of the reaction term in eqn(62) calculated from the reaction jump and the heat of reaction.

System	reaction term calculated from reaction jump	heat of reaction
K - H	1 ± 0.45	$1 \pm 0,75$
Na- H	1 ± 0.47	$1 \pm 0,66$

Similar values of the heat of reaction were found by Boyd, Vaslow and Lindenbaum(33).

Concerning the values of Helfferich's interdiffusion coefficent, values of D_H/D_{Na} and D_H/D_{Na} smaller than those found in solutions must be taken into account. Similar results have been found by Dickel and Franke (34) in investigations of isotonic osmosis.

Finally an example presented in fig. (14) should demonstrate, that cases are possible where $dE_r = 0$. In the exchange RH⇄RLi,

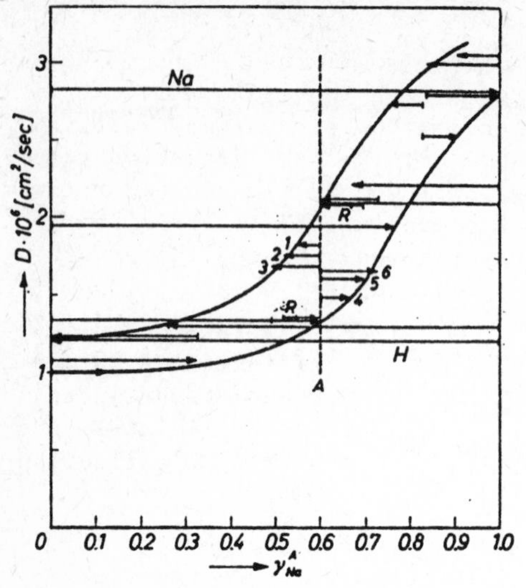

Fig. 13. Interdiffusion coeff. of RH ⇄ RNa.
Arrowhead: end-concentration
Crossline: initial-concentr.
Abscissa: mole fraction of Na˙

the branches of the forward and the backward reaction touch each other at the point γ = 0.6. This effect is connected with a minimum of water sorption in this concentration range(35).

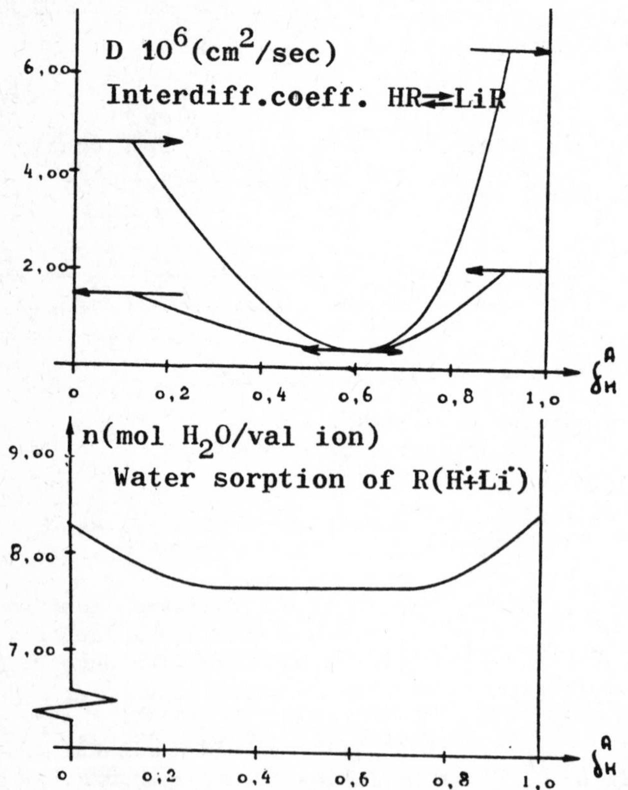

Fig. 14. Interdiff. coefficient of RH ⇄ RLi and water-sorption of this system in dependence of the mole fraction of H˙.

8. DISCUSSION

The circumstance that the phenomenological theories used in the treatment of membrane phenomena didn't have the desired result, gave rise to search for a more general principle. Since there was no doubt about the application of the Nernst-Planck equation resp. the electrochemical potentials, we proceeded to a much more rigorous foundation of these potentials by introducing the theory of potentials into the treatment of the transport phenomena. As the theory of potentials involves the important problem of the boundary conditions and because few attention was paid to this problem up to date, only, it laid near at hand to start our investigations by using the method of the variation of the endpoint. This method was the starting point of Hamilton's famous theory of eiconal. This was the base of the fields of extremals which led immediately to the Hilbert's independence theorem as used in our investigations.

Though the first investigations concerning the ion exchange demonstrated the applicability of this method, it was necessary to set up first the flux equations, based on this theory. In this way we have found the isotonic osmosis. Whilst the first experiments showed a linear relation between the flux of water and the ion fluxes, with increasing fluxes however, deviations of the linearity occured (fig. 15), (34). This is not surprisingly. According to the conception of the thermodynamical transport theory, the linearity should be restricted to the neighbourhood of the equilibrium, whilst deviations must be expected with increasing distance from equilibrium. Fortunately, reasons which will be discussed in the following gave rise to further investigations which led to the discovery of the invariance of the inversion of the isotonic osmosis (see fig. 6).

These reasons had their origine in the concept of Hamilton's field theory which was taken as a base of our flux equations. Whilst Hamilton considered the propagnation of the light, Hilbert persuaded the idea, to explain the gravitation as a consequence of the expanding universe. Though this problem could not be solved up to date, it is possible, however, to explain the electrochemical potentials as a consequence of the expansion of an electrolyte from a box of high concentration to a box of low concentration. The advantage of this concept is, that dynamical processes are taken into consideration only, and no equilibrium states. Therefrom result the following differences between the thermodynamical conception and the idea of the theory of fields of extremals.

Expanding the potential introduced according to Hamilton's theory in a Taylor series, in the case of an extremum, all derivatives of higher order vanish. This was proved by Weistrass, who had introduced the field into the calculus of variations. The remainder of this Taylor series is called the excess function of Weierstrass.

An expansion in a Taylor series is applied in the thermodynamical treatment of transport processes, also. In this case all processes are restricted to the neighourhood of the equilibrium and in this case only, the linear relations are applicable. Otherwise, non-linear terms must be applied.

The value of such a conception was demonstrated in our first investigations of isotonic osmosis. In this case, the invariance of the inversion of isotonic osmosis would never have been predicted by using the non-linear statement. In contrast to the theorem of least dissipation of energy which is based on a fundamental principle, there is no base, which justifies results, originating from non-linear thermodynamics.

In table 2 we have compiled the differences between the theory of linear systems and the theory of the fields of extremals once more.

Table 2 Comparison between the Theory of Linear Systems and the Theory of the Fields of Extremals.

Fields of Extremals		Linear Systems
	Balance of Energy	
Entropy Production		Entropy Production (Dissipation of Energy)
	Balance of Forces	
Vanishing of the sum of specific Forces		No conception
	Flux Conditions for the System of Reference i	
J_i = const;(arbitrary value)		Generally $J_i = 0$
	Boundary Conditions	
Closed resp. Compensated Systems must be regarded, where $J_i = 0$ at the Boundaries		Generally open Systems are regarded.
	Onsager Relations	
Following a posteriori from the Field Theory		A priori supposed

Finally we emphasize that we have never intended to make a new theory of the membrane transport. Rather it was the purpose of our investigation to prove whether a transport process proceeds in such a way that at every moment and at any position the Onsager's principle of least dissipation of energy is fulfilled. The methods necessary to solve this question have been developed by the mathematicians in the last century, allready. As far as we have performed investigations, all phenomena found, could be explained in this way without using arbitrary presumptions.

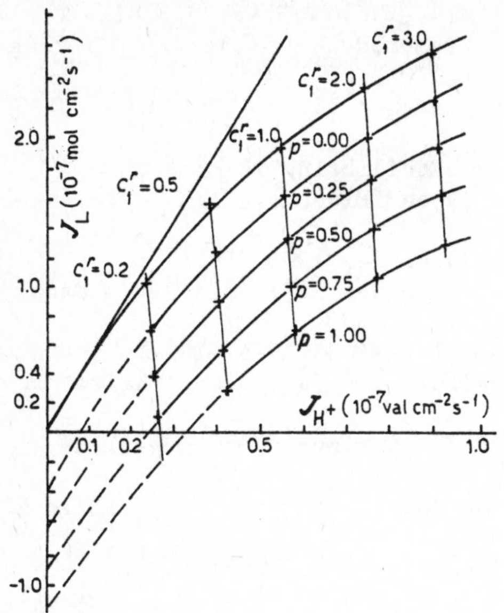

Fig.15.Water flux J_L in iso= tonic osmosis of HCl/NaCl vs. ion flux J_H.

Parameter:Pressure p and concentration c_H^r on the right hand side.

References

1. Boyd, G.E., Adamson, A.W., and Myers, L.S., Jr., J.Amer.Chem. Soc. 67, 2836 (1947)
2. See "The Analytical Theory of Heat by Joseph Fourier. Translated, with notes, by Alexander Freeman, Dover Publications, Inc., New York (1955).
3. Helfferich, F. and Plesset M.S. ,J.Chem.Phys.28,418(1958)
4. I.Gyarmati, Non-equilibrium Thermodynamics,Springer Verlag Berlin-Heidelberg-New York (1970)
5. G.Dickel and G.Backhaus, J.Chem.Soc.Faraday Trans.2,74,115, 124 (1978)
6. G.Dickel, Electro-Mechanical Equilibrium in Membranes, Topics in Bioelectrochemistry and Bioenergetics Vol.4.(1981). John Wiley and Sons Ltd.
7. K.S.Spiegler, Trans.Faraday Soc. 54, 1408 (1958)
8. P.Meares, D.G.Dawson, A.H.Sutton and J.F.Thain,Ber.Bunsenges. Phys.Chem.71, 765 (1967)
9. L.Onsager, Phys.Rev. 37,405 (1931); 38,2265 (1931)
10. O.Bolza, Lectures on the Calculus of Variations,Chelsea Publ. Co., New York
11. D.Hilbert, Ges.Werke Bd.3.S.32. Springer Verlag Berlin-Heidelberg-New York (1970)
12. G.Dickel and B.Pitesa, J.Chem.Soc., Faraday Trans.II,77,441 (1981)
13. I.Prigogine, Etude thermodynamique des Phénomènes Irréversibles(Thesis) Dunod, Paris and Desoer, Liege (1947)
14. G.Schmid, Z.Elektrochem., 54, 424 (1950); 55,229,295,684 (1951); 56,35,181 (1952) Ber.Bunsenges., Phys.Chem. 71,778 (1967)
15. P.Meares, J.F.Thain and D.G.Dawson, Membranes Vol.1, Chapter 2, edited by G.Eisenman, Marcel Dekker, Inc.,New York(1972)
16. R. Schloegl, Stofftransport durch Membranen.Steinkopf-Verlag Darmstadt, (1969)
17. G.Dickel and H.Hoenig, Z.Phys.Chem.(Frankfurt/Main) 90, 198, 1974)
18. H.S.Harned, B.B.Owen, Physical Chemistry of Electrolytic Solutions, Reinhold, Pub. Co., New York (1950)
19. G.Dickel, G.Backhaus and E.Hauerwaas, Z.Phys.Chem., 258, (1978)
20. R.Kretner, H.Hoenig and G.Dickel, Z.Phys.Chem.(Frankfurt/Main) lo6, 330 (1977)
21. G.Dickel and R.Kretner, Z. Phys.Chem.(Frankfurt/Main) 118,161(1979)
22. A.L.Hodgkin and B.Katz. J.Physiolog.London 108,37(1949)
23. F.Helfferich, J.Phys. Chem. 66,39 (1962)
24. G.Dickel and R.Kretner, J.Chem. Soc., Faraday Trans.2 74, 2225 (1978)
25. W.Guggenheim, J.Phys.Chem.33, 842 (1929); 34,1540(1930)
26. G.Dickel, to be published

27. R.S.De Groot, Thermodynamics of Irreversible Processes. North Holland Publ. Co., Amsterdam (1958)
28. D.Enskog, Physik. Z. 12,56,553(1911). Ann.Physik 38, 731 (1912)
29. I.W.R.Hamilton, Irish Transactions 15-17(1828-30)
30. G.Dickel, to be published
31. P.Fries, Dissertation München,(1951)
32. G.Dickel and A.Meyer,Z.Electrochem.Ber.der Bunsenges. 57,901(1953)
33. G.E.Boyd,F.Vaslow and S.Lindenbaum,J.Phys.Chem.68,590(1964)
34. G.Dickel u.W.Franke,Z.Phys.Chem.(Frankfurt/Main)80,190(1972)
35. D.Fiederer,Dissertation München (1971)

WATER AND SALT TRANSPORT IN TWO CATION-EXCHANGE MEMBRANES

by

A. Berg*, T. S. Brun**, A. Schmitt*** and K. S. Spiegler****
University of California, Berkeley and Michigan Technological University, Houghton, U.S.A.

1. ABSTRACT

Accurate measurements of the isothermal transport of ions and of water across a synthetic ion-exchange membrane were made. A feedback method was used, in which electrolyte transport is determined by the amounts of electrolyte and water respectively that have to be added to and/or withdrawn from the half-cells separated by the membrane so as to keep the solution concentrations constant. This automatic feedback is actuated by conductivity probes.

Four groups of experiments, leading to the measurement of eight independent transport coefficients, could be readily made in the "concentration-clamp" apparatus developed in our Laboratory, viz. (a) electric conductivity, (b) hydraulic permeability and streaming potential, (c) electromigration and electroosmosis, and (d) osmosis, dialysis, and membrane potential. /The ninth parameter ("degree of hyperfiltration") cannot be measured with sufficient accuracy in the existing "concentration-clamp" apparatus./ In systems of membranes separating two dilute sodium-chloride solutions, in which silver-silver chloride electrodes are immersed, two reciprocity relations could be checked.

It was found that these two reciprocities hold for a cation-exchange membrane (CL-25T, Tokuyama Soda Co., Tokuyama City, Japan); the highest solution concentration was 0.5N NaCl. Comparison of the conductance coefficients to those of C-103 (American Machine and Foundry Co., Stamford, Conn.), measured previously, indicates higher permeability characteristics for CL-25T.

* Present address: Israel Desalination Engineering, P.O. Box 18041, Tel Aviv, Israel

** Present address: Department of Chemistry, University of Bergen, Bergen, Norway

*** Present address: Centre de recherches sur les macromolécules, 6 rue Boussingault, 67803 Strasbourg, France

**** To whom correspondence should be addressed: Department of Mechanical Engineering, University of California, Berkeley, CA 94720, U.S.A.

2. INTRODUCTION

While ion-exchange processes were originally used for the treatment of very dilute solutions, many applications for the treatment of concentrated solutions have been developed in recent years. In these situations, the mass-transfer bottlenecks are located in the solid, rather than the liquid phase. Therefore, the development of quantitative models for ion-exchange kinetics requires knowledge about the conductance characteristics of ions and solvent in the solid phase. A useful approach towards this aim is the study of transport characteristics of these species, and of their interactions in solid ion-exchange membranes.

Many different transport processes and related phenomena can be observed in membrane-solution systems, e.g., ion migration, electroosmosis, diffusion and self-diffusion, osmosis, hydraulic flow, hyperfiltration (reverse osmosis) or ultrafiltration, streaming potential and streaming current, and membrane potentials (also called "membrane concentration potentials"). It is important to correlate all these phenomena so as to avoid a very large number of unnecessary measurements. Such correlation is often possible [Meares, 1976] since all these phenomena are determined by the ease of migration of the different species across the membrane. Important correlations have been made and summarized even before high-capacity ion-exchange membranes became commercially available [Sollner, 1950, 1976].

For practical purposes, it is often important to determine the interactions between membrane permeants. The aim of this research was the development of precise experimental methods for this purpose, and also of methods for the calculation of Onsager conductance coefficients at specific salt concentrations.

3. THEORY

3.I. Linear Transport Equations

We consider the isothermal system

Ag/AgCl | chloride solution, c' | membrane | chloride solution, c" | Ag/AgCl

There are three types of particles which can migrate in the membrane, viz. cations (+), Cl^- (-) and water (w). Three equations relating the fluxes, J (mole $cm^{-2}sec^{-1}$), to the generalized forces, F (wattsec $mole^{-1}$), are necessary to characterize the transport properties of the system.

The most fundamental set of transport equations is based on the classical concepts of non-equilibrium thermodynamics /Staverman, 1952/ originally developed by Onsager /1931/ and applied by him to diffusion in liquids /1945/. Previous experimental work justifies the description of the fluxes as sums of the products of generalized forces and conductance coefficients, L_{ij}:

$$J_i = \sum_j L_{ij} F_j \tag{1}^*$$

From a practical viewpoint it is essential to look for transport descriptions with reasonably constant conductance coefficients (i.e., conductance coefficients which are reasonably independent of the forces). Only in this case does a single set of L_{ij} coefficients characterize a given membrane-solution system, i.e., enable us to predict the fluxes from knowledge of the forces and/or vice-versa.

For ion and water transport in membranes under the influence of differences of pressure, concentration and/or electric potential, the generalized force acting on particle i is

$$F_i = -(\bar{v}_i \Delta p + \Delta\mu_i^c + z_i \mathfrak{F} \Delta\phi) \tag{2}$$

where Δp (dekabar) is the difference of pressure (1 dekabar ≡ 10 bar ≈ 9.87 atm. = 1 MNewton m^{-2}, also called 1 megapascal; this pressure unit is used in order to minimize the use of conversion factors in our system of units - see also List of Symbols). Δp, $\Delta\mu_i^c$ (wattsec mol^{-1}) and $\Delta\phi$ (volt) are the differences of pressure, of the "concentration-dependent

*The list of symbols with units is in Section 6 of this report.

part of the chemical potential" /i.e., $\left(\frac{\partial \mu_i}{\partial c_i}\right)_{p,T} \Delta c_i = RT\Delta \ln a_i$/ and the electrical potential across the membrane respectively. Hence the original set of transport equations (which we call the "Onsager set") is

$$J_+ = -L_{++}(\bar{v}_+\Delta p + \Delta\mu_+^c + \mathfrak{F}\Delta\phi) - L_{+-}(\bar{v}_-\Delta p + \Delta\mu_-^c - \mathfrak{F}\Delta\phi) - L_{+w}(\bar{v}_w\Delta p + \Delta\mu_w^c) \quad (3)$$

$$J_- = -L_{-+}(\bar{v}_+\Delta p + \Delta\mu_+^c + \mathfrak{F}\Delta\phi) - L_{--}(\bar{v}_-\Delta p + \Delta\mu_-^c - \mathfrak{F}\Delta\phi) - L_{-w}(\bar{v}_w\Delta p + \Delta\mu_w^c) \quad (4)$$

$$J_w = -L_{w+}(\bar{v}_+\Delta p + \Delta\mu_+^c + \mathfrak{F}\Delta\phi) - L_{w-}(\bar{v}_-\Delta p + \Delta\mu_-^c - \mathfrak{F}\Delta\phi) - L_{ww}(\bar{v}_w\Delta p + \Delta\mu_w^c) \quad (5)$$

Onsager's theoretical work, based on statistical mechanics, led to the conclusion that the conductance coefficients, L_{ij}, satisfy the reciprocity condition

$$L_{ij} = L_{ji} \quad (6)$$

Onsager's work is of general nature and covers many more phenomena than transport in membranes, e.g., heat conduction and chemical reactions. In general, the range of validity of linear laws of the type of eq. (1) is limited; for instance, for most chemical reactions the linear approximation is reasonably adequate only very close to chemical equilibrium. If this were generally true for transport in membranes also, the linear equations would therefore be only of very limited use; for many transport phenomena, however, the range of approximate linearity extends far beyond equilibrium /Chartier, Gross and Spiegler, 1975; Silver, 1977/. For instance, it has been known for many decades that the ion flux in electrolyte solutions is proportional to the applied electric voltage over a considerable voltage range, provided proper electrodes and stirring devices are used ("Kohlrausch's law"). Also self-diffusion of particles follows a strictly linear relationship between flux and force even in solutions and gases which are very far from isotopic equilibrium. Fick's law of diffusion can be shown to be a linear law in the sense of eq. (1) /Barrer, 1956; Chartier, Gross and Spiegler, 1975/.

The practical application of the Onsager set of transport equations /eqs. (3)-(5)/ is hampered by the lack of knowledge of the partial ionic volumes v_+ and v_- (as opposed to the partial molar volumes of water and electrolyte which can be unequivocally defined from macroscopic measurements and often found in tables of properties of electrolyte solutions). Moreover, the electrical potential difference across the membrane, $\Delta\phi$, appears in two of the three forces, whereas the potential difference actually measured, $\Delta\phi_-$, is that between two Ag/AgCl electrodes placed at some distance from the membrane. Therefore previous investigators have performed certain transformations of the Onsager set of equations to make it directly amenable to evaluation of the experimental results [Duncan, 1962]. We shall discuss one transformation [Michaeli and Kedem, 1961; Katchalsky and Curran, 1966] which we found useful for the evaluation of our past experiments and which we also used in the present work.* We shall then demonstrate how we developed a method for the evaluation of our transport measurements, based on these "M-K" equations.

3.II. The Michaeli-Kedem Transformation [Katchalsky and Curran, 1966, p. 151; Michaeli and Kedem, 1961; Kedem and Katchalsky, 1963]

This transformation leads to three fluxes, J_s, J_w and $i/\mathfrak{F}$, which are all uniform in the steady state. Moreover the three conjugated forces, $(-\Delta\mu_s)$, $(-\Delta\mu_w)$ and $\Delta\tilde{\mu}_{el}$ $[= -\mathfrak{F}\Delta\phi_- + (\bar{v}_{AgCl} - \bar{v}_{Ag})\Delta p]$ can be readily measured and/or found in thermodynamic tables. For this reason we have come to the conclusion that this set of equations is the most practical for the evaluation of our past and present transport experiments, and that the L-conductance coefficients in this matrix are the most practical for characterizing the transport properties of the membrane. It should be noted that we consider this type of characterization as an important intermediate step in the friction coefficient method for membrane characterization [Spiegler, 1958; Richardson, 1970 and 1971; Meares, Thain and Dawson, 1972], which is not treated here.

The starting point for this transformation is the original (Onsager) set of flux equations (3)-(5) which is rewritten, using the symbol $\Delta\tilde{\mu}_i$ for the difference between the total potentials of component i in the right and left solution respectively [Spiegler, 1974]:

---

*This transformation is referred to as the "Michaeli-Kedem" ("M-K") transformation in this report. Some elements of this transformation were anticipated in an earlier paper [Lorimer, Boterenbrod and Hermans, 1956].

$$\Delta\tilde{\mu}_i \equiv \Delta\mu_i + z_i\mathcal{F}\Delta\phi = \bar{v}_i\Delta p + \Delta\mu_i^c + z_i\mathcal{F}\Delta\phi \tag{7}$$

Since the net charge of water, z_w, is zero, $\Delta\tilde{\mu}_w = \Delta\mu_w$.

With this notation, the set of original Onsager flux equations is

$$J_+ = L_{++}(-\Delta\tilde{\mu}_+) + L_{+-}(-\Delta\tilde{\mu}_-) + L_{+w}(-\Delta\mu_w) \tag{8}$$

$$J_- = L_{-+}(-\Delta\tilde{\mu}_+) + L_{--}(-\Delta\tilde{\mu}_-) + L_{-w}(-\Delta\mu_w) \tag{9}$$

$$J_w = L_{w+}(-\Delta\tilde{\mu}_+) + L_{w-}(-\Delta\tilde{\mu}_-) + L_{ww}(-\Delta\mu_w) \tag{10}$$

The total potential of the salt component, $\Delta\tilde{\mu}_s$, is equal to the sum of the total potentials of the constituent ions. Also, since salt carries no net charge, $z_s = 0$, and hence $\Delta\tilde{\mu}_s = \Delta\mu_s$:

$$\Delta\tilde{\mu}_s = \Delta\tilde{\mu}_+ + \Delta\tilde{\mu}_- \tag{11}$$

We substitute $\Delta\mu_s - \Delta\tilde{\mu}_-$ for $\Delta\tilde{\mu}_+$ in eqs. (8)-(10) and then $-\mathcal{F}\Delta\phi_- + v_{Cl}\Delta p$ for $\Delta\tilde{\mu}_-$ [Krämer and Meares, 1969; Spiegler, 1974]. The result is the following set of equations:

$$J_+ = J_s = L_{++}(-\Delta\mu_s) + (L_{++} - L_{+-})(-\mathcal{F}\Delta\phi_- + v_{Cl}\Delta p) + L_{+w}(-\Delta\mu_w) \tag{12}$$

$$J_- = L_{-+}(-\Delta\mu_s) + (L_{-+} - L_{--})(-\mathcal{F}\Delta\phi_- + v_{Cl}\Delta p) + L_{-w}(-\Delta\mu_w) \tag{13}$$

$$J_w = L_{w+}(-\Delta\mu_s) + (L_{w+} - L_{w-})(-\mathcal{F}\Delta\phi_- + v_{Cl}\Delta p) + L_{ww}(-\Delta\mu_w) \tag{14}$$

Finally, we calculate $i/\mathcal{F}$ from eqs. (12) and (13):

$$i/\mathcal{F} = J_+ - J_- = (L_{++} - L_{-+})(-\Delta\tilde{\mu}_s) + (L_{++} - L_{+-} - L_{-+} + L_{--})(-\mathcal{F}\Delta\phi_- + v_C$$

$$+ (L_{+w} - L_{-w})(-\Delta\mu_w) \tag{15}$$

Summarizing, the transformed set of flux equations (12)-(14) is:

$$J_s = L_{ss}(-\Delta\mu_s) + L_{sw}(-\Delta\mu_w) + L_{se}(-\mathcal{F}\Delta\phi_- + v_{Cl}\Delta p) \quad (16)$$

$$J_w = L_{ws}(-\Delta\mu_s) + L_{ww}(-\Delta\mu_w) + L_{we}(-\mathcal{F}\Delta\phi_- + v_{Cl}\Delta p) \quad (17)$$

$$i/\mathcal{F} = L_{es}(-\Delta\mu_s) + L_{ew}(-\Delta\mu_w) + L_{ee}(-\mathcal{F}\Delta\phi_- + v_{Cl}\Delta p) \quad (18)$$

"Modified M-K set of flux equations"

where the meaning of the new L conductance coefficients in terms of those of the original set, eqs. (3)-(5), is shown in Table 1. Note that when reciprocity [eq. (6)] is satisfied in the original set, reciprocity also prevails in the transformed one. We conclude that in eqs. (16)-(18) fluxes and forces are properly conjugated.

The transport coefficients, L, of the original "Onsager-type" equations [eqs. (3)-(5)] or even those in the transformed equations (16)-(18) give little direct insight into the transport processes, because they are strongly concentration-dependent. Moreover, this dependence limits their use to very small concentration intervals across the membrane [Foley, Klinowski and Meares, 1974]. Therefore, methods have been developed to describe transport across membranes in terms of transport parameters less dependent on the concentrations of the migrating species than the conventional L-coefficients.

First, it is desirable to invert the matrix of equations (3)-(5), i.e., to represent the generalized forces, F_i, in terms of the fluxes, J_i, rather than vice versa. In this manner <u>resistance coefficients</u>, R_{ij}, are obtained, instead of <u>conductance coefficients</u>, L_{ij}:

$$F_+ = -(\bar{v}_+\Delta p + \Delta\mu_+^c + \mathcal{F}\Delta\phi) = R_{++}J_+ + R_{+-}J_- + R_{+w}J_w \quad (19)$$

$$F_- = -(\bar{v}_-\Delta p + \Delta\mu_-^c - \mathcal{F}\Delta\phi) = R_{-+}J_+ + R_{--}J_- + R_{-w}J_w \quad (20)$$

Table 1

Relations between Conductance Coefficients in Onsager-Type Flux Equations (3)-(5) and Transformed Set (I. Michaeli and O. Kedem), eqs. (16)-(18) in this Report

$$L_{ss} = L_{++}$$

$$L_{ww} = L_{ww}$$

$$L_{ee} = L_{++} - L_{+-} - L_{-+} + L_{--}$$

$$L_{sw} = L_{+w}$$

$$L_{ws} = L_{w+}$$

$$L_{se} = L_{++} - L_{+-}$$

$$L_{es} = L_{--} - L_{-+}$$

$$L_{we} = L_{w+} - L_{w-}$$

$$L_{ew} = L_{+w} - L_{-w}$$

All coefficients have the units mole^2 joule^{-1} cm^{-2} sec^{-1}.

$$F_w = -(\bar{v}_w \Delta p + \Delta\mu_w^c) = R_{w+}J_+ + R_{w-}J_- + R_{ww}J_w \qquad (21)$$

For once, it is more logical to add resistive terms if the generalized force is desired; just as in electric networks, the total voltage across a series of resistors is more simply expressed in terms of sums of resistances, rather than in terms of sums of conductances /the RJ terms on the right sides of equations (19)-(21) represent additive terms of entropy creations due to the interactions of the migrating species (each entropy-creation term multiplied by the constant temperature), which is not true of the right-hand terms of equations (3)-(5)/ /Oster, Perelson and Katchalsky, 1971; Chartier, Gross, and Spiegler, 1975, p. 81/. Moreover, the six resistance coefficients, R_{ij} ($i \neq j$), which appear on the right sides of equations (19)-(21) are numerically equal to the mutual friction coefficients ζ_{ij} /Richardson, 1971, eq. (8); Meares, Thain and Dawson, 1972, p. 67/ which are amenable to physical interpretation /Spiegler, 1958; Meares, 1976/.

3.III. The "Additivity Rule" for Flows and Potential Differences Across Membranes /Krämer and Meares, 1969/

Consider a mass-transport experiment as schematically described in Figure 1 /Spiegler and Kedem, 1966/. Although the figure shows hyperfiltration (reverse osmosis) as the transport process, for the sake of illustration, the following remarks about the additivity rule refer to mass flow across the membrane in general, not necessarily flow induced by a pressure gradient. The membrane polymer is assumed to be homogeneous along the horizontal coordinate.

We replace the real membrane by a series sequence of many infinitesimally thin membrane elements, separated by thin solution compartments of uniform concentration, and we look at one such membrane element.

We consider the steady state. Concentrations c_s' and c_s'' in the terminal solutions are maintained constant ("clamped"). The components in the (hypothetical) solution compartments are assumed to be equilibrated with the adjacent membrane faces.

If we imagine the insertion of electrodes reversible to one of the salt ions in the terminal compartments, as well as in one of the intermediate solution compartments (salt concentration c_1), the total electric potential drop between the terminal electrodes, $\Delta E_{c' \to c''}$, equals the sum of the potential drops $\Delta E_{c' \to c_1}$ and $\Delta E_{c_1 \to c''}$. Since all potential drops are independent of the membrane thickness, the additivity rule can be stated as follows: "Consider three identical membranes, (1) one bracketed by solutions of concentrations c" and c_1; (2) one by c_1 and c"; (3) one by c' and c". The potential drop across the third is equal to the sum of the potential drops across the first two."

A similar additivity rule holds also for the mass flows, i.e., the flow across the third membrane equals the sum of the flows across the

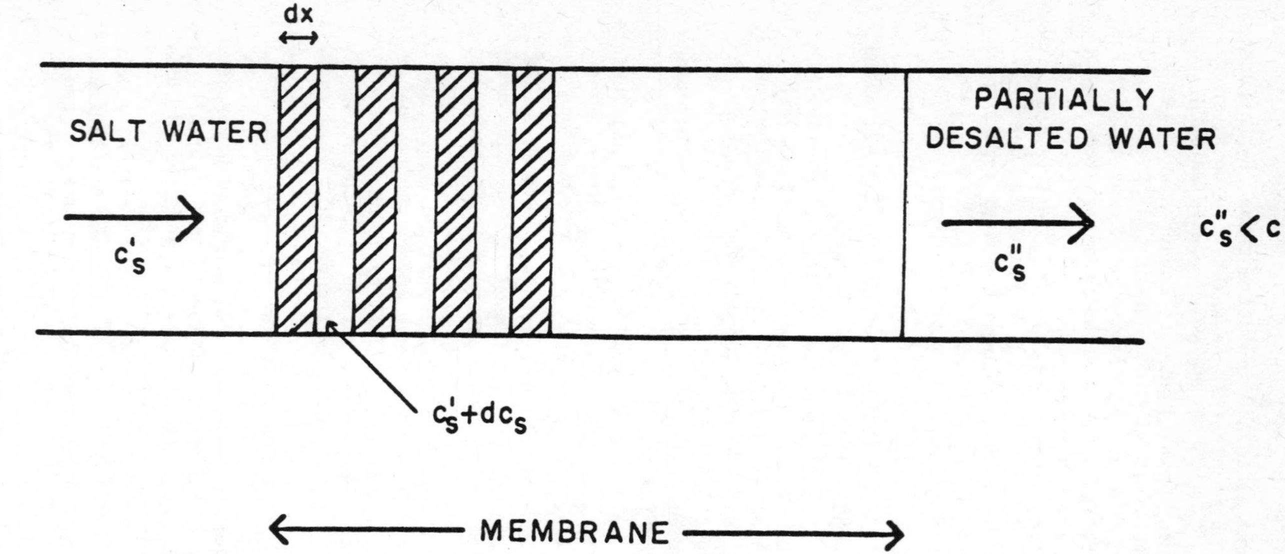

Figure 1. Schematic Representation of Hyperfiltration Membrane. Membrane is broken down into differential elements, separated by <u>uniform</u> solution segments which are in equilibrium with the two <u>adjacent</u> membrane faces. In a thought experiment, the solution elements are imagined to be infinitesimally narrow and continuously stirred. Therefore, their resistance to mass transfer is negligible [Spiegler and Kedem, 1966].

first and the second, as shown in the following.

According to Ohm's law:

$$i = \Delta E/R' = \kappa\Delta E/\Delta\ell \qquad (22)$$

where i is the current density, ΔE the electric potential across a resistor of unit cross-section and electrical resistance R', κ the electric conductivity and Δℓ the length of the resistor. Similarly the mass flux, J, across a membrane equals:

$$J = k\frac{\Delta\tilde{\mu}}{\Delta\ell} \qquad (23)$$

where $\tilde{\mu}$ is the electrochemical potential of the species considered; k is a conductivity coefficient of this species, which often depends on $\tilde{\mu}$.

In differential form, eq. (23) becomes:

$$J = k(\tilde{\mu})\frac{d\tilde{\mu}}{d\ell} \qquad (24)$$

In the steady state, J is uniform. Therefore, we obtain by integration of (24) across a membrane of thickness Δℓ, bracketed by solutions in which the electrochemical potentials of the migrating species are $\tilde{\mu}'$ and $\tilde{\mu}''$ respectively

$$\left(J_{c'\to c''}\right)\Delta\ell = \int_{\tilde{\mu}'}^{\tilde{\mu}''} k(\tilde{\mu})\,d\tilde{\mu} \qquad (25)$$

By the laws of calculus, the latter integral equals:

$$\left(J_{c'\to c''}\right)\Delta\ell = \int_{\tilde{\mu}'}^{\tilde{\mu}_1} k(\tilde{\mu})d\tilde{\mu} + \int_{\tilde{\mu}_1}^{\tilde{\mu}''} k(\tilde{\mu})d\tilde{\mu} \qquad (26)$$

where $\tilde{\mu}_1$ is the potential of the migrating species in a solution of intermediate concentration c_1.

Consider now two identical membranes of equal thickness, $\Delta\ell$, one bracketed by solutions c' and c_1, the second by c_1 and c". By the reasoning which led to eq. (25), the flux of a migrating species through the first membrane is:

$$J_{c'\to c_1} = (1/\Delta\ell) \int_{\tilde{\mu}'}^{\tilde{\mu}_1} k(\tilde{\mu})d\tilde{\mu} \qquad (27)$$

and through the second:

$$J_{c_1\to c''} = (1/\Delta\ell) \int_{\tilde{\mu}_1}^{\tilde{\mu}''} k(\tilde{\mu})d\tilde{\mu} \qquad (28)$$

Adding equation (28) to (27), and comparing the result to (26) we see that:

$$J_{c'\to c''} = J_{c'\to c_1} + J_{c_1\to c''} \qquad (29)$$

In other words, if the flux of a species is measured when the membrane is interposed between solutions c' and c_1, and also when the same membrane is interposed between solutions c_1 and c", one can predict the flux which would be obtained if the same membrane were interposed between c' and c", by just adding the results of the two measurements. (It is important to make all measurements after a steady state has been reached.) In this manner a composite curve may be prepared yielding $J_{c\to c''}$ vs. c at a constant c". The tangent of this curve is defined as the <u>permeability coefficient</u> (of the respective components), g_c:

$$g_c \equiv (\partial J/\partial c)_{c''} \qquad (30)$$

The permeability coefficient depends on both terminal concentrations c' and c" respectively. In a series of measurements to test the additivity rules, and thus the underlying assumptions about membrane homogeneity, one terminal concentration is held constant, and the other is varied in the different experiments.

If J is plotted as a function of the chemical potential of one of the solution components, similar considerations apply. In that case we define conductivity coefficients of the respective components, g_μ :

$$g_\mu \equiv \partial J/\partial(\Delta\mu^c) \qquad (31)$$

3.IV. Calculation of Transport Coefficients from Experimental Data

In the strict sense, the Michaeli-Kedem ("M-K") transport equations, (16)-(18), apply to the total system between the electrodes, i.e., the membrane plus the two adjacent solutions. Since we are primarily interested in the transport coefficients of the membrane proper, we inquire first which of the fluxes and forces would be appreciably affected if we progressively reduced the thickness of the solution layers in series with the membrane.

Since the solutions are well-stirred, concentration gradients in them are negligible. Thus the generalized forces $-\Delta\mu_s$ and $-\Delta\mu_w$, which represent the chemical potential differences between the solutions near the electrodes, are identical with the corresponding differences across the membranes only. The electrical potential difference, $\Delta\phi_-$, depends on the magnitude of the current density, i. When the latter is zero, $\Delta\phi_-$ is independent of the position of the electrodes, and hence the experimental $\Delta\phi_-$, measured at a finite distance, can be substituted for the corresponding potential difference of electrodes in very close vicinity of the membranes. When the current density is finite, however, and kept constant, moving the electrodes towards the membrane will decrease the absolute value of $\Delta\phi_-$, since the ohmic drop through the solutions is decreased. The ohmic drop (volt) is:

$$\text{Ohmic drop} = i(\rho' z' + \rho'' z'') \qquad (32)$$

where the ρ's are the specific resistances of the solutions and the z's the distances between the membrane surfaces and the electrode surfaces in the two solutions, respectively.

Therefore, if we want to calculate transport coefficients for the membrane only, we use the value corrected for infinitesimally small distance from the membrane:

$$(\Delta\phi_-)_{\text{across membrane}} = (\Delta\phi_-)_{\text{measured}} + i(\rho' z' + \rho'' z'') \qquad (33)$$

When the current is negligible (e.g., in measurements of streaming potentials or membrane potentials), it is seen from eq. (33) that the measured potential differences are independent of the position of the electrodes.

In a thought experiment one can reduce all the L-coefficients to a system in which the electrodes are so close to the membrane surfaces that the mass and ion-transfer resistances of the remaining thin solution layers are entirely negligible. The L-coefficients calculated for this system may be considered as the transport coefficients for the membrane.

Four groups of experiments, leading to the measurement of eight independent transport properties, can be readily made in the concentration-clamp apparatus, viz. (a) electric conductivity, (b) hydraulic permeability and streaming potential, (c) electromigration and electroosmosis, and (d) osmosis, dialysis, and membrane potential. /The ninth property ("degree of hyperfiltration") cannot be measured with sufficient accuracy in the existing concentration-clamp apparatus/. Hence it is possible to test the range of applicability of two out of the following three reciprocity relations:

$$L_{we} = L_{ew} \quad (34)$$

$$L_{se} = L_{es} \quad (35)$$

$$L_{sw} = L_{ws} \quad (36)$$

Two reciprocity relations, viz. (34) and (35), were indeed tested experimentally over a limited range and found to hold. The third, eq. (36), was assumed to hold also.

The conductance coefficients, L_{xy} , were calculated from the experimental results in the following manner:

Measurement of membrane conductivity, κ

This measurement is performed with identical solutions on the two sides of the membrane ($\Delta\mu = 0$), and the (diagonal) conductance coefficient, L_{ee} , calculated from eq. (18).

L_{ee} When the solutions are of the same concentration and at the same pressure, the algebraic sum of the two Ag/AgCl electrode potentials (i.e., of the potential drops between each silver metal and the solution in contact with it) is zero. Therefore in this case $\Delta\phi_- = \Delta\phi$. Since the conductivity of the membrane is defined as

$$\kappa = (-i/\Delta\phi)\underline{d} \quad (37)^*$$

it follows from eq. (18) that

$$(L_{ee})_{\Delta\mu=0} = -i/(\Delta\phi_-\mathcal{F}^2) = -i/(\Delta\phi\mathcal{F}^2) = \kappa/(\mathcal{F}^2\underline{d}) = (\mathcal{F}^2\rho\underline{d})^{-1} \quad (38)$$

In the determination of the membrane resistance (per unit active area), $\rho\underline{d}$, the solution resistance is eliminated by extrapolation to zero solution thickness. Hence $(L_{ee})_{\Delta\mu=o}$ is a transport constant of the membrane alone. In this measurement, platinized platinum electrodes carrying alternating current were used, but for $\Delta\mu = o$ this makes no difference, since in this case the electrical potential difference between these electrodes is the same as $\Delta\phi_-$.

L_{se} Dividing eq. (16) by (18) for $\Delta\mu = 0$, we obtain

$$L_{se}/L_{ee} = [J_s/(i/\mathcal{F})]_{\Delta\mu=o} \equiv \bar{t}_+ \quad (39)$$

Hence

$$(L_{se})_{\Delta\mu=o} = \bar{t}_+(L_{ee})_{\Delta\mu=o} \quad (40)$$

---

*The negative sign derives from the convention that for positive potential difference ($\phi'' > \phi'$), the current direction (taken as the flow direction of positive carriers) is negative.

Measurements of hydraulic permeability and streaming potential

The hydraulic permeability of the membrane, L_p, is the volume flux through the membrane per unit of applied pressure. In its measurement, both sides of the membrane are exposed to the same solution ($\Delta\mu_s^c = 0$; $\Delta\mu_w^c = 0$). No electric current flows ($i = 0$), i.e., if there are electrodes near the two membrane faces, they are either not connected at all or the electric potential difference between them, the negative value of which is termed the "streaming potential differential" ["S.P.D.," eq. (42)], is measured with a voltmeter of negligible conductance. In other words:

$$L_p \equiv -(J_v/\Delta p)_{\Delta\mu^c=0,i=0} \tag{41}$$

$$(SPD) \equiv [\partial(-\Delta\phi_-)/\partial(\Delta p)]_{\Delta\mu^c=0,i=0} \tag{42}$$

Using these definitions, we apply the general transport equations (16) and (17) respectively to the special conditions $\Delta\mu^c = 0$, $i = 0$. When $\Delta\mu_i^c = 0$, the chemical potential differences reduce to pressure terms only [Katchalsky and Curran, 1966, eq. (10-6)]:

$$\Delta\mu_i = \bar{v}_i \Delta p \tag{43}$$

where $\bar{v}_i$ is the partial molal volume of component i. Keeping (41)-(43) in mind, the transport equations (16) and (17) become:

$$L_{ws}\bar{v}_s + L_{ww}\bar{v}_w = -[L_{we}\mathfrak{F}(SPD) + (L_p/\bar{v}_w) + L_{we}v_{Cl^-}] \equiv A_3 \tag{44}$$

$$L_{es}\bar{v}_s + L_{ew}\bar{v}_w - L_{ee}[\mathfrak{F}(SPD) + v_{Cl^-}] = 0 \tag{45}$$

An approximation has been made in substituting eq. (43) in eqs. (16) and (17), and in using the condition $i = 0$, to derive eqs. (44) and (45), namely,

$$J_v = J_w\bar{v}_w + J_s\bar{v}_s \approx J_w\bar{v}_w \tag{46}$$

which means that the total volume flux through the membrane consists primarily of the water flux, $J_w\bar{v}_w$. This approximation is often made (see, for instance, Katchalsky and Curran, 1966).

The derivations of eqs. (44) and (45) represent two steps in the calculation of the eight conductance coefficients, L_{xy}.

Measurements of electromigration-electroosmosis; L_{se} and L_{we}

The transport numbers of the positive ions, and of water, t_+ and t_w, respectively, are determined in these experiments. From these transport numbers, the conductance coefficient, L_{se} , can be calculated from eq. (40). L_{we} is calculated in the following manner:*

Dividing eq. (17) by (18) for $\Delta\mu = 0$, we obtain

$$L_{we}/L_{ee} = [\bar{J}_w/(i/\mathcal{F})]_{\Delta\mu=o} \equiv \bar{t}_w \tag{47}$$

Hence

$$(L_{we})_{\Delta\mu=o} = \bar{t}_w(L_{ee})_{\Delta\mu=o} \tag{48}$$

Measurements of osmosis-dialysis-membrane potential

In these measurements, solutions of different salt concentration in the two half-cells bracket the membrane. The flow of water (osmosis) and of salt (dialysis) between the half-cells is measured, and so is the potential difference, ϕ_- , between two silver-silver chloride electrodes in the half-cells. While this measured potential difference, ϕ_- , is used directly in the following equations for calculation of the conductance coefficients, it is well to keep in mind that the potential difference, ϕ , between the two solutions (which is simply ϕ_- corrected

---

*It should be noted that (in the absence of the concentration-clamp mechanism) in the membrane/(chloride solution) system described here, the flow of positive ions through the membrane would be numerically equal to the loss and gain respectively of salt in the two compartments, because electroneutrality causes each positive ion to be balanced by a chloride ion removed or produced respectively by the electrode reaction: $J_+ = J_s$.

for the electrode potentials) is often called the "membrane potential" [see, for instance, Spiegler and Wyllie, 1968].

To use the experimental results, it is useful to derive transport equations for systems with infinitesimally small concentration gradients across the membrane, and to conclude from the results obtained with finite concentration gradients what transport phenomena to expect for the infinitesimal gradients. [This is possible by means of a plotting procedure described later; the plotting procedure, in turn, is based on the "additivity rule".]

To calculate the differences between the chemical potentials of the salt in the two half-cells from the concentration, we use the Gibbs-Duhem relationship:

$$(\partial\mu_w^c/\partial\mu_s^c)_{p,T} = -c_s/c_w \tag{49}$$

We now apply the transport equation (18) to the measurement of the electric potential difference between the two Ag/AgCl electrodes in a membrane/solution system with infinitesimal concentration (and hence chemical-potential) differences. The presumption is that one concentration remains constant, while the other varies. We define $d\mu \equiv d(\Delta\mu) = d(\mu'' - \mu')$. In those experiments with membrane C-103 in which the solutions separated by the membrane were not of equal concentration, c_s' was held constant, while c_s'' was varied. Hence, in these cases $d\mu = d\mu''$. On the other hand, in similar experiments with membrane Cl-25T, c'' was held constant, while c' was varied. The evaluation method remained the same, but in these cases $d\mu = -d\mu'$.

To calculate the conductance coefficients L_{ew}, L_{es}, L_{ws} and L_{ss} from the experimental results, we first divide eq. (18) by $[-d(\Delta\mu_s)]$, go to the limit of infinitesimal concentration differences ($\Delta\mu^c \rightarrow d\mu^c$), and then substitute the concentration ratio c_s/c_w for $-(d\mu_w^c/d\mu_s^c)$ [eq. (49)].

$$L_{es} - L_{ew}(c_s/c_w) + L_{ee}\mathfrak{F}[\partial(\Delta\phi_-)/\partial(\Delta\mu_s)]_{p,T} =$$

$$L_{es} - L_{ew}(c_s/c_w) + L_{ee}\mathfrak{F}(\partial\phi_-/\partial\mu_s^c)_{p,T} = 0 \quad [\text{For } i=0;\ p,T \text{ uniform}] \tag{50}$$

Similarly, we obtain from eq. (17):

$$L_{ws} - L_{ww}(c_s/c_w) = -[\mathfrak{F}L_{we}(\text{PDC}) + (\partial J_w/\partial\mu_s^c)_{p,T}] \equiv A_6 \quad [p,T \text{ uniform}] \tag{51}$$

Here the measured potential-difference coefficient has been designated as (PDC):

$$(\text{PDC}) \equiv (\partial\phi_-/\partial\mu_s^c)_{p,T} \tag{52}$$

Although eq. (51) is derived in a formal manner from eq. (17), it is worthwhile to prove eq. (51) from the "additivity rule", so as to better illustrate the physical meaning of the different terms:

Consider three situations, in which the same membrane separates three different pairs of solutions, viz.

c' |Membrane| c"	Flux J_1	(p,T uniform)
c" |Membrane| c" + dc	Flux J_2	(p,T uniform)
c' |Membrane| c" + dc	Flux J_3	(p,T uniform)

Then, by the additivity rule:

$$J_3 = J_1 + J_2 \tag{53}$$

We define

$$dJ \equiv J_3 - J_1 \tag{54}$$

Also

$$J_3 - J_1 = J_2 \tag{55}$$

Apply eq. (17) to the second situation shown above:

$$dJ_w = L_{ws}(-d\mu_s^c) + L_{ww}(-d\mu_w^c) + L_{we}[-\mathfrak{F}d(\Delta\phi_-) + v_{Cl}dp] \tag{56}$$

Since p is uniform, dp = 0. Divide by $-d\mu_s^c$, and then substitute for $(\partial\mu_w^c/\partial\mu_s^c)_{p,T}$ from eq. (49):

$$-\left(\frac{\partial J_w}{\partial\mu_s^c}\right)_{p,T} \equiv g_{\mu,w} = L_{ws} + L_{ww}\left(\frac{\partial\mu_w^c}{\partial\mu_s^c}\right)_{p,T} + L_{we}\left(\mathfrak{F}\frac{\partial\phi_-}{\partial\mu_s^c}\right) = L_{ws} - L_{ww}\left(\frac{c_s}{c_w}\right) + L_{we}\mathfrak{F}(\text{PDC}) \tag{57}$$

Upon rearrangement, this equation, derived from the additivity rule, is indeed identical to eq. (51).

In analogy to the procedures leading to eq. (51), we obtain from eq. (16):

$$L_{ss} - L_{sw}(c_s/c_w) + \mathfrak{F}L_{se}(\text{PDC}) + (\partial J_s/\partial\mu_s)_{p,T} = 0 \tag{58}$$

To obtain an expression for L_{ew} in terms of measured parameters, we first multiply eq. (50) by $(-\bar{v}_s)$:

$$-L_{es}\bar{v}_s + L_{ew}(\bar{v}_s c_s/c_w) - \mathfrak{F}L_{ee}\bar{v}_s(\text{PDC}) = 0 \tag{59}$$

Add to this eq. (45):

$$L_{ew}[\bar{v}_w + (\bar{v}_s c_s/c_w)] - L_{ee}[\mathfrak{F}(\text{SPD}) + \mathfrak{F}\bar{v}_s(\text{PDC}) + v_{Cl}] = 0 \tag{60}$$

Solve for L_{ew}:

$$L_{ew} = \frac{L_{ee}}{\bar{v}_w + (\bar{v}_s c_s/c_w)}[\mathfrak{F}(\text{SPD}) + \mathfrak{F}\bar{v}_s(\text{PDC}) + v_{Cl}] \tag{61}$$

To obtain an expression for L_{es}, again in terms of measured parameters, we multiply eq. (50) by $(\bar{v}_w c_w / c_s)$:

$$(L_{es}\bar{v}_w c_w/c_s) - L_{ew}v_w + L_{ee}\mathfrak{F}(\bar{v}_w c_w/c_s)(PDC) = 0 \tag{62}$$

Add eq. (45):

$$L_{es}[(\bar{v}_w c_w/c_s) + \bar{v}_s] + L_{ee}[(\bar{v}_w c_w/c_s)\mathfrak{F}(PDC) - \mathfrak{F}(SPD) - v_{Cl}] = 0 \tag{63}$$

Multiply by (c_s/c_w) and solve for L_{es}:

$$L_{es} = \frac{-L_{ee}}{\bar{v}_w + (\bar{v}_s c_s/c_w)}[-\mathfrak{F}(c_s/c_w)(SPD) + \mathfrak{F}\bar{v}_w(PDC) - v_{Cl}(c_s/c_w)] \tag{64}$$

An expression for L_{ws} in terms of measured parameters is obtained by first multiplying eq. (44) by $c_s/(\bar{v}_w c_w)$:

$$L_{ws}[\bar{v}_s c_s/(\bar{v}_w c_w)] + L_{ww}(c_s/c_w) = A_3 c_s/(\bar{v}_w c_w) \tag{65}$$

Add eq. (51) and solve for L_{ws}:

$$L_{ws} = \frac{A_3(c_s/c_w) + \bar{v}_w A_6}{\bar{v}_w + (\bar{v}_s c_s/c_w)} \tag{66}$$

To obtain an expression for the conductance coefficient $\underline{L_{ww}}$ in terms of measured parameters, we multiply eq. (51) by $(-\bar{v}_s)$:

$$-L_{ws}\bar{v}_s + L_{ww}(\bar{v}_s c_s/c_w) = -A_6\bar{v}_s \tag{67}$$

Add eq. (44) and solve for L_{ww}:

$$L_{ww} = \frac{A_3 - \bar{v}_s A_6}{\bar{v}_w + (c_s\bar{v}_s/c_w)} \tag{68}$$

Finally, L_{ss} is calculated from eq. (58), assuming that the reciprocity relation $L_{sw} = L_{ws}$, eq. (36) is valid:

$$L_{ss} = L_{ws}(c_s/c_w) - \mathfrak{F}L_{se}(PDC) - (\partial J_s/\partial\mu_s)_{p,T} = L_{ws}(c_s/c_w) - \mathfrak{F}L_{se}(PDC) - g_{\mu,s} \quad (69)$$

Summary of sequence for the calculation of the conductance coefficients

The sequence of calculation of the coefficients L_{xy}, from measured and tabulated data, is as follows: (1) L_{ee} is calculated from the membrane conductivity by use of eq. (38); (2) L_{se} is calculated from the cation transport number by eq. (40); (3) L_{we} is calculated from the water transport number by eq. (48); (4) L_{ew} is calculated from streaming-potential and membrane-potential determinations by eq. (61); (5) L_{es} is determined by using the measurements which appear in eq. (64); (6) L_{ws} is calculated from eq. (66); (7) L_{ww} is calculated from eq. (68); (8) L_{ss} is calculated from eq. (69).

In these calculations for L_{ws} and L_{ww}, there appears A_3 [eq. (44)], that contains only parameters which can be measured in experiments at uniform concentrations. As for A_6 [eq. (51)], data on the water conductivity coefficient, $g_{\mu,w} \equiv (\partial J_w/\partial\mu_s^c)_{p,T}$, are necessary. The calculation of L_{ss} [eq. (69)] necessitates data on $g_{\mu,s} \equiv (\partial J_s/\partial\mu_s^c)_{p,T}$, and for all equations containing the potential-difference coefficient, (PDC), data on $(\partial\phi_-/\partial\mu_s)_{p,T}$ [eq. (52)] are necessary. All three of these derivatives with respect to μ_s are concentration-dependent [as is the streaming potential, (SPD)]. They are found by plotting the respective property against μ_s (at constant pressure and temperature), and taking tangents of these curves at the desired concentrations (which define the desired μ_s's). These curves are prepared by making use of the "additivity rules".

V. Magnitude of the potential-difference coefficient, (PDC)

The electric potential of a Ag/AgCl electrode in a 1-1 chloride solution can be expressed in terms of solution properties by application of elementary thermodynamics of electrolyte solutions [see, for instance, Spiegler and Wyllie, 1968]:

$$(\phi_-)_{elde} = \text{constant} - (RT/\mathfrak{F})\ln a_{Cl^-} \quad (70)$$

where a_{Cl^-} is the mean ionic activity of the ions in the chloride solution.

The mean ionic activity is related to the chemical potential of the electrolyte [ibid., eq. (11)] by:

$$\mu_s^c = C + 2RT\ln a_{Cl^-} \tag{71}$$

where the constant C is independent of the concentration.

Consider the system

Ag/AgCl|NaCl solution, c'|Membrane|NaCl solution, c"|Ag/AgCl

at uniform pressure and temperature. In a series of experiments, c' is held constant and c" is varied. In this case, the potential difference between the electrodes is

$$\Delta\phi_- = C' + \Delta\phi - (RT/\mathfrak{F})\ln a''_{Cl^-} \tag{72}$$

where $\Delta\phi$ is the membrane potential and C' is a constant, independent of the concentration c''_s.

In principle, $\Delta\phi$ can vanish. For instance, this would be the case for two KCl solutions bracketing an entirely non-permselective ("NPS") membrane. In this case $\Delta\phi_-$ is the difference between the electrode potentials $\Delta(\phi_-)_{elde}$

$$(\Delta\phi_-)_{NPS,KCl} = C' - (RT/\mathfrak{F})\ln a''_{Cl^-} = \Delta(\phi_-)_{elde} \tag{73}$$

and, from eqs. (71) and (73),

$$(PDC)_{NPS,KCl} \equiv (\partial\Delta\phi_-/\partial\mu_s^c)_{NPS,KCl} = -1/2\mathfrak{F} = (PDC)_{elde} \tag{74}$$

On the other hand, if the membrane is ideally cation-selective ("ICS") the membrane potential, $(\Delta\phi)_{ICS}$, is of the same magnitude and direction as $\Delta\phi_-$ [ibid., eq. (68)]. Hence we obtain from eqs. (72) and (74):

$$(PDC)_{ICS} = -1/2\mathfrak{F} + (PDC)_{NPS,KCl} = -1/\mathfrak{F} \tag{75}$$

It should be noted that for a perfectly anion-selective ("IAS") membrane, the membrane potential $(\Delta\phi)_{IAS} = -(\Delta\phi)_{ICS}$. Hence from eqs. (74) and (75):

$$(PDC)_{IAS} = (PDC)_{elde} + \Delta\phi_{IAS} = (PDC)_{elde} + (1/2\mathfrak{F}) = 0 \tag{76}$$

It is seen that the potential-difference coefficient, (PDC), is a quantitative indicator of the permselectivity, as is $\bar{t}_+$.

4. EXPERIMENTAL DETAILS

4.I. The "Concentration-Clamp" Method

This method was devised for the measurement of transport in membranes under steady-state conditions. This implies that the concentrations of the solutions separated by the membrane remain constant. To achieve this aim the P.I. and his collaborators devised the system shown in Fig. 2 [Zelman et al., 1971; Spiegler et al., 1979]. In both solutions bracketing the membrane, a conductivity cell is inserted in the system. The resistance of the cell solution is compared to that of a reference resistor by a very accurate impedance comparator. The output of this comparator is then amplified with a phase-sensitive amplifier to open or close a relay. At the "salt-donor" side (right side in Figure 2) this relay activates an automatic buret which pushes a concentrated NaCl solution into the system until the reference concentration is reached again. At the "salt-acceptor" side of the membrane the relay actuates a pump which introduces distilled water into this cell compartment. In the measurements performed with <u>membrane C-103</u> the pump circulated the contents of a reservoir connected to the "enriched" half-cell by solution tubing and valves, as shown in Fig. 2. This reservoir of known volume is of the same size as the test cell; its contents were continuously agitated by a magnetic stirrer. It is originally filled with deionized water, and its contents serve to dilute the solution in the "enriched" half-cell when needed. A small portion of the solution in the half-cell passes through the reservoir while most of the solution is recirculated without change. Solenoid valves remain open until the reservoir has released enough water to lower the concentration to that of a reference cell. The concentration of the solution in the reservoir was continuously being monitored by measurement of its electric conductivity. A simular dilution method had been used by Delmotte and Chanu (1973) [see also Delmotte (1975)]

In the experiments with <u>membrane CL-25T</u>, a somewhat different flow scheme, designed and put into operation by A. Berg, was used, as shown in Fig. 3. A small amount of the solution circulated by the pump in the reservoir circuit was bled into the "salt-acceptor" compartment for dilution of the solution in that compartment until the original solution concentration was restored. The amount of water necessary for this purpose was measured by determination of the volume increase of the solution in that compartment into an overflow pipet (Fig. 3). The valve system shown in Fig. 3 included a check valve which prevented the flow of salt solution into the reservoir. Consequently, the amount of salt transported into the "salt-acceptor" compartment was calculated from the increase of the solution volume, rather than from the increase of the electrical conductivity of the solution in the reservoir. In this "modified" concentration-clamp cell, the salt-donating compartment was also equipped with an overflow pipet. The salt and water transport from this compartment to the "salt-acceptor" compartment could be calculated from

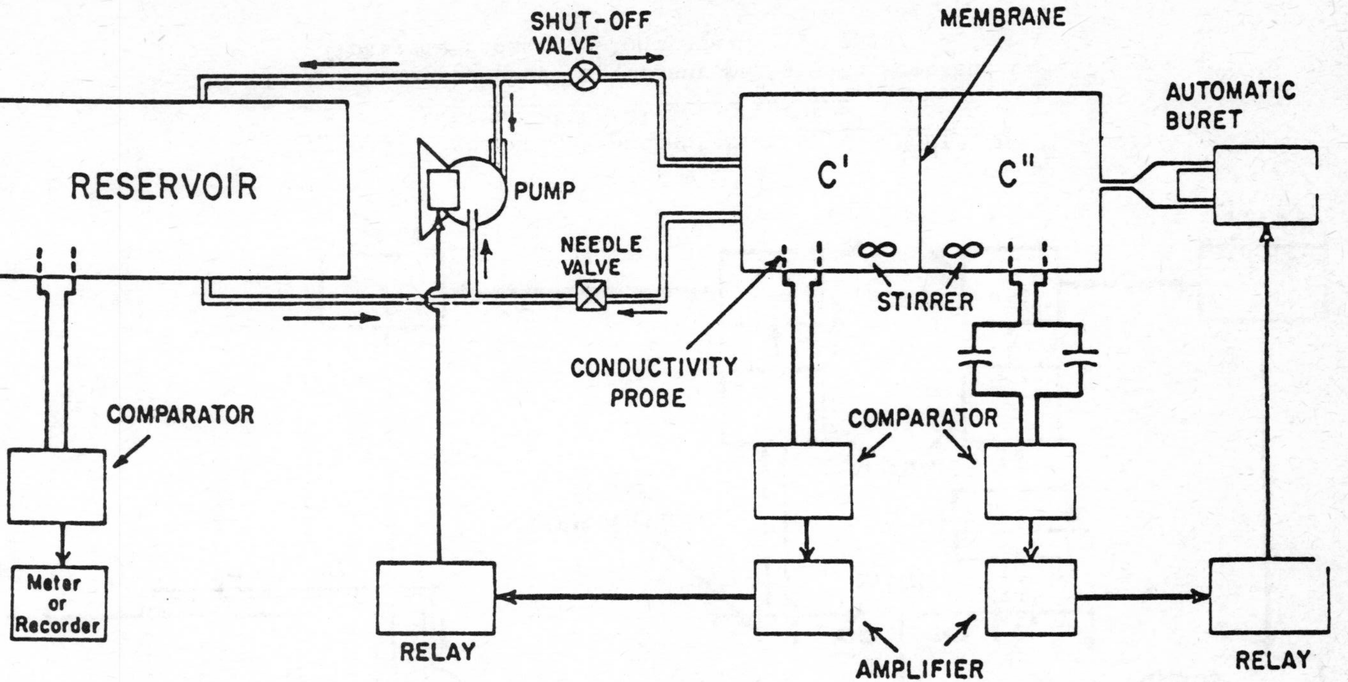

Figure 2. Schematic Representation of "Second-Generation Concentration-Clamp" Apparatus [Spiegler et al., 1979]

This flow scheme was used for measurement with membrane C-103. (American Machine and Foundry Co., Stamford, Conn.)

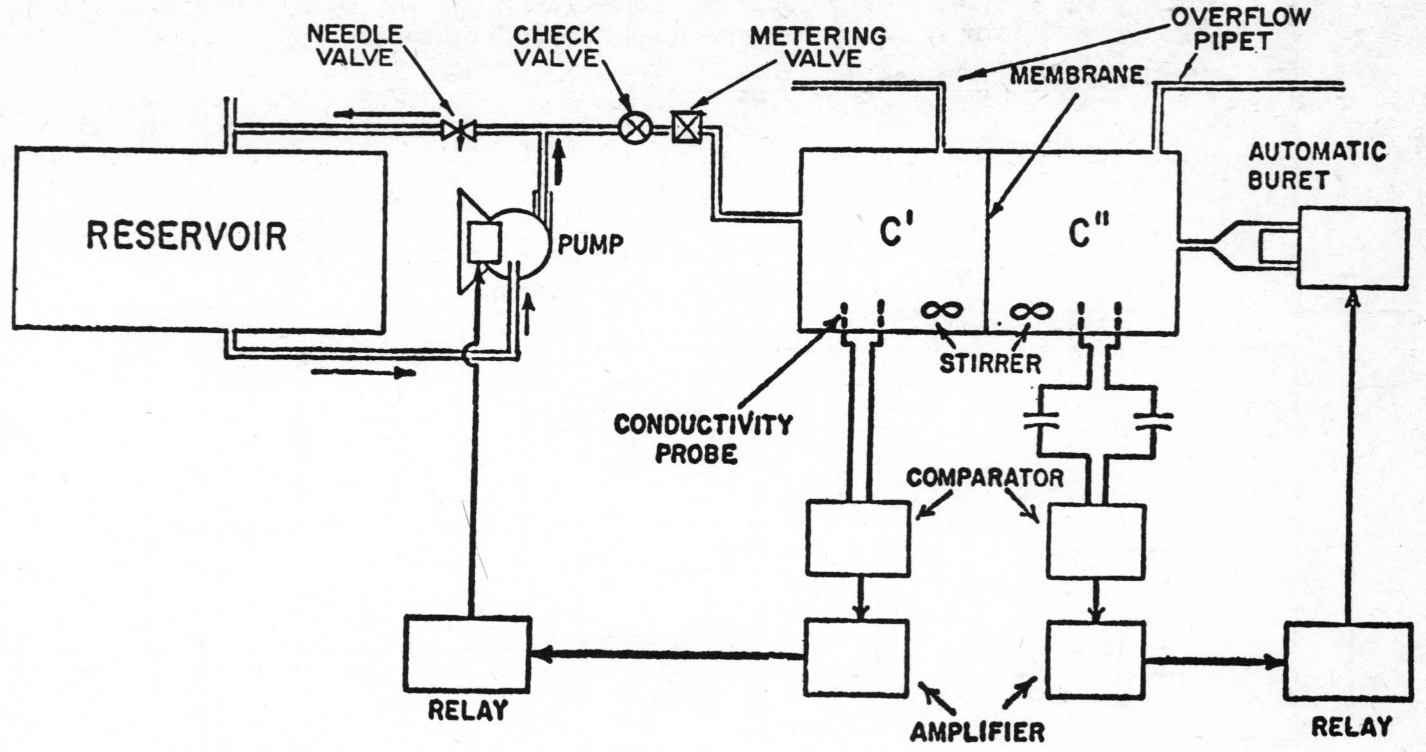

<u>Figure 3. Schematic Representation of the MODIFIED "Concentration-Clamp" Apparatus</u>

This flow scheme was used for measurements with membrane CL-25T. (Tokuyama Soda Co., Tokuyama City, Japan)

the difference between the amount of salt and water injected by the automatic buret, and the amount of water and salt in this overflow pipet of the "salt-donating" compartment.

When all driving forces for the transport process are chosen sufficiently small, so that no side-reactions occur (e.g. pH changes at the electrodes or membrane/solution interfaces in electromigration-electroosmosis experiments), this system keeps the concentration at either side of the membrane within 0.02%.

The design of the concentration-clamp apparatus, specifying in detail the components used, and the experimental procedure are presented in a detailed publication [Zelman, Kwak, Leibovitz and Spiegler, 1971; see also Halary, 1979] except that in the first-generation "concentration-clamp" apparatus, salt removal from the solution in the "enriched" half-cell was accomplished by a demineralizing column. The present method uses dilution, as was done in an earlier modification of the original method [Delmotte and Chanu, 1973]. The cell and associated electronics are shown in Figures 4-6. Sodium chloride was the electrolyte in all experiments.

4.II. Measurement of the Electrical Conductance of the Membrane

Because the electrical conductance of the membrane has to be determined accurately, and bearing in mind that similar cell and membrane geometrics are required in all experiments, the cell shown in Fig. 7 was designed, using parts of the "concentration-clamp" transport cell. It incorporates concepts of cells previously used for the measurement of the resistance of separators [Spiegler, 1966; Guillou, Guillou and Buvet, 1969]. The a.c. electrical resistance of the cell was measured as a function of the distance between the electrodes, with and without the membrane. The cell consists of two interchangeable hollow cylinders and two end parts, clamped together by steel rods. "Buna N" rubber gaskets were used as seals between the various parts.* All parts were machined of "Lexan" (General Electric Co., Plastic Div., Pittsfield, Mass.), a transparent polycarbonate plastic. The cylinders have outer and inner diameters of 4.5 and 2.0 inches respectively. The end parts are 1.25 inch thick. Two platinized platinum electrodes of 1.90 inch diameter can be moved forwards and backwards in the cell by means of a screw-driven mechanism, such that the electrodes remain parallel to the plane of the membrane; displacements to one-thousandth of an inch can be monitored with two precise dial-indicators (No. 25-441, L. S. Starret Co., Athol, Mass.), which are mounted on both sides of the cell.** The moving part of the gauge touches a metal strip fixed to the moving electrode. This method of measuring the distance traveled by

* In the measurements of the conductance of CL-25T membrane, "Teflon" sheets were used.
**Not shown in Figure 7.

Figure 4. "Concentration-Clamp" Cell and Associated Equipment

Figure 5. Electronics for "Concentration-Clamp" Cell Feedbacks

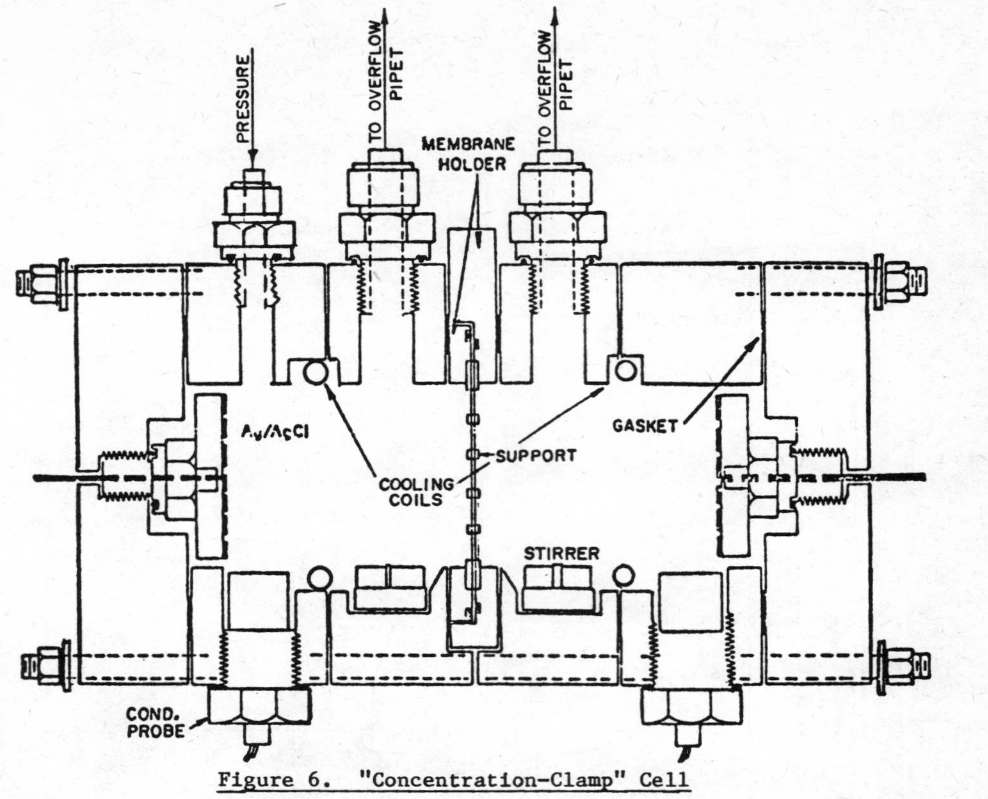

Figure 6. "Concentration-Clamp" Cell

This cell is an adaptation of the cell used in previous work [Zelman, Kwak, Leibovitz and Spiegler (1971); Spiegler (1979)]. The membrane-support plate had 130 circular perforations of nominal diameter 0.28 cm. The diameter of the circular membrane in the holder was 7.8 cm. $\mathcal{A}$=8.12cm^2.

Not shown here are two thermistors, located behind the overflow pipets, and drain valves (behind the conductivity probes). Connectors to reservoir loop and automatic buret are located behind the left and right stirrers respectively and not shown here. This cell was used in the experiments with membrane CL-25T.

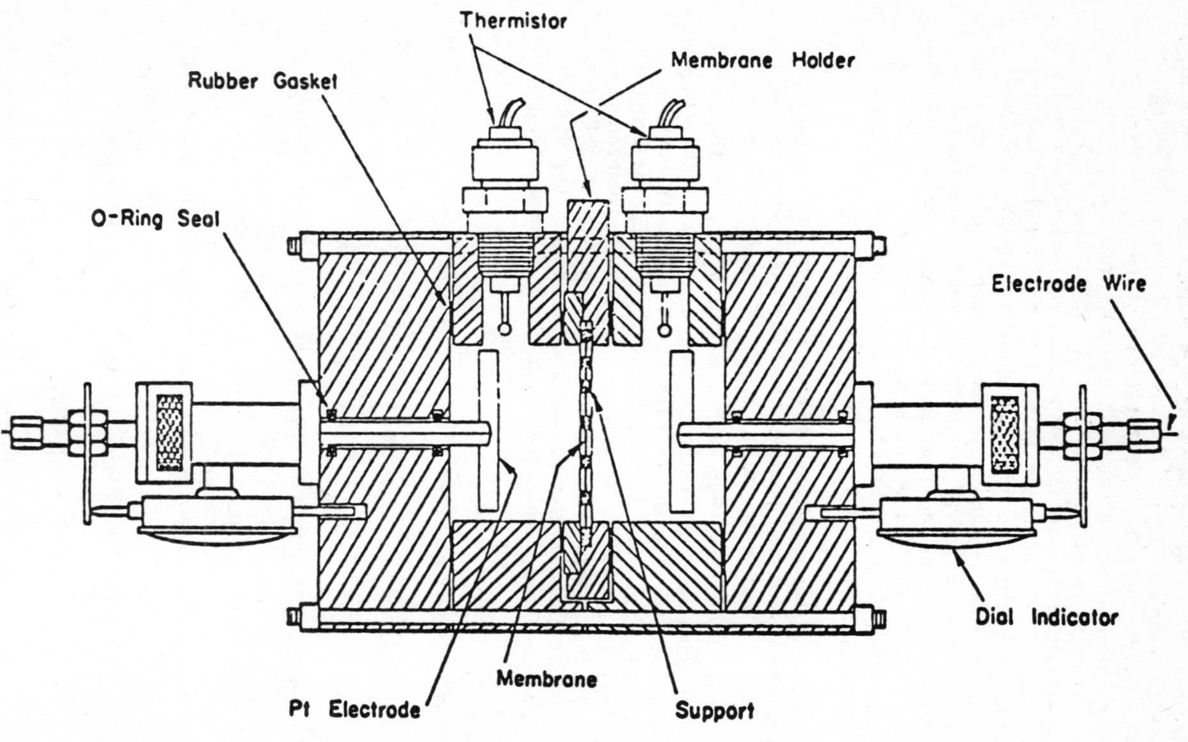

Figure 7. Cell for Measurement of Membrane Conductance

Inner diameter: 2" (= 5.08 cm)

the electrode was used in the experiments with membrane C-103, whereas for measurements with membrane CL-25T the distance traveled by an electrode was determined from the number of rotations of the screw (Fig. 6) moving the electrode (0.050 inches for each complete rotation of this 0.5"-20NF screw). One electrode usually remains stationary close to the membrane, but not in direct contact with it. The other is moved along the cylinder axis. The membrane is held in the membrane holder by O-ring seals. This membrane holder, the membrane support and the membrane itself are the same as used with the transport cell. The resistance between the electrodes is compared to a reference resistance by means of a 1605-AH grounded impedance comparator (General Radio Co., West Concord, Mass.). Two Veco glass-embedded thermistors (No. 47A13 Victory Engineering Corp., Springfield, N.H.) monitor the temperature of the solutions on both sides of the membrane.

The determination of membrane resistivity is based on a substitution method, i.e., the resistance between the electrodes is measured (a) with the membrane in the holder (result: $R_{tot,m}$ ohm), and (b) with the membrane removed from the holder, but with a membrane ring on the periphery of the two perforated membrane supports [Zelman, Kwak, Leibovitz and Spiegler, 1971, p. 685], but not in the path of the electric current between the electrodes (Fig. 7). In this manner, the supports are held at the same distance, $\underline{d}$, as in the presence of a complete membrane between them [$\underline{d}$(cm) is the membrane thickness]. Thus a body of solution of the same geometrical dimensions replaces the membrane after its removal. The solution resistance measured under these circumstances is $R_{tot,s}$.

From Ohm's law, the two measured resistances are:

$$R_{tot,m} = (\rho_s D'/S) + (\rho_m d/\mathcal{A}) + (2\rho_s \underline{d}'/\mathcal{A}) \quad (77)$$

and

$$R_{tot,s} = (\rho_s D'/S) + (\rho_s d/\mathcal{A}) + (2\rho_s \underline{d}'/\mathcal{A}) \quad (78)$$

where ρ_m and ρ_s (ohm cm) are the membrane and solution resistivities respectively, D'(cm) the sum of the distances between electrodes and membrane (cm), S the cross-sectioned area of the cell bore (cm^2), and $\mathcal{A}$ (cm^2) the exposed area of the membrane,* i.e., the area of the calendaring holes in the membrane support. $\underline{d}'$ (cm) is the thickness of the perforated support plate.

Subtracting eq. (77) from (78) and solving for ρ_m we obtain:

$$\rho_m = \rho_s - [\mathcal{A}(\delta R)/\underline{d}] \quad (79)$$

where $\delta R \equiv R_{tot,s} - R_{tot,m}$ is the difference between the two resistance

* For the experiments with membrane CL-25T, $\mathcal{A} = 8.12\ cm^2$.

measurements. Note that ρ_m is calculated from the difference, δR, between the two straight lines R_{tot} vs. D*, not from the absolute values of $R_{tot,s}$ and $R_{tot,m}$.

The solution resistivity can be found in electrochemical tables, but we thought it was preferable for this purpose to measure it under the same conditions as the other resistance measurements. This was done by measuring the resistance, R, of the cell filled with solution, at different electrode distances, z (cm), thus determining (dR/dz). The solution resistivity is

$$\rho_s = S(dR/dz) \tag{80}$$

Substituting eq. (80) in (79) we obtain an expression for the membrane resistivity, which was used in the evaluation of the data. **

$$\rho_m = S(dR/dz) - [\mathcal{A}(\delta R)/\underline{d}] \tag{81}$$

The measurements of the conductance of membrane Cl-25T were made with the membrane inserted in the membrane holder, as were many previous measurements of the P.I.'s group. Because the insertion of the membrane with its perforated supports is time-consuming and gives rise to temperature changes which decay quite slowly, a modified method was used for the measurement of the electrical conductivity of membrane C-103. In this method, the membrane holder was modified for rapid insertion or withdrawal of the membrane without supports. The membrane conductance data for C-103 were obtained by use of the modified membrane holder. The basic assumption was that, on the average, the electrical conductance per unit area of the membrane exposed in the modified membrane holder is the same as in the original holder.

4.III. Summary of Differences between Measurements with C-103 and CL-25T Membranes respectively

The experiments in this reporting period were all performed with membrane CL-25T. For comparison, results obtained earlier with membrane C-103 are also summarized and quoted in this report. The methods used in both stages of this research were similar in principle, but in accordance with the experience gained in the course of this research, and the different nature of the two membranes studied, the procedures varied in some respects, described under the following paragraphs (a) and (b).

---

* or plots of R_{tot} vs. z, where z is any length coordinate indicating the position of the movable electrode [Spiegler, 1966].

**A minor correction was applied, to account for the deviation of the cell geometry from the ideal cylindrical configuration.

(a) Concentration-clamp cells (Figs. 2 and 3). The "modified" apparatus shown in Fig. 3, was used for CL-25T, whereas previous experiments, with C-103, had been performed according to the scheme described in Fig. 2. [Spiegler, 1979]

(b) Conductance cell (Fig. 7). For membrane CL-25T, the complete membrane holder, including perforated membrane supports, was used. Changes of the electrode distance were determined from the number of revolutions of the screw moving the non-stationary electrode. Membrane C-103 was inserted in the cell without supports. Changes of the electrode distance were measured by means of distance gauges with dial indicators.

5. RESULTS and DISCUSSION

5.I. Equilibrium Properties of Membrane

The equilibrium properties of CL-25T membrane samples, which had been conditioned by several cycles of alternating treatment with 0.1 M HCl and 0.1 M NaCl solutions, were determined by T. S. Brun and are shown in column 1 of Table 2. Some of the properties of this kind of membrane are also listed in the manufacturer's catalog, and are also tabulated here for comparison.

A circular membrane specimen (diameter 7.8 cm) was cut from the sheet supplied by the manufacturer. It was placed into the membrane holder which, in turn, was inserted in the "concentration-clamp" cell (Figure 6), in which all transport measurements were performed. The electric conductance was determined in the cell shown in Figure 7. The results are shown in Figures 8-20.

The transport parameters determined from those Figures by the method described in Section 3 are tabulated in Tables 3-5.

5.II. Electrical Resistance Measurements

According to eqs. (77)-(79), the resistance between the electrodes (Figure 7) is expected to vary linearly with the axial position of the movable electrode, and the membrane resistance can be determined from the vertical distance of plots of resistance in the presence and absence of the membrane respectively. Figures 8-10 show such plots for three different solution concentrations bracketing the membrane.

5.III. Electromigration-electroosmosis

In these experiments, electric current was passed through the "concentration-clamp" cell (Figure 6); the membrane was bracketed by solutions of equal concentrations. These concentrations were maintained constant by the feedback mechanisms. The salt transport was determined by mass balance in both compartments. The results were close, as seen in Figures 11-13. The average of the salt transport determined from salt balances in the two compartments respectively was used in the subsequent calculations. In the present modification of the "concentration-clamp" method, the water transport could be determined only from the water balance in the "salt-donating" compartment (the right cell compartment in Figure 3). The water transport at different times is also shown in Figures 11-13, which describe three electromigration-electroosmosis experiments performed with solutions of three different concentrations respectively.

Table 2

Equilibrium Properties of Membrane

CL-25T (Tokuyama Soda Co.), Na-form

	Determined in this work	From manufacturer's "Neosepta Ion Exchange Membranes" catalog
Exchange capacity, eq/g dry membrane	1.8×10^{-3}	1.5 - 1.8
Water content, g/g dry membrane	0.391	0.25 - 0.35
Thickness, cm	0.0152	0.015 - 0.017
Non-exchange salt, mole/g wet membrane when equilibrated		
with 0.1 M NaCl solution	0.22×10^{-5}	
with 0.5 M NaCl solution	2.4×10^{-5}	

The manufacturer lists a Mullen burst strength of 3-5, and an area resistance of 2.2-3.0 ohm cm^2 for a membrane equilibrated with 0.5 M NaCl solution at 25°C.

The membrane samples used in this research were supplied by the manufacturer; they are reinforced by a plastic mesh.

Exposed membrane area, $\mathcal{A} = 8.12 cm^2$.

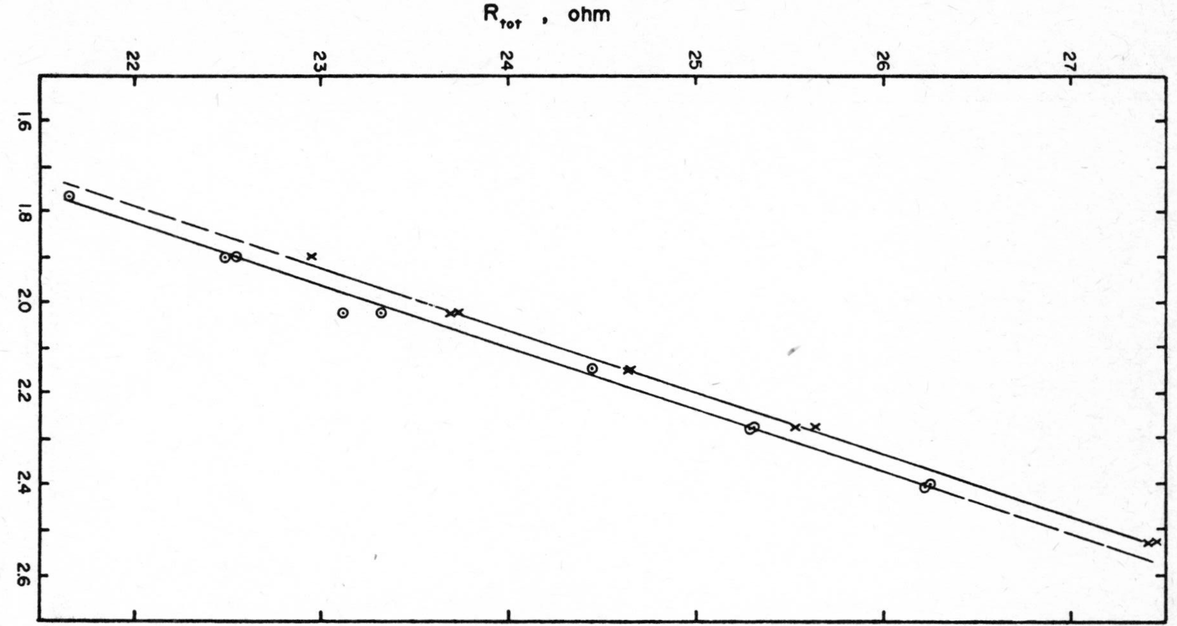

Figure 8. Resistance vs. Axial Length Coordinate in Conductance Cell (Fig. 7) (Experiment 22)

Membrane CL-25T (Tokuyama). NaCl solution. 25°C. $c_s = 0.07 \times 10^{-3}$ mol cm^{-3}.

(a) Without membrane $R_{s,tot} = 7.325z_1 + 8.598$ ohm (b) With membrane $R_{m,tot} = 7.193z_1 + 9.197$

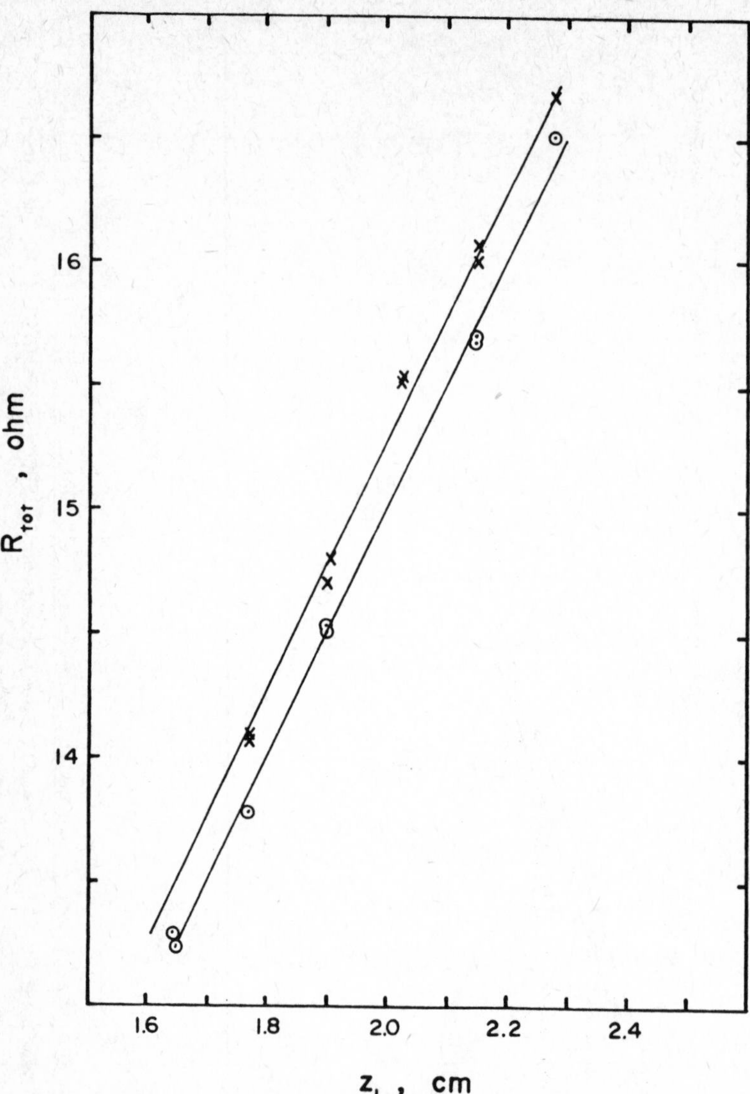

Figure 9. Resistance vs. Axial Length Coordinate in Conductance Cell (Fig. 7) (Experiment 21)

Membrane CL-25T (Tokuyama). NaCl solution. 25°C.

$c_s = 0.1 \times 10^{-7}$ mol cm^{-3}.

(a) Without membrane $R_{s,tot} = 5.037z_1 + 4.931$

(b) With membrane $R_{m,tot} = 5.110z_1 + 5.076$

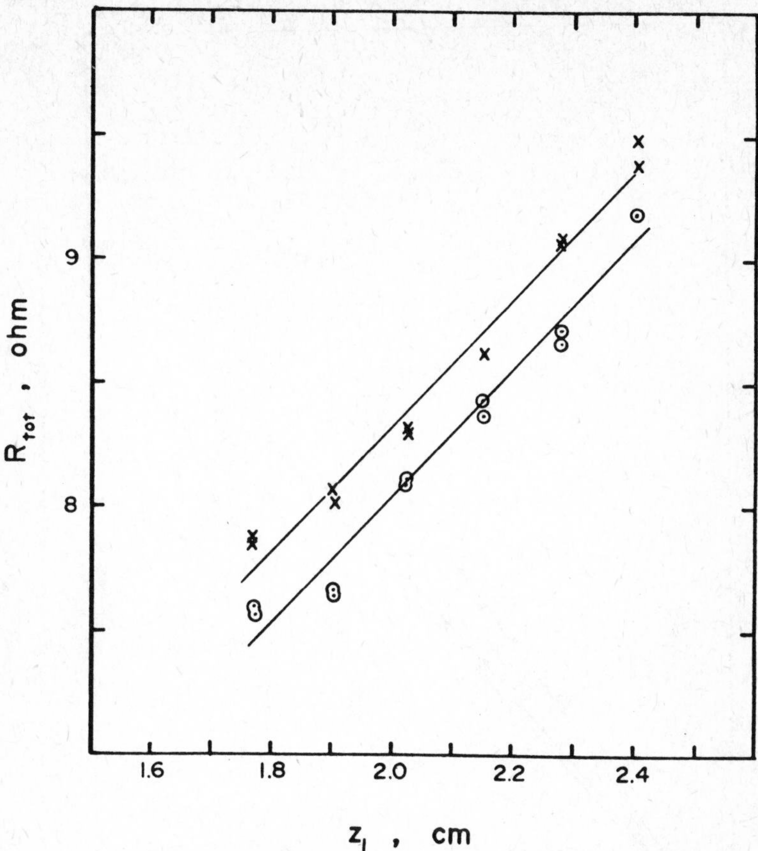

Figure 10. Resistance vs. Axial Length Coordinate in Conductance Cell (Fig. 7) (Experiment 23)

Membrane Cl-25T (Tokuyama). NaCl solution. 25°C.

$c_s = 0.2 \times 10^{-3}$ mol cm^{-3}.

(a) Without membrane $R_{s,tot} = 2.574z_1 + 2.894$

(b) With membrane $R_{m,tot} = 2.537z_1 + 3.264$

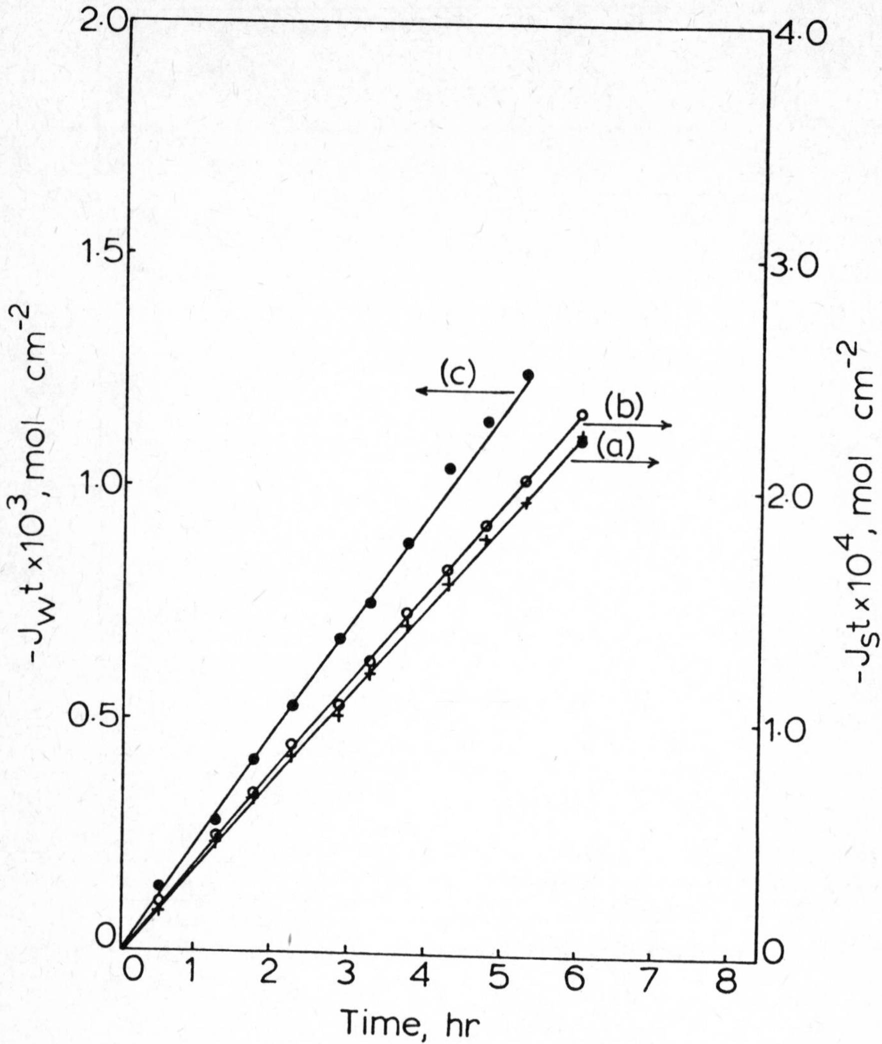

Figure 11. Electromigration-Electroosmosis $c_s' = 0.07$ M $= c_s''$

Membrane CL-25T (Tokuyama). NaCl solutions. $i = 0.9975$ mamp cm^{-2}.

(a) + Cumulative salt transport calculated from mass balance in salt-receiving compartment
(b) o " " " " " " " salt-donating "
(c) • " water " " " " " " " "

(a) $J_s = -1.034 \times 10^{-8}$ (b) $J_s = -1.079 \times 10^{-8}$ (c) $J_w = -6.517 \times 10^{-8}$ mol cm^{-2}sec^{-1}

(Transport data normalized for unit exposed area.) 25°C.

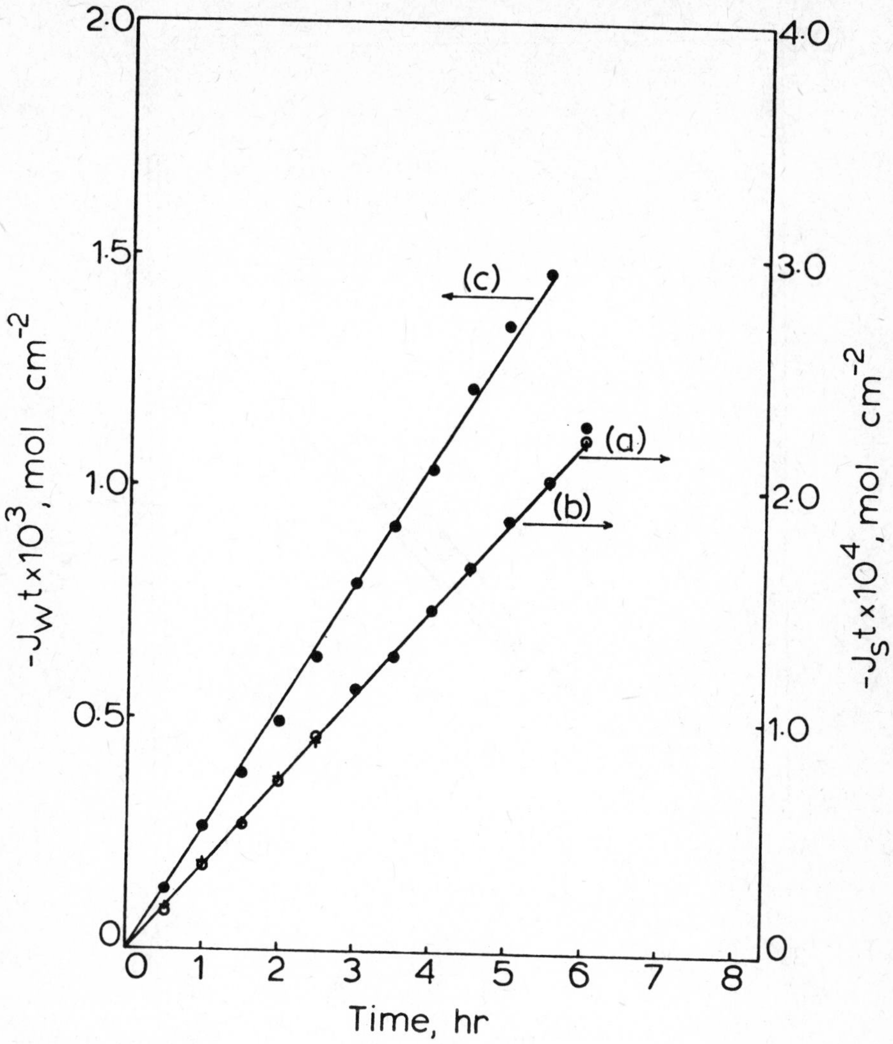

Figure 12. Electromigration-Electroosmosis $c_s' = 0.1$ M $= c_s''$

Membrane CL-25T (Tokuyama). NaCl solutions. $i = 0.9977$ mamp cm^{-2}

(a) + Cumulative salt transport calculated from mass balance in salt-receiving compartment
(b) o " " " " " " " salt-donating "
(c) • " water " " " " " " "

(a) $J_s = -1.023 \times 10^{-8}$ (b) $J_s = -1.028 \times 10^{-8}$ (c) $J_w = -7.323 \times 10^{-8}$ mol cm^{-2}sec^{-1}
(Transport data normalized for unit exposed area.) 25°C.

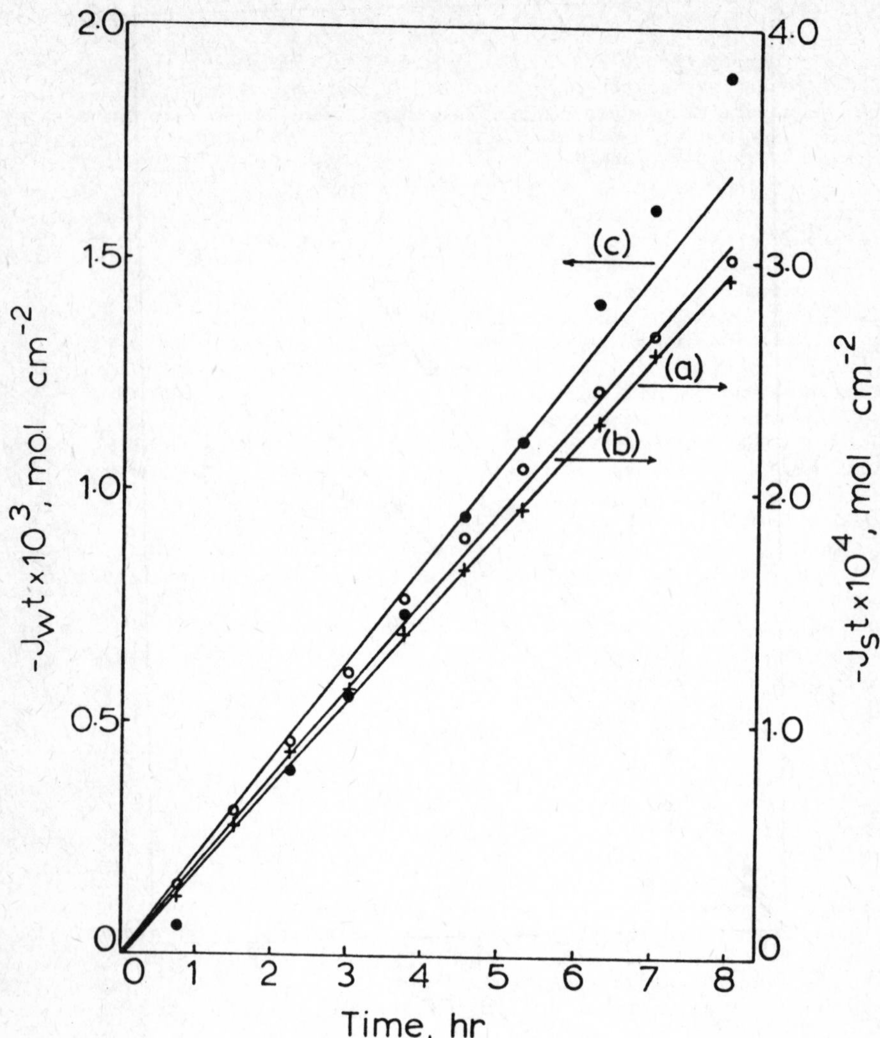

Figure 13. Electromigration-Electroosmosis $c_s' = 0.2$ M $= c_s''$

Membrane CL-25T (Tokuyama). NaCl solutions. $i = 0.9977$ mamp cm^{-2}.

(a) + Cumulative salt transport calculated from mass balance in salt-receiving compartment
(b) o " " " " " " " "
(c) • " water " " " " " " salt-donating "
" "

(a) $J_s = -1.020 \times 10^{-8}$ (b) $J_s = -1.059 \times 10^{-8}$ (c) $J_w = -5.781 \times 10^{-8}$ mol cm^{-2}sec^{-1}

(Transport data normalized for unit exposed area. $\mathcal{A} = 8.12$ cm^2) 25°C

It is seen that despite the relatively large amounts of salt transported across the membrane, the plots are linear, i.e. the rate of both salt and water transport remained constant. This demonstrates that the feedback mechanisms maintained the solution concentrations constant, rather than permitting salt-depletion and build-up to occur in the "salt-donating" and "salt-receiving" compartments respectively, with concomitant changes in the transport rates. In other words, the feedback mechanisms maintained steady-state conditions, rather than continuous changes of transport rates, as would occur in a closed electrodialysis cell in the absence of the "concentration clamp" mechanism.

The parameter termed "cumulative salt transport", plotted on the ordinates of Figures 11-13, actually represents the transport of Na^+ from the "salt-donating" to the "salt-receiving" compartment. Because the Ag/AgCl electrode in the "salt-donating" compartment binds, in the form of AgCl, one equivalent of Cl^- for each equivalent of Na^+ electromigrating through the membrane, the mols of NaCl lost from the original solution in the "salt-donating" compartment are numerically equal to the equivalents of Na^+ transported through the membrane. [This number of mols of NaCl is replaced in the solution by injection of salt solution from the automatic buret (Fig. 3)]. Similarly, since the Ag/AgCl electrode in the "salt-receiving" compartment releases one equivalent Cl^- for each equivalent Na^+ which electromigrates through the membrane, the mols NaCl gained by the solution in the "salt-receiving" compartment numerically equal the equivalents of Na^+ which electromigrated through the membrane. Hence the salt transport, plotted on the ordinates, numerically equals the sodium-ion transport.

5.IV. Dialysis-Osmosis

The salt concentration in the "salt-donating" compartment was $c_s'' = 0.5 \times 10^{-3}$ mol cm^{-3} in all the osmosis-dialysis experiments described in Figures 14-17. The salt concentration in the "salt-receiving" compartments, c_s', was different (and always lower than c_s'') in the four experiments described in these Figures. In all cases, the "concentration-clamp" feedback mechanism held the concentrations in the two compartments constant during the course of the experiments in which spontaneous salt transport ("dialysis") took place from the "salt-donating" to the "salt-receiving" compartment*, while spontaneous water transport ("osmosis") took place simultaneously in the opposite direction.

In the previous embodiments of the "concentration-clamp" cell [Zelman, Kwak, Leibovitz and Spiegler, 1971; Spiegler, 1979], the fluxes of salt and water could be calculated from mass balances in both the "salt-donating" and the "salt-receiving" compartments. Satisfactory agreement

---

* right to left in Figure 3.

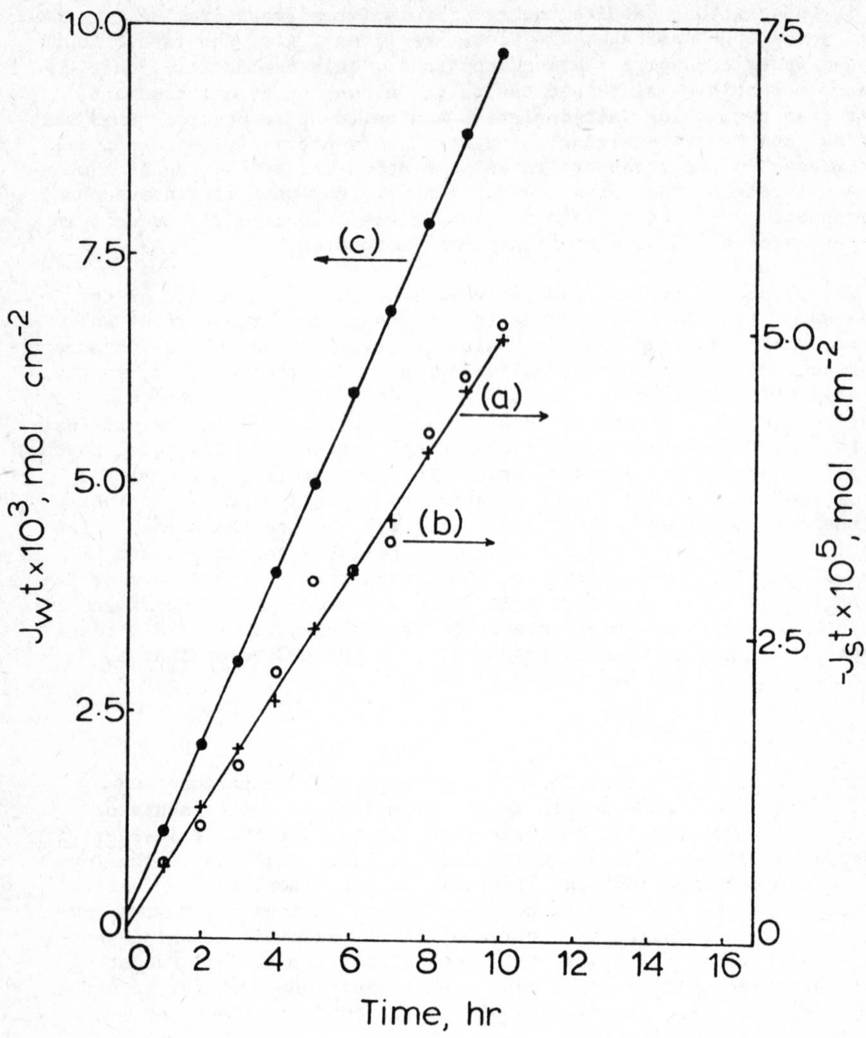

<u>Figure 14. Dialysis-Osmosis. 0.05 - 0.5 M NaCl. 25°C.</u>

Membrane CL-25T (Tokuyama). NaCl solutions.

(a) + Cumulative salt transport calculated from mass balance in salt-<u>receiving</u> compartment
(b) o " " " " " " " " salt-donating "
(c) ● " water " " " " " " " "

(a,b) J_s = -1.389×10^{-9} mol cm^{-2}sec^{-1} (c) J_w = $+2.708 \times 10^{-7}$ mol cm^{-2}sec^{-1}

(Transport data normalized for unit exposed area. = 8.12 cm^2)

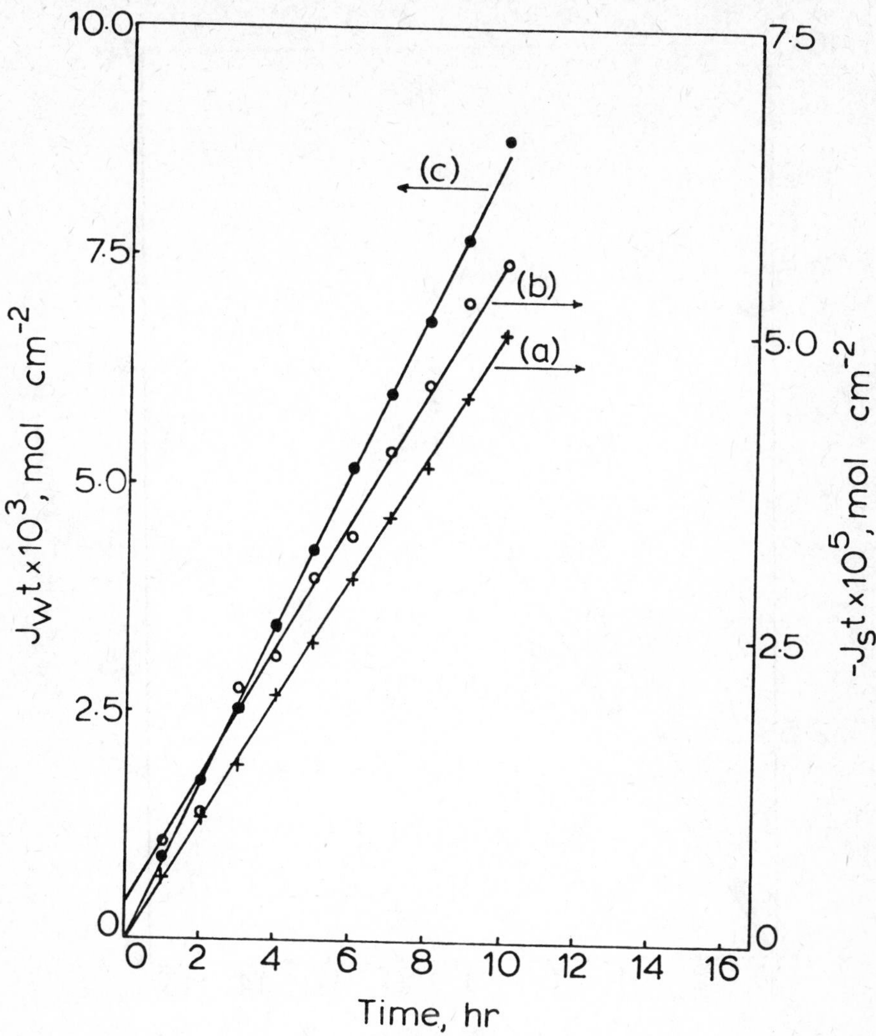

Figure 15. Dialysis-Osmosis. 0.07 - 0.5 M NaCl. 25°C.

Membrane CL-25T (Tokuyama). NaCl solutions.

(a) + Cumulative salt transport calculated from mass balance in salt-receiving compartment
(b) o " " " " " " " " salt-donating "
(c) ● " water " " " " " " " "

(a) $J_s = -1.381 \times 10^{-9}$ (b) $J_s = -1.480 \times 10^{-9}$ (c) $J_w = +2.377 \times 10^{-7}$ mol cm^{-2}sec^{-1}

(Transport data normalized for unit exposed area.)

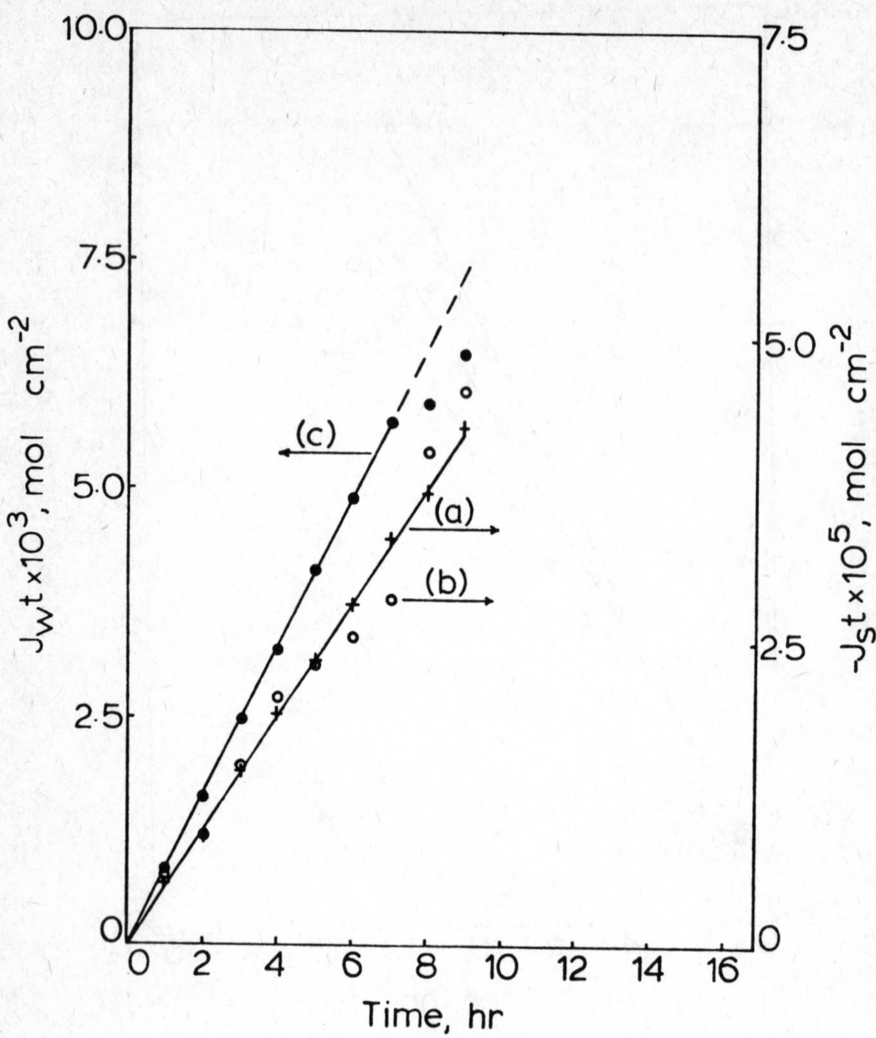

Figure 16. Dialysis-Osmosis. 0.1 - 0.5 M NaCl. 25°C.

Membrane CL-25T (Tokuyama). NaCl solutions.

(a) + Cumulative salt transport calculated from mass balance in salt-<u>receiving</u> compartment
(b) o " " " " " " " " salt-<u>donating</u> "
(c) • " water " " " " " " " "

(a,b) $J_s = -1.297 \times 10^{-9}$ mol cm^{-2}sec^{-1} (c) $J_w = +2.275 \times 10^{-7}$ mol cm^{-2}sec^{-1}

(Transport data normalized for unit exposed area.)

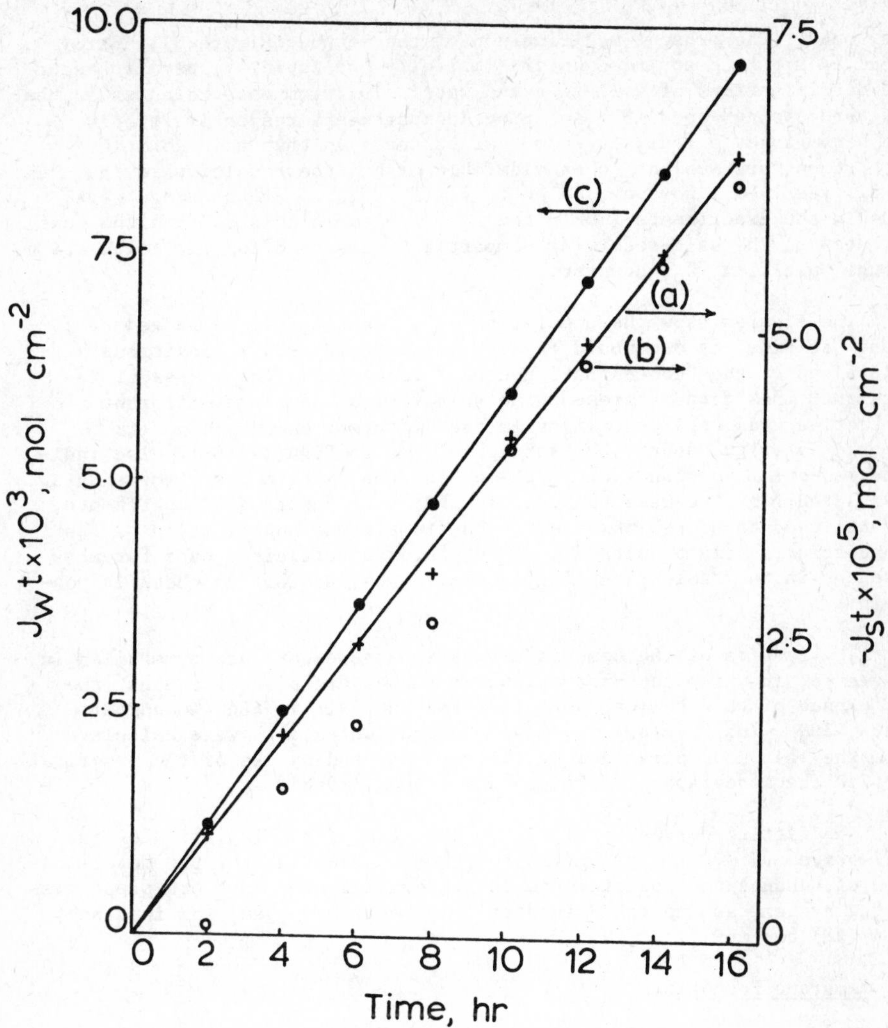

Figure 17. Dialysis-Osmosis. 0.2 - 0.5 M NaCl. 25°C.

Membrane CL-25T (Tokuyama). NaCl solutions.

(a) + Cumulative salt transport calculated from mass balance in salt-receiving compartment
(b) o " " " " " " " " salt-donating "
(c) ● " water " " " " " " salt-donating "

(a) $J_s = -1.117 \times 10^{-9}$ (b) $J_s = -1.094 \times 10^{-9}$ (c) $J_w = +1.667 \times 10^{-7}$ mol cm^{-2}sec^{-1}

(Transport data normalized for unit exposed area.)

was found. The present modification of the method (Figure 3), introduced by Mr. Berg for reasons of simplicity and rapidity, permits accurate determinations of salt flux and water flux from mass balances in the "salt-receiving" and "salt-donating" compartments respectively only. The salt fluxes calculated from mass balances in the "salt-donating" compartment are subject to considerable error; the results were included in Figures 14-17, however, so as to verify that no gross errors were made in the experiments. Only the salt fluxes calculated from the mass balances in the "salt-receiving" compartment were used for the calculation of the salt flux, J_s, however.

The Figures show the cumulative salt transport to be indeed proportional to time, as one would expect, since steady-state conditions are maintained by the "concentration-clamp" mechanism. The slopes of the straight lines found represent the salt fluxes under the different concentration gradients prevailing in the different experiments. As for osmotic water transport, the straight lines in Figures 14-17 also indicate essential constancy of J_w throughout the experiments. These data, calculated from the mass balances in the "salt-donating" compartments, are believed to be reliable, although the present modification of the "concentration-clamp" method does not furnish sufficient data for mass balances in the "salt-receiving" compartments, so that no check is possible.

The results of the osmosis-dialysis measurements are summarized in Figure 18, in which the salt and water fluxes are plotted against the difference of the chemical potential of the salt in the two compartments, $\Delta\mu_s^c \equiv (\mu_s^c)'' - (\mu_s^c)'$. These chemical potentials were calculated from the salt concentrations by the methods, and by use of the numerical data in the monograph of Robinson and Stokes (1968).

The fitting curves in Figure 18 were used for calculation of the derivatives $\partial J_s/\partial\Delta\mu_s^c$ and $\partial J_w/\partial\Delta\mu_s^c$ which are essential for the determination of conductance coefficients characterizing the transport properties of the system, as described in detail in Section 3 [see, for instance, eqs. (51) and (69)].

5.V. Membrane Potentials

The relationship between the electric potential difference between the Ag/AgCl electrodes and the difference of the chemical potentials of the salt in the two solutions bracketing the membrane is shown in Figure 19. From elementary solution thermodynamics, proportionality is expected for a perfectly permselective membrane (see, for instance, Spiegler and Wyllie, 1968). The slope of the straight line is expected to be $1/\mathfrak{F}$ [eq. (75)], i.e. 1.03×10^{-5} eq. coul^{-1}. The slope found from Figure 19 is 0.96×10^{-5} eq. coul^{-1}. The difference between these values is primarily due to the non-ideal permselectivity of the membrane ($t_+ < 1$).

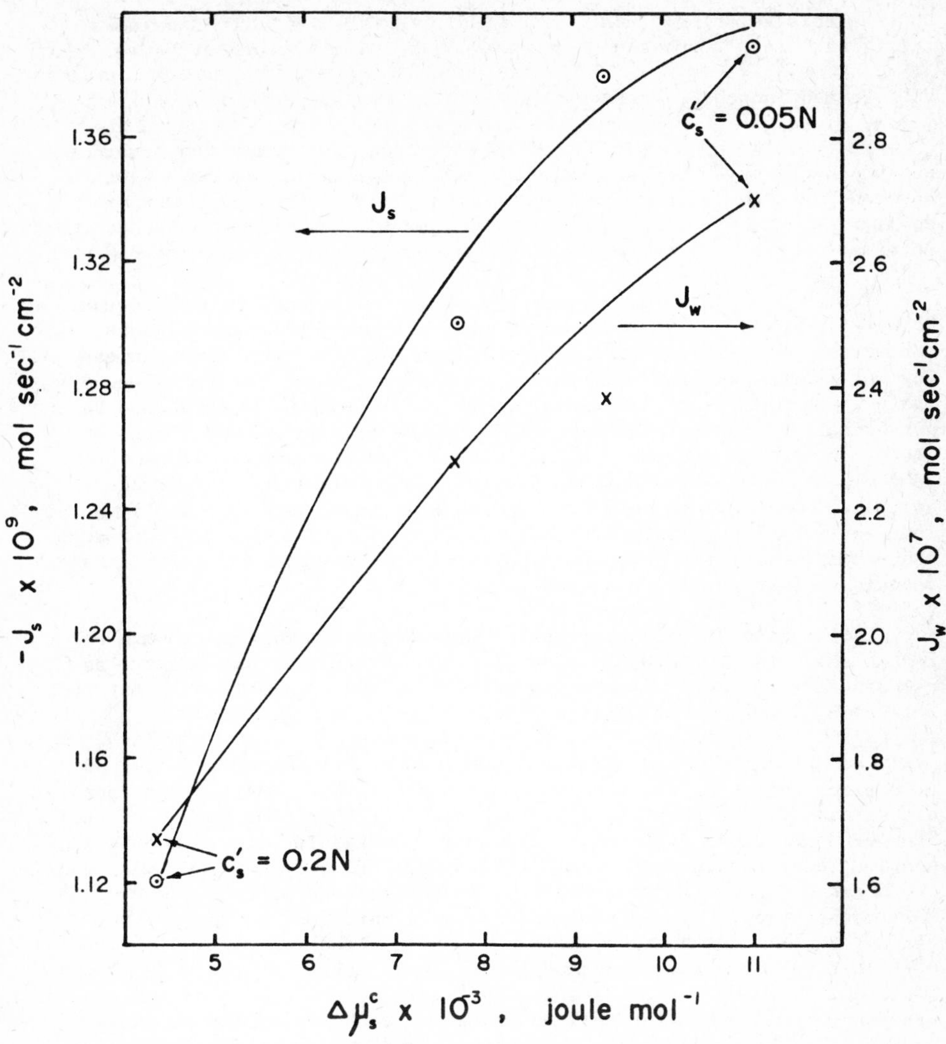

Figure 18. Dialysis-Osmosis

Salt and Water Fluxes vs. Salt Chemical Potential Difference

Membrane CL-25T (Tokuyama). NaCl solutions. $c_s'' = 0.5 \times 10^{-3}$ mol cm^{-3}. 25°C.

The following polynomials were used for curve-fitting:

$$-J_s = 2.173 \times 10^{-22}(\Delta\mu_s^c)^3 - 1.076 \times 10^{-17}(\Delta\mu_s^c)^2 + 1.667 \times 10^{-13}(\Delta\mu_s^c) + 5.750 \times 10^{-10}$$

$$J_w = -7.885 \times 10^{-16}(\Delta\mu_s^c)^2 + 2.785 \times 10^{-11}(\Delta\mu_s^c) + 5.976 \times 10^{-8}$$

Despite this relatively small difference, CL-25T may be considered a fairly efficient "membrane electrode" in dilute solutions of NaCl.

5.VI. Permeation under Pressure

The results of a hydraulic-flow experiment, in which the membrane was bracketed by 0.2 M NaCl solutions are shown in Figure 20. Because the concentration changes in the solutions were known to be extremely small, the "concentration-clamp" feedback mechanisms were considered unnecessary for this experiment, and were consequently disconnected.

It is seen that after a period of relatively rapid flow (Sections "a" and "b" of Figure 20) a lower flow rate was reached which remained constant for several hours. (Section "c" of Figure 20) The hydraulic permeability was calculated from this part of the curve, which later flattened.* There is an element of arbitrariness in this choice, but it should be noted that the construction materials of the cell are not suitable for operation at high pressures when the hydraulic permeability, and the concomitant salt-concentration changes, can be measured with greater accuracy. Such experiments have indeed been performed in a steel vessel long ago [McKelvey, Spiegler and Wyllie, 1959], and techniques for such measurements are available as a result of the development of the reverse-osmosis process [Dresner and Johnson, 1980].

In the present work, hydraulic-permeability measurements were performed only with 0.2 M NaCl solution. The hydraulic permeability, as calculated from the straight-line portion of the curve shown in Figure 20, was 6.07×10^{-6} cm^4 watt sec^{-2}.

The streaming potential was determined from the sudden change of electrical potential observed when the pressure was suddenly increased or decreased. The pressure difference between the compartments ranged from 0.073 to 0.075 Megapascal (547 to 562 mm Hg) in different experiments. The streaming potential difference, "SPD", [eq. (42)], was 1.067, 1.075 and 1.152×10^{-3} volt Mpa^{-1} at concentrations 0.07, 0.1 and 0.2 M NaCl respectively. These measurements were performed at low pressure drops, for the reason stated above. (The feedback mechanisms were not operated for these measurements.) The influence of Δp on (SPD) was not studied in the "concentration-clamp" cell, but a series of systematic measurements in a smaller cell confirmed that the streaming potential, ϕ, (volt) is proportional to the pressure difference between the compartments ($c_s = 10^{-4}$mol cm^{-3}). The streaming-potential differential in the small cell, 1.105×10^{-3} volt Mpa^{-1}, agreed fairly well with the corresponding value measured in the "concentration-clamp" cell. By

---

*An unexplained steep rise of the rate occurred later ("d" in Figure 20). This part of the flow, which might have been caused by a leak, was disregarded.

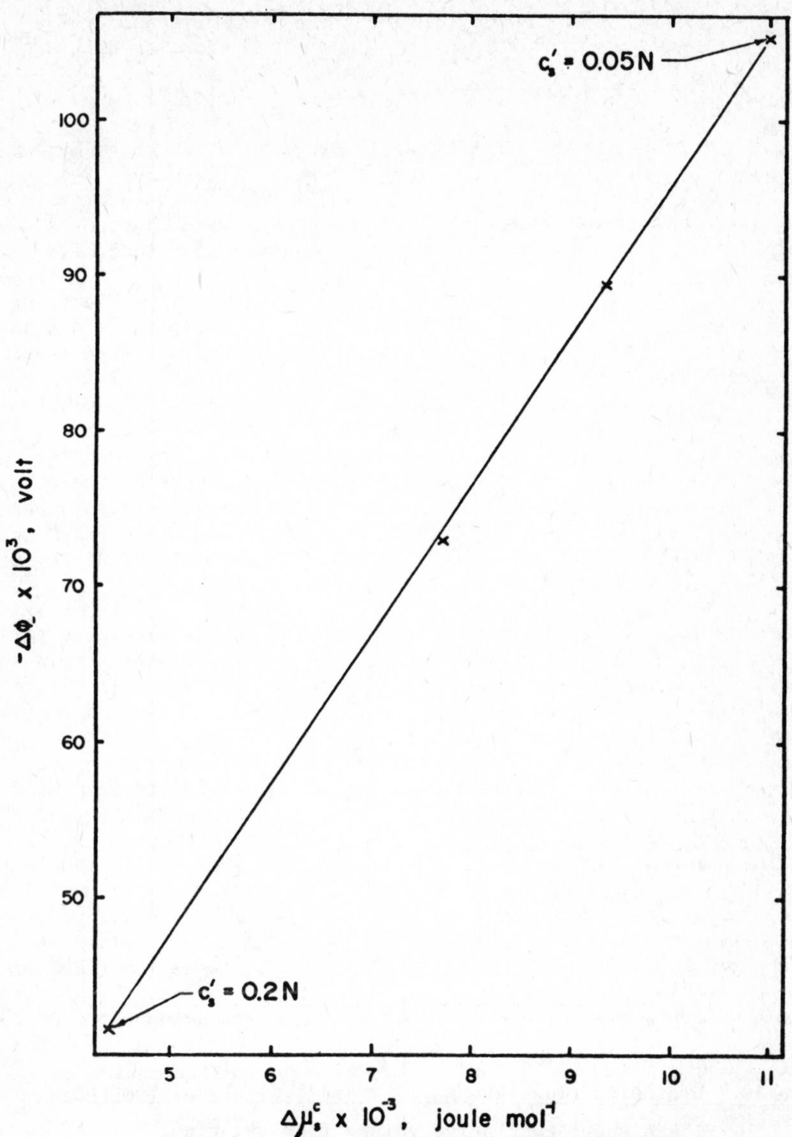

Figure 19. Electrode-Potential Difference vs. Chemical-Potential Difference (Dialysis Experiments)

Membrane CL-25T (Tokuyama). NaCl solutions. $c_s'' = 0.5 \times 10^{-3}$ mol cm^{-3}. 25°C. The following line was curve-fitted to the experimental points:

$$-\Delta\phi_- = 9.578 \times 10^{-6} \Delta\mu_s + 3.147 \times 10^{-4} \text{volt}$$

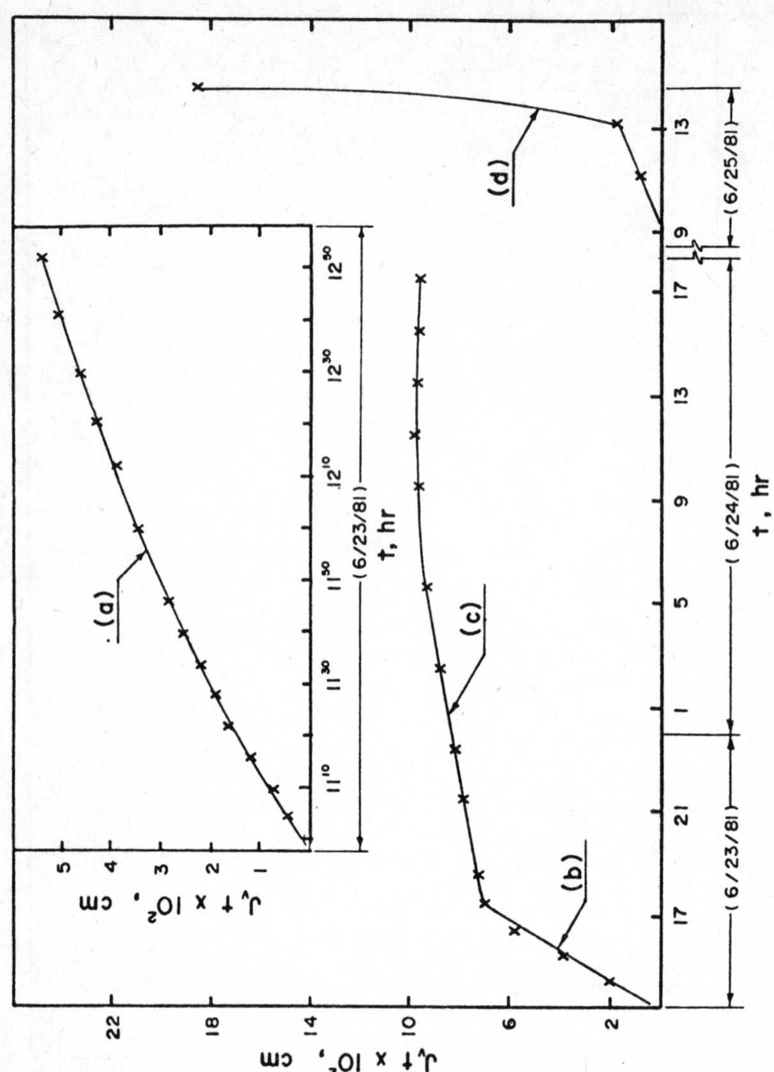

Figure 20. Hydraulic Flow Through Tokuyama Soda CL-25T Membrane.

Figure shows cumulative volume flow vs. time.

Solution: NaCl 0.2×10^{-3} mol cm^{-3}. $25^{\circ}C$.

(a) p" = 1 atm, p' = 1.799 atm, p = -0.0810 Mpa
(b) p" = 1 atm, p' = 1.805 atm, p = -0.0816 Mpa
(c) p" = 1 atm, p' = 1.904 atm, p = -0.0916 Mpa

(Transport data normalized for unit exposed area. $\mathcal{A}$ = 8.12cm^2)

Saxén's law (see, for instance, Spiegler and MacLeish, 1981), this corresponds to an electroosmotic transport of 6.4 mol water Faraday^{-1}, which is in fair agreement with the water transport number, $t_w = 7.08$ mol Faraday^{-1}, found in the electromigration-electroosmosis experiments (0.1 N NaCl) in this work.

It is also of interest to compare this parameter with a rule of thumb for prediction of the electroosmotic water transport from the chemical composition of the membranes, which states that the amount of water transported in electroosmosis is about one half of the amount of water per exchange site, determined by chemical analysis [Spiegler, 1956; Paterson, 1976]. * For membrane CL-25T, the latter ratio is (Table 2): (0.391 g H_2O/g dry membrane/ 1.81 x 10^{-3} eq. fixed charge/g dry resin) x (1/18) mol H_2O/g H_2O = 12.0 mol H_2O/eq. fixed charge. If one half of this water is transported in electroosmosis, one would expect t_w to be 6.0, compared to 6.4 predicted from the streaming potential differential and 7.08 mol H_2O/ Faraday found by direct transport measurement in this work.

5.VII. Summary of Experimental Data

Table 3 summarizes experimental data necessary for calculation of the conductance coefficients, L. Some equilibrium data, necessary for this purpose, have also been added, namely the partial molal volumes of NaCl and water. For comparison, a table with data for a different membrane (AMF C-103), taken from a previous research [Spiegler, 1979] is also presented here (Table 4).

5.VIII. Conductance Coefficients, L

The conductance coefficients, L, [eqs. (16)-(18)], which characterize the transport properties of the membrane in the given solution were calculated by the method described in Section 3. (A step-by-step procedure for these calculations is outlined on p. 20). These coefficients are listed in Table 5. For comparison, data previously determined [Spiegler, 1979] for membrane AMF C-103 are shown in Table 6.

It is seen that in both membranes, L_{ee}, which is proportional to the electrical conductivity of the membrane [eq. (38)] increases with increasing solution concentration, but in terms of percentage change the membrane conductivity varies much less than the conductivity of the equilibrating solution. This behavior is as expected for high-capacity ion-exchange materials, since, at the dilute solution concentrations considered here, most of the electric current is carried by the counterions of the membrane, rather than by the "Donnan invasion" of electrolyte from the solution into the membrane [Helfferich, 1962]. The

* This rule of thumb seems to hold for systems consisting of high-capacity synthetic-resin ion-exchange membranes and dilute alkali halide solutions.

Table 3

Data for Calculation of the Transport Coefficients

(Membrane CL-25T)

c_s mol cm^{-3} x 10^3	c_w mol cm^{-3} x 10^3	$\bar{V}_s$ $cm^3 mol^{-1}$	$\bar{V}_w$ $cm^3 mol^{-1}$	ρ_m ohm cm	$\bar{t}_+$	$\bar{t}_w$	(SPD) volt Mpa^{-1} x 10^3	L_p $watt^{-1}$ cm^4 sec^{-2} x 10^6	$-(PDC)x10^5$ volt $watt^{-1}$ $sec^{-1} mol^{-1}$	$-g_{\mu,s}$ * $watt^{-1} cm^{-2}$ $sec^{-2} mol^2$ x 10^{15}	$g_{\mu,w}$ ** $watt^{-1} cm^{-2}$ $sec^{-2} mol^2$ x 10^{12}
0.07	55.29	17.36	18.08	296	1.00	6.30	1.067	6.13	0.958	22.3	13.1
0.1	55.23	17.49	18.08	248	0.989	7.08	1.075	6.12	0.958	39.6	15.7
0.2	55.15	17.87	18.078	206	0.986	5.59	1.152	6.07	0.958	84.9	20.9

* $g_{\mu,s} \equiv (\partial J_s / \partial \Delta\mu_s^c)_{p,T}$ eq. (31)

** $g_{\mu,w}$ $(\partial J_w / \partial \Delta\mu_s^c)_{p,T}$ " "

Table 4 [Spiegler, 1979]

Data for Calculation of the Transport Coefficients
(Membrane: AMF C-103)

c_s (mole cm^{-3} x 10^3)	c_w (mole cm^{-3} x 10^3)	$\bar{v}_s$ (cm^3 $mole^{-1}$)	$\bar{v}_w$ (cm^3 $mole^{-1}$)	L_{ee} ($joule^{-1}$ $mole^2$ cm^{-2} sec^{-1}) x 10^{11}	$\bar{t}_+$	$\bar{t}_w$	(SPD) (volt dek-$abar^{-1}$) x 10^3	L_p ($watt^{-1}$ cm^4 sec^{-1}) x 10^6	$\left\|\frac{\partial\phi_-}{\partial\mu_s^c}\right\|_{\Delta p=0}$ (volt $watt^{-1}$ sec^{-1} $mole^{-1}$) x 10^5 (PDC)	$\left(\frac{-dJ_s}{d\mu_s}\right)_{p,T}$ ($watt^{-1}$ $mole^2$ cm^{-2} sec^{-2}) x 10^{15}	$\left(\frac{dJ_w}{d\mu_s}\right)_{p,T}$ ($watt^{-1}$ $mole^2$ cm^{-2} sec^{-2}) x 10^{12}
0.01	55.32	16.894	18.08	0.90	1.0	7.0	1.53	2.50	1.0	1.7	1.30
0.025	55.30	17.050	18.08	1.28	0.99	7.24	1.49	2.36	1.0	3.2	2.26
0.05	55.275	17.232	18.08	1.50	0.98	7.33	1.45	2.24	1.0	7.36	4.03
0.10	55.225	17.491	18.08	1.64	0.975	7.33	1.40	2.06	0.98	25	8.27
0.25	55.072	18.008	18.077	1.85	0.965	7.24	1.33	1.80	0.97	122	17.5
0.50	54.82	18.626	18.074	2.22	0.965	7.04	1.27	1.50	0.96	310	28.3
1.00	54.31	19.525	18.062	2.95	0.96	6.85	1.20	0.86	0.914	933	42.1

Table 5

Computed "M-K" Conductance Coefficients (watt^{-1}mol^{2}cm^{-2}sec^{-2})

[Equations (16)-(18)]

Membrane: CL-25T (Tokuyama Soda Co.)

Solution Concentration (mol cm^{-3}) x 10^3	L_{ss}x10^{10}	L_{ww}x10^{8}	L_{ee}x10^{10}	L_{ws}x10^{9}	L_{es}x10^{10}	L_{se}x10^{10}	L_{ew}x10^{9}	L_{we}x10^{9}	Reciprocity Index %*	
									$(RI)_s^e$	$(RI)_w^e$
0.07	0.225	1.96	0.239	0.151	0.225	0.234	0.135	0.150	-5.9	-11.1
0.1	0.2650	1.99	0.285	0.207	0.267	0.282	0.162	0.202	-5.8	-24.7
0.2	0.3211	1.97	0.342	0.227	0.324	0.338	0.208	0.191	-4.2	+ 8.1

* $(RI)_s^e \equiv 100(L_{es} - L_{se})/L_{es}$

$(RI)_w^e \equiv 100(L_{ew} - L_{we})/L_{ew}$

Table 6

Computed "M-K" Conductance Coefficients (watt^{-1} mol ^{2} cm^{-2} sec^{-2})

[Equations (16)-(18)]

Membrane: AMF C-103

Solution Concentration (mol · cm^{-3}) x 10^3	$L_{ss} \times 10^{10}$	$L_{ww} \times 10^{8}$	$L_{ee} \times 10^{10}$	$L_{ws} \times 10^{9}$	$L_{es} \times 10^{10}$	$L_{se} \times 10^{10}$	$L_{ew} \times 10^{9}$	$L_{we} \times 10^{9}$	Reciprocity Index %*	
									$(RI)^e_s$	$(RI)^e_w$
0.01	0.087	0.810	0.090	0.064	0.087	0.090	0.065	0.063	-3.5	+3.6
0.025	0.120	0.787	0.128	0.095	0.124	0.127	0.090	0.093	-2.3	-2.9
0.05	0.143	0.759	0.150	0.117	0.146	0.147	0.102	0.110	-0.9	-7.6
0.10	0.154	0.707	0.164	0.135	0.157	0.160	0.107	0.120	-1.8	-12.0
0.25	0.176	0.629	0.185	0.171	0.178	0.179	0.114	0.134	-0.1	-18.0
0.50	0.222	0.542	0.222	0.223	0.217	0.214	0.128	0.156	+1.4	-22.0
1.0	0.312	0.362	0.295	0.287	0.289	0.283	0.158	0.202	+2.1	-28.0

* $(RI)^e_s \equiv 100(L_{es} - L_{se})/L_{es}$

$(RI)^e_w \equiv 100(L_{ew} - L_{we})/L_{ew}$

variation of L_{ww} with the solution concentration, c_s, is also only moderate over a considerable range of c_s-values. The potential-difference coefficient, PDC, is practically independent of c_s as expected for membranes separating dilute solutions (see, for instance, Spiegler and Wyllie, 1968), and there is also little variation of the transport numbers, $\bar{t}_s$ and $\bar{t}_w$, with c_s. The streaming-potential differential is known to vary little with c_s for dilute equilibrating solutions [McKelvey, Spiegler and Wyllie, 1959]. Because of the relative smallness of the variation of all these parameters with the solution concentration (in dilute solutions), as demonstrated by the data in Tables 3 and 4, L_{se} [eq. (40)], L_{we} [eq. (48)] and L_{ew} [eq. (61)] are expected to vary only moderately with c_s, and this is indeed confirmed by the data in Tables 5 and 6.

It is seen that the cation transport number in the membrane, $\bar{t}_+$, is close to unity over the whole concentration range, but slightly decreasing with increasing solution concentration, as expected for cation-exchange membranes [Helfferich, 1962]. The water transport number in the membrane, $\bar{t}_w$, is related to the streaming-potential differential, (SPD), as a result of a reciprocity relation. This is best seen from the near-equality of L_{we} and L_{ew} (Tables 5 and 6). It is well known [McKelvey, Spiegler, and Wyllie, 1959; Chartier, Gross and Spiegler, 1975, eq. (9-15)] that this equality amounts to a prediction of the electroosmotic flow from the streaming potential differential, or vice-versa (Saxén's law):

$$\left(\frac{\Delta\phi}{\Delta p}\right)_{i=o,\Delta\mu=o} = -\left(\frac{J_v}{i}\right)_{\Delta p=o,\Delta\mu=o} \tag{82}$$

In tables 5 and 6 the discrepancy between two reciprocal conductance coefficients, L_{xy} and L_{yx} respectively, is expressed in percent, viz. as the Reciprocity Index, $(RI)^x_y$.

It is seen that for L_{es} and L_{se} there is good agreement over the whole concentration range, confirming that the conductance coefficients computed from transport numbers determined from (a) careful membrane-potential measurements and from (b) direct ion-transport measurements are the same. [Scatchard, 1953; Spiegler and Wyllie, 1968]. This assumption has been made long ago for many membranes [Sollner, 1950] although the relatively small contribution of the water transport to the membrane potential between dilute electrolyte solutions has not always been duly taken into account.

The streaming potential-electroosmosis reciprocity [eq. (82), "Saxén's law"] should lead to equality of the transport coefficients L_{ew} and L_{we}. Previous experiments (not published in the conventional literature) with the "concentration-clamp" apparatus [Spiegler and coll., 1972] had demonstrated reciprocity for a different specimen of the membrane C-103, when the solution concentration was 0.1N. At that time, the results for this single concentration were 0.177 and 0.188 x 10^{-9}watt^{-1}mol^{2}cm^{-2}sec^{-2} for

L_{ew} and L_{we} respectively, i.e., agreement within six percent. The results in Tables 5 and 6 also show fair agreement at low concentrations of the equilibrating solutions (up to 0.025 M), but significant systematic discrepancies at higher concentrations. These discrepancies are yet unexplained. At much higher pressures for the streaming-potential measurements, the validity of Saxén's law for a system consisting of a commercial cation-exchange membrane and 0.1 M sodium chloride solution had been proven before [McKelvey, Spiegler and Wyllie, 1957 and 1959; Demisch and Pusch, 1979].

The validity of the latter reciprocity relation [eq. (82)] which makes it possible to predict the electroosmotic water transport from the streaming potential is by no means merely of academic interest. This relation made it possible to predict that in certain modified cellulose-acetate membranes as used in many reverse-osmosis installations, the electroosmotic water transport should be almost one order of magnitude higher than found for the high-capacity ion-exchange membranes studied in this research (Table 3) and a preceding one (Table 4) because the streaming potential differential, (SPD), is almost one order of magnitude higher [Minning and Spiegler, 1976]. In other words, it was predicted that each Faraday would cause the transfer of not $\bar{t}_w = 7$, but of 50-60 water molecules through the membrane, depending on the solution concentration. This was indeed found to be the case, and a method for the electroosmotic backwash of modified-cellulose membranes was worked out [Spiegler, 1980; Spiegler and MacLeish, 1981]. This method is based on the validity of the reciprocity relation $L_{ew} = L_{we}$.

While it is not easy to draw immediate conclusions on the transport mechanism from the phenomenological conductance coefficients, L_{xy}, several important points emerge from consideration of the L-values in Tables 5 and 6 and their relation to the physical situation:

First, the coefficients, related to a larger extent to the solvent than to the salt, are larger [$L_{ew} > L_{es}$; $L_{ww} >> L_{ws}$; L_{sw} ($= L_{ws}$) $>> L_{ss}$]. This is probably due to the much higher concentration of water than of salt in the ion-exchange membrane, for the coefficients L_{ij} increase in general with increasing c_j; while the flux equations derived in the friction model [Spiegler, 1958] are not identical with equations (16)-(18) used here, some insight into the physical meaning of the L_{xy}-coefficients, and in particular the above conclusion about their variation with c_j (borne out by the results) may be obtained from that reference; see eqs. (17) or (32) of the reference. This property of the set of L-coefficients thus probably reflects the exclusion of salt from the ion-exchange membrane ("Donnan effect"), which also explains why L_{ee} changes only relatively little for large variations of c_s.

Second, two of the three reciprocity relations were tested by the results of our experiments, and confirmed for the conditions of these measurements.

Third, the numerical calculations showed the large dependence of all L-coefficients on the membrane <u>conductance</u>. In fact, membrane conductance emerges as a decisive transport characteristic. The implication of this conclusion for future experiments is the emphasis on refinements of the technique of membrane conductance measurements.

Comparison of the "diagonal" conductance coefficients. L_{ss}, L_{ww} and L_{ee}, for membrane CL-25T (Table 5) with those for AMF C-103 (Table 6) shows that these coefficients are higher for CL-25T. The latter membrane may therefore be considered somewhat more permeable than the first.

It may be concluded that the methods described in Section 3 for calculating a set of L-coefficients from specified transport measurements lead to the unequivocal characterization of systems of the type: solution/membrane/solution, provided only that these specified measurements can be performed with sufficient accuracy. The "concentration-clamp" apparatus represents a significant step towards the attainment of this experimental objective.

6. LIST OF SYMBOLS

$\mathcal{A}$	effective surface area of membrane, cm^2
Λ_3, Λ_6	parameters defined in equations (44) and (51), respectively
c	solution concentration, mol cm^{-3}
$\bar{c}$	concentration in membrane, mol (cm^3 membrane)$^{-1}$
d	membrane thickness, cm
D'	sum of distances between electrodes and adjacent membranes, cm
F	generalized force (e.g.,$-\Delta\tilde{\mu}$), joule mol $^{-1}$
g_μ	conductivity coefficient, $watt^{-1} cm^{-2} mol^2 sec^{-2}$ [eq. (31)]
i	electric current density, amp cm^{-2}
J_i	flux of component i, mol cm^{-2} sec^{-1}
J_v	volume flux, cm sec^{-1}
L_{ij}	conductance coefficient in "Onsager set of transport equations" (3)-(5). i, j = +, -, w; mol ^{2} $joule^{-1}$ cm^{-2} sec^{-1}
L_p	hydraulic permeability, $watt^{-1}$ cm^4 sec^{-1}
L_{ss}, L_{ww}, L_{ee}, etc.	conductance (transport) coefficients in "Michaeli-Kedem set of transport equations" (16)-(18), mol ^{2} $joule^{-1}$ cm^{-2} sec^{-1} = $watt^{-1}$ mol ^{2} cm^{-2} sec^{-2}
(PDC)	potential-difference coefficient, volt mol $joule^{-1}$, [eq. (52)]
(PD)	membrane potential, volt
p	pressure, Mpascal (=dekabar)*
R	gas constant, 8.31 joule mol $^{-1}$ (°K)$^{-1}$

*1 dekabar = 1 joule cm^{-3} = 10 bar = 9.8692 atm = 1 MNewton m^{-2} = 1 Mpascal

S	cross-section of membrane resistance cell, cm^2
(SPD)	streaming-potential differential, volt $dekabar^{-1}$ [eq. (42)]
T	temperature, °K
t	time, sec
$\bar{t}_i$	transference number of component i, relative to the membrane
$\bar{v}_i$	partial molal volume of component i, mol cm^{-3}
v_{Cl}	$\equiv \bar{v}_{AgCl} - \bar{v}_{Ag}$, $cm^3\ mol^{-1}$
z	length coordinate, cm
z_1	length of 2" diameter cylinder of solution between electrodes in conductance cell (Fig. 6). [This length does not include the thickness of the support plates.]
z_i	valency of particle i, equiv mol^{-1} (positive for cations, negative for anions, zero for water)
$\mathcal{F}$	Faraday's constant, 0.965×10^5 coul eq^{-1}
$\Delta\phi$	electric potential difference across membrane, volt*
$\Delta\phi_-$	potential difference between two Ag/AgCl (anion-reversible) electrodes in solutions adjacent to membrane, volt
κ	conductivity, $ohm^{-1}\ cm^{-1}$
ρ	resistivity, ohm cm
$\tilde{\mu}_i$	total potential of component i, joule mol^{-1}
μ_i	chemical potential of component i (includes pressure-volume term), joule mol^{-1}
μ_i^c	"concentration-dependent part" of chemical potential of component i, joule mol^{-1}

*Often measured with "Luggin capillaries"

Subscripts

+	cation
-	anion (Cl^- in this report
m	membrane
s	electrolyte (NaCl in this report). The equations refer to an electrolyte composed of a cation of valency +1 and an anion of valency -1 ("1-1" electrolyte)
w	water

Sign Conventions

(1) Positive direction is from left to right. (2) Fluxes from left to right are counted positive. (3) The operator, Δ, for finite differences refers to the value on the right (double primes) minus the value on the left (single primes), as does conventionally the differential operator, d.

Driving forces are of the general form ($-d\tilde{\mu}/dz$). Thus positive values of the driving force ($-d\tilde{\mu}/dz$) > 0, lead to positive fluxes. For example, Ohm's law is

$$I = (1/\rho) \underbrace{(\quad -\Delta\phi \quad)}_{\text{"Driving force"}}$$

and Fick's law

$$J_i = D_i \underbrace{(\quad -\Delta c_i \quad)}_{\text{"Driving force"}}$$

7. REFERENCES

Brun, T. S. and Vaula, D., Ber. Bunsenges 71, 824 (1967).

Chartier, P., M. Gross and K. S. Spiegler, Applications de la thermodynamique du non-équilibre, Herman, Paris, 1975.

Demisch, H-U and Pusch, W., J. Coll. Interfac. Sci. 69, 247 (1979).

Delmotte, M., Approche dissipative des transport membranaires en biologie, Dr. Sci. thesis, University of Paris VII, 1975.

Delmotte, M., and J. Chanu, Electrochimica Acta 18, 963 (1973).

Dresner, L. and Johnson, J. S., Jr., Chapter 8 in "Principles of Desalination", 2nd ed., K. S. Spiegler and A. D. K. Laird eds., Academic Press, New York, 1980.

Duncan, B. C., J. Res. National Bureau Standards 66A, 83 (1962).

Foley, T., J. Klinowski and P. Meares, Proc. Roy. Soc. London A336, 327 (1974).

Guillou, M., D. Guillou and R. Buvet, in "Membranes à perméabilité sélective," Editions du Centre National de la Recherche Scientifique, Paris (1969), p. 131.

Halary, J. L., "Relations entre la structure et les phénomènes de transport dans les membranes de diacétate de cellulose pour osmose inverse", Ph.D. thesis, University of Paris, 6, Dec. 18, 1979.

Halary, J. L., Noël, C. and Monnerie, L., "Analysis of Transport Phenomena in Cellulose-Acetate Membranes," submitted to J. Polym. Sci., 1977.

Helfferich, F. G., Ion Exchange, McGraw-Hill, New York, 1962.

Katchalsky, A. and P. Curran, Non-Equilibrium Thermodynamics for Biophysicists Harvard University Press, Boston, Mass., 1966.

Kedem, O. and A. Katchalsky, Trans. Farad. Soc. 59, 1918 (1963).

Krämer, H. and P. Meares, Biophys. J. 9, 1006 (1969).

Lorimer, J. W., E. I. Broterenbrod and J. J. Hermans, in "Membrane Phenomena," Disc. Faraday Soc. 21, 141 (1956).

Mackay, D. and P. Meares, Trans. Farad. Soc. 55, 1221 (1959).

McKelvey, J. G., K. S. Spiegler and M. R. J. Wyllie, J. Electrochem. Soc. 104, 387 (1957).

McKelvey, J. G., K. S. Spiegler and M. R. J. Wyllie, Chem. Engin. Progr. 55 (No. 24), 199 (1959).

Meares, P., J. F. Thain and D. G. Dawson, "Transport Across Ion-Exchange Membranes; The Frictional Model of Transport," Chapter 2 in Membranes, G. Eisenman, ed., Marcel Dekker, Inc., New York, 1972.

Meares, P., ed., Membrane Separation Processes, Elsevier, Amsterdam, 1976.

Michaeli, I. and O. Kedem, Trans. Farad. Soc. 57, 1185 (1961).

Minning, C. P. and K. S. Spiegler, in Charged Gels and Membranes, Part I, E. Sélégny, ed., D. Reidel Publ. Co., Dordrecht (Holland) and Boston, Mass., 1976.

Onsager, L., Phys. Rev. 37, 405 (1931), ibid., 38, 2265 (1931).

Onsager, L., Ann. New York Acad. Sci. 46, 24 (1945).

Oster, G., A. Perelson and A. Katchalsky, Nature 234, 393 (1971).

Paterson, R., in Biological and Artificial Membranes and Desalination of Water (Proceedings of the Study Week at the Pontifical Academy of Sciences, Vatican City), R. Passino, ed., Elsevier, 1976.

Pusch, W. and H. J. Wolff, Rev. Sci. Instr. 45, 1403 (1974).

Richardson, I. W., Bull. Math. Biophys. 32, 237 (1970); J. Membrane Biol. 4, 3 (1971).

Robinson, R. A. and H. Stokes, "Electrolyte Solutions", 2nd ed. (revised), Butterworth's, London (1968).

Scatchard, G., J. Amer. Chem. Soc. 75, 2883 (1953).

Silver, R. S., Physics Letters 63A, 73 (1977).

Sollner, K., J. Electrochem. Soc. 97, 139C (1950).

Sollner, K., in Biological and Artificial Membranes and Desalination of Water, R. Passino, ed., Elsevier, Amsterdam, 1976.

Sollner, K., "The Early Developments of the Electrochemistry of Polymer Membranes", in Charged Gels and Membranes, E. Sélégny, ed., D. Reidel Publ. Co., Dordrecht-Holland, 1976.

Spiegler, K. S., Trans. Faraday Soc. 54, 1408 (1958).

Spiegler, K. S., J. Electrochem. Soc. 113, 161 (1966).

Spiegler, K. S. and O. Kedem, Desalination 1, 311 (1966).

Spiegler, K. S., "Study of Membrane-Solution Interfaces by Electrochemical Methods", United States Department of the Interior, Office of Saline Water Research and Development Report No. 353, Superintendent of Documents, Washington, D.C. (1968).

Spiegler, K. S. and M. R. J. Wyliie, "Electrical Potential Differences," Chapter 7 in Physical Techniques in Biology, 2nd ed., Vol. II, D. H. Moore, ed., Academic Press, New York, 1968.

Spiegler, K. S. and collaborators, Study of Permeability Characteristics of Membranes, Combined Quarterly Reports Nos. 15 and 16 to the Jet Propulsion Laboratory, Pasadena, Calif., January 15, 1972.

Spiegler, K. S., Desalination 15, 135 (1974).

Spiegler, K. S. and J. H. MacLeish, J. Membr. Sci. 8, 173 (1981).

Spiegler, K. S. and collaborators, "Measurement and Interpretation of Ion and Water Flows Across Membranes", Final Report, Grant ENG75-21038, National Science Foundation, 1979.

Spiegler, K. S., "Backwashing Reverse-Osmosis and Ultrafiltration Membranes by Electro-osmosis", United States Patent 4,231,865, Nov. 4, 1980.

Staverman, J., Trans. Faraday Soc. 48, 176 (1952).

Teorell, T., "The Development of the Modern Membrane Concepts and the Relations to Biological Phenomena," in Charged Gels and Membranes, E. Sélégny, ed., D. Reidel Publ. Co., Dordrecht, Holland, 1976.

Zelman, A., J. C. T. Kwak, J. Leibovitz and K. S. Spiegler, Experientia Suppl. 18, 676 (1971). |Also listed as "Biological Aspects of Electrochemistry", G. Milazzo, ed., Birkhaeuser Verlag, Basel, 1971.|

ACKNOWLEDGEMENTS

This paper represents a summary of different phases of work on the "concentration-clamp" method, performed under Contract No. 952109, Jet Propulsion Laboratory, Pasadena, California, U.S.A., as well as Grants 14-34-0001-8539, Office of Water Research and Technology, U.S. Department of the Interior, and ENG 75-21038, (U.S.) National Science Foundation.

The authors express their gratitude to all these sponsors to whom they have reported this work. They thank Dr. J. Leibovitz for many helpful discussions and his very constructive suggestions, and to Ruth Jenkins Moore for her assistance in the early stages of this work.

Thanks are also due to the following institutions who granted different forms of supplemental support to the investigators so as to enable them to participate in this work during their sabbatical leaves in the United States: Israel Desalination Engineering (A.B.); University of Bergen (T.S.B.); and University of Strasbourg (A.S.).